Integrated Optics

Springer
Berlin
Heidelberg
New York
Barcelona
Budapest
Hong Kong
London
Milan
Paris
Santa Clara
Singapore
Tokyo

Robert G. Hunsperger

Integrated Optics
Theory and Technology

Fourth Edition

With 195 Figures and 17 Tables

 Springer

Professor ROBERT G. HUNSPERGER

Department of Electrical Engineering
University of Delaware
140 Evans Hall
Newark, DE 19716, USA

The first three editions appeared as Vol. 33 of Springer Series in Optical Sciences

ISBN 3-540-59481-7 4th edition Springer-Verlag Berlin Heidelberg New York
ISBN 3-540-53305-2 3rd edition Springer-Verlag Berlin Heidelberg New York

CIP data applied for

Die Deutsche Bibliothek – CIP-Einheitsaufnahme
Hunsperger, Robert G.:
Integrated optics: theory and technology / Robert G.
Hunsperger. – 4., ed. – Berlin; Heidelberg; New York;
Barcelona; Budapest; Hong Kong; London; Mailand; Paris;
Tokyo: Springer, 1995
   ISBN 3-540-59481-7

© Springer-Verlag Berlin Heidelberg 1982, 1984, 1991, 1995
Printed in Germany

The use of general descriptive names, registered names, trademarks, etc. in this publication does not imply, even in the absence of a specific statement, that such names are exempt from the relevant protective laws and regulations and therefore free for general use.

Typesetting: Best-set Typesetter Ltd., Hong Kong

SPIN: 10503856   31/3144/SPS – 5 4 3 2 1 0 – Printed on acid free paper

To my wife, Elizabeth

Preface

Once again it has become necessary to produce a new edition in order to update material provided in earlier editions and to add new descriptions of recently emerging technology. All of the chapters have been revised to include new developments, and to incorporate additional literature references.

In the past few years there has been a vast expansion of worldwide telecommunications and data transmission networks. In many localities fiber-to-the-home and integrated services digital networks (ISDN) have become a reality. Many people are now logging-on to the Internet and the World Wide Web. The growth of these networks has created a strong demand for inexpensive, yet efficient and reliable, integrated optic components such as signal splitters, couplers and multiplexers. Because of this demand, there has been a great deal of work recently on devices made using polymers and glasses. Descriptions of these components have been added to the book in the appropriate chapters.

A number of new practice problems have been added, and an updated booklet of problem solutions is available. The supplementary series of video-taped lectures described in the preface to earlier editions continues to be available. Inquires regarding these materials should be sent directly to the author.

The author wishes to thank Mrs. Barbara Westog, who helped with the organization of new material and typed the revisions.

Newark, July 1995   R.G. Hunsperger

Preface to the Third Edition

The field of integrated optics is continuing to develop at a rapid pace, necessitating the writing of this third edition in order to update the material contained in earlier editions. All of the chapters have been revised to reflect the latest developments in the field and a new chapter has been added to explain the important topic of newly invented quantum well devices. These promise to significantly improve the operating characteristics of lasers, modulators and detectors.

The trend of telecommunications toward the use of single mode systems operating at the longer wavelengths of 1.3 and 1.55 μm has been explained and documented with illustrations of recently developed devices and systems. In this regard, broader coverage of GaInAsP devices and optical integrated circuits has been provided, and the new growth techniques of molecular beam epitaxy (MBE) and metal-organic chemical vapor deposition (MOCVD) have been described. The extensive development of hybrid optical integrated circuits in lithium niobate has also been described. Notably, this progress has led to the production of the first commercially available optical integrated circuits.

A number of new practice problems have been added. An updated booklet of problem solutions is available, and the supplementary series of videotaped lectures described in the preface to the first edition has been expanded and updated. Inquiries regarding these materials should be sent directly to the author.

The author wishes to thank Mr. Garfield Simms, who has generated the artwork for a number of the new illustrations which appear in this edition, and Mrs. Barbara Westog, who typed the revisions.

Newark, January 1991   R.G. Hunsperger

Preface to the Second Edition

Our intent in producing this book was to provide a text that would be comprehensive enough for an introductory course in integrated optics, yet concise enough in its mathematical derivations to be easily readable by a practicing engineer who desires an overview of the field. The response to the first edition has indeed been gratifying; unusually strong demand has caused it to be sold out during the initial year of publication, thus providing us with an early opportunity to produce this updated and improved second edition.

This development is fortunate, because integated optics is a very rapidly progressing field, with significant new research being regularly reported. Hence, a new chapter (Chap. 17) has been added to review recent progress and to provide numerous additional references to the relevant technical literature. Also, thirty-five new problems for practice have been included to supplement those at the ends of chapters in the first edition. Chapters 1 through 16 are essentially unchanged, except for brief updating revisions and corrections of typographical errors.

Because of the time limitations imposed by the need to provide an uninterrupted supply of this book to those using it as a course text, it has been possible to include new references and to briefly describe recent developments only in Chapter 17. However, we hope to provide details of this continuing progress in a future edition.

The author wishes to thank Mr. Mark Bendett, Mr. Jung-Ho Park, and Dr. John Zavada for their valuable help in locating typographical errors and in developing new problems for this edition.

Newark, December 1983   R.G. HUNSPERGER

Preface to the First Edition

This book is an introduction to the theory and technology of integrated optics for graduate students in electrical engineering, and for practicing engineers and scientists who wish to improve their understanding of the principles and applications of this relatively new, and rapidly growing field.

Integrated Optics is the name given to a new generation of opto-electronic systems in which the familiar wires and cables are replaced by light-waveguiding optical fibers, and conventional integrated circuits are replaced by optical integrated circuits (OIC's). In an OIC, the signal is carried by means of a beam of light rather than by an electrical current, and the various circuit elements are interconnected on the substrate wafer by optical waveguides. Some advantages of an integrated-optic system are reduced weight, increased bandwidth (or multiplexing capability), resistance to electromagnetic interference, and low loss signal transmission.

Because of the voluminous work that has been done in the field of integrated optics since its inception in the late 1960's, the areas of fiber optics and optical integrated circuits have usually been treated separately at conferences and in textbooks. In the author's opinion, this separation is unfortunate because the two areas are closely related. Nevetheless, it cannot be denied that it may be a practical necessity. Hence, this book includes an overview of the entire field of integrated optics in the first chapter, which relates the work on optical integrated circuits to progress in fiber-optics development. Specific examples of applications of both fibers and the OIC's are given in the final chapter. The remaining chapters of the book are devoted to the detailed study of the phenomena, devices and technology of optical integrated circuits.

This book is an outgrowth of a graduate level, single-semester course in integrated optics taught first at the University of Southern California in 1975 and later at the University of Delaware. The course has also been produced as a series of 20 color videotaped lectures, which can be used along with this book for self-study of the subject. A booklet of solutions to the problems given at the end of the chapters is also available. Inquiries regarding these supplementary materials should be sent directly to the author.

The author wishes to thank those persons who have contributed to making this book a reality. In particular, the critical comments and constructive suggestions provided by Dr. T. Tamir throughout the preparation of the manuscript have been most helpful. The continuing support and encouragement of Dr. H. Lotsch are also greatly appreciated. The competent and efficient typing of the manuscript by Mrs. Anne Seibel and Miss Jacqueline Gregg has greatly facilitated timely publication.

Newark, April 1982   R.G. HUNSPERGER

Contents

1. Introduction

The transmission and processing of signals carried by optical beams rather than by electrical currents or radio waves has been a topic of great interest ever since the early 1960s, when the development of the laser first provided a stable source of coherent light for such applications. Laser beams can be transmitted through the air, but atmospheric variations cause undesirable changes in the optical characteristics of the path from day to day, and even from instant to instant. Laser beams also can be manipulated for signal processing, but that requires optical components such as prisms, lenses, mirrors, electro-optic modulators and detectors. All of this equipment would typically occupy a laboratory bench tens of feet on a side, which must be suspended on a vibration-proof mount. Such a system is tolerable for laboratory experiments, but is not very useful in practical applications. Thus, in the late 1960s, the concept of "integrated optics" emerged, in which wires and radio links are replaced by light-waveguiding optical fibers rather than by through-the-air optical paths, and conventional electrical integrated circuits are replaced by miniaturized optical integrated circuits (OIC's).

For a historical overview of the first years of integrated optics, the reader is referred to the books edited by *Tamir* [1.1] and *Miller* et al. [1.2]. During the later years of the 1970s, several factors combined to bring integrated optics out of the laboratory and into the realm of practical application; these were the development of low-loss optical fibers and connectors, the creation of reliable cw GaAlAs and GaInAsP laser diodes, and the realization of photolithographic microfabrication techniques capable of submicron linewidths. In the 1980s, optical fibers largely replaced metallic wires in telecommunications, and a number of manufacturers began production of optical integrated circuits for use in a variety of applications. In the 1990's, the incorporation of optical fibers into telecommunications and data-transmission networks has been extended to the subscriber loop in some systems. This provides an enormous bandwidth for multichannel transmission of voice, video and data signals. Access to worldwide communications and data banks has been provided by computer networks such as the Internet. We are in the process of developing what some have called the "Information superhighway." The implementation of this technology has provided continuing impetus to the development of new integrated optic devices and systems.

Because of the very broad scope of the field of integrated optics, it is common practice to consider optical fiber waveguides and optical integrated circuits as two separate areas of study, even though they are closely related. For example, the first conference on integrated optics, per se, sponsored by the Optical Society of America, in February 1972 [1.3] included sessions on both optical fibers and on optical integrated circuits. However, by February 1974, when the second conference of this series [1.4] was held, only two papers were presented on optical fibers [1.5, 6]. Subsequent conferences in this biennial series have included only papers on OIC's. In this book, we will concentrate mainly on optical integrated circuits, but we consider first the advantages of a combined fiber-optic OIC system in order to put the subject matter in proper perspective.

The integrated optics approach to signal transmission and processing offers significant advantages in both performance and cost when compared to conventional electrical methods. In this chapter, we consider those advantages in order to generate an understanding of the motivating force behind the development of integrated optics. We also examine the basic question of which substrate materials are most advantageous for the fabrication of OIC's, and whether a hybrid or a monolithic approach should be used.

1.1 Advantages of Integrated Optics

To consider the advantages of a fiber-optic OIC system as compared to its electrical counterpart, we show in Fig. 1.1 a hypothetical fiber-optic OIC system for optical communications that can be used to illustrate many of the special advantages of the integrated optic approach. In this system, the transmitter and receiver are each contained on an OIC chip, and the two are interconnected by means of an optical fiber waveguide. The elemental devices

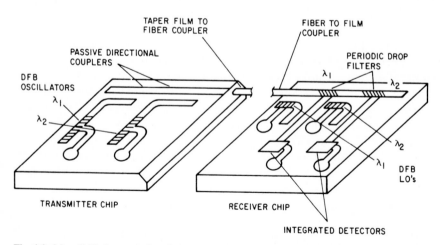

Fig. 1.1. Monolithic integrated optic system for optical communications

of the system will be explained in detail in later chapters, but, for now, let us consider only their general functions. The light sources are integrated laser diodes of the distributed feedback (DFB) type, emitting at different wavelengths λ_1 and λ_2. Only two diodes are shown for simplicity, but perhaps hundreds would be used in a practical system. Since the light emitted by each laser is at a different wavelength, it travels via an essentially independent optical "carrier" wave within the waveguide, so that many signals can be transmitted simultaneously, or "multiplexed", by the optical fiber. In the receiver, these signals can be separated by wavelength selective filters and routed to different detectors. Additional laser diodes may be used in the receiver as local oscillators (LO) for heterodyne detection of the optical signals. Let us now consider the advantages of an optical fiber interconnector like that shown in Fig. 1.1.

1.1.1 Comparison of Optical Fibers with Other Interconnectors

For many years, the standard means of interconnecting electrical subsystems, including integrated circuits, has been either the metallic wire or the radio link through the air. The optical fiber waveguide has many advantages over either of these conventional methods. The most important of these have been listed in Table 1.1 for easy reference, and they are discussed further below.

In modern electronic systems, such as those found in aircraft or in groundbased communications systems, it is often necessary to run bundles of wires over considerable distances. These wires can act as receiving antennas, in which extraneous signals are generated by induction from the electromagnetic fields that surround the wire. These fields may be, for example, stray fields from adjacent wires, radio waves in the surrounding environment, or gamma radiation released during a nuclear explosion. In such applications as airborne radar, missile guidance, high-voltage power line fault sensing, and multichannel telecommunications, it is critically important that the system continue to operate normally in the presence of severe electromagnetic interference

Table 1.1. Comparative evaluation of optical interconnectors

Advantages
Immunity from electromagnetic interference (EMI)
Freedom from electrical short circuits or ground loops
Safety in combustible environment
Security from monitoring
Low-loss transmission
Large bandwidth (i.e., multiplexing capability)
Small size, light weight
Inexpensive, composed of plentiful materials

Major disadvantage
Cannot be used for electrical power transmission

(EMI). Metallic wires can, of course, be shielded, as in the case of coaxial cables, but the metallic shield adds weight, is costly, and produces parasitic capacitance that limits the frequency response or the bandwidth. The optical fiber waveguide has inherent immunity to most forms of EMI, since there is no metallic wire present in which current can be induced by stray electromagnetic coupling. In addition, it is easy to exclude undesired light waves by covering the fiber (or fiber bundle) with an opaque coating. "Cross talk", or interference between the signals carried on adjacent optical fibers in a bundle, is also minimal because each waveguiding core of the fibers is surrounded by a relatively thick cladding, through which the fields of the guided optical wave cannot penetrate. The immunity that guided optical waves have to EMI is sufficient reason, of itself, to prefer the use of optical fiber interconnectors instead of wires or cables in many applications. However, as we shall see, there are also quite a few other advantages to be gained by their use.

Unlike metal wires, optical fibers do not allow flow of electrical current, so that electrical short circuits cannot occur. Thus, optical fibers can be bound into a tight bundle and can be routed through metallic conduits without concern for electrical insulation. Optical fiber interconnections are particularly useful in high-voltage applications, such as transmitting telemetry data from, and control signals to, power transmission lines and switchgear. In this case, the insulating properties of the optical fiber itself eliminate the need for expensive isolation transformers. Since no spark is created when an optical fiber breaks, these fibers can also be used advantageously in combustible, or even explosive, environments.

In military and other high-security applications, optical fibers provide more immunity from *tapping* or monitoring than do wires or radio links. This is because there is no electromagnetic field extending outside of the optical fiber from which a signal can be tapped by electromagnetic induction in a pickup loop or antenna. In order to tap a signal from an optical fiber waveguide, one must somehow pierce the light-confining cladding of the fiber. This is very difficult to do without disrupting the light-wave transmission within the fiber to an extent which can be detected, thus easily permitting appropriate countermeasures.

Perhaps the most important advantage of fibers is that they can be used for low-loss transmission of optical signals over 100 kilometer-length paths with bandwidths greater than 20 Gigabit/s at the present state-of-the-art [1.7]. The loss can be reduced to less than 2 dB/km, even in relatively inexpensive, commercially available fibers. In the laboratory, fibers with attenuation of less than 0.2 dB/km have been demonstrated [1.8]. The losses in fibers are relatively independent of frequency, while those of competitive interconnectors increase rapidly with increasing frequency. For example, the data shown in Fig. 1.2 indicate losses in twisted-pair cable, which is commonly used as an interconnector in avionic systems, increase substantially at modulation frequencies above about 100 kHz. Coaxial cables are useful for transmission over relatively short paths, at frequencies up to about 100 MHz, even though losses

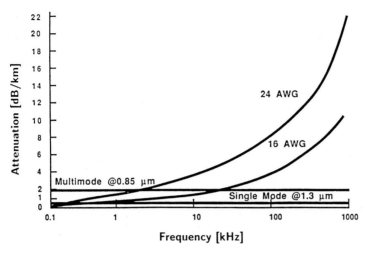

Fig. 1.2. Comparison of attenuation in optical fiber with that in twisted-pair cable

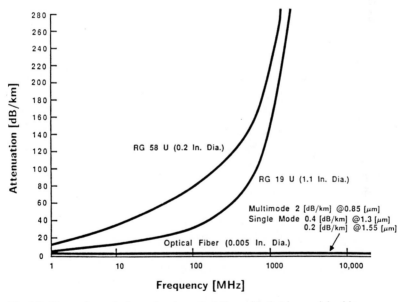

Fig. 1.3. Comparison of attenuation in optical fiber with that in coaxial cable

are large, but above that frequency losses become excessive, as shown in Fig. 1.3. By comparison, attenuation in fibers is insignificant even at frequencies up to 10 GHz. The maximum frequency at which fibers can be used to transmit a signal is limited not by attenuation, per se, but rather indirectly by the phenomenon of dispersion.

The least expensive optical fibers that are readily available today are multimode fibers, in which the light waves propagate simultaneously in a number of different optical modes. Since each of these modes has a different group velocity, a pulse of light traveling along a multimode fiber is broadened. Transformation of this time-domain pulse broadening to the frequency domain results in a corresponding bandwidth product of about 200 MHz km for currently available fibers. Modal dispersion can be avoided, of course, by using a single-mode fiber, in which the core diameter is made very small ($\leqq 10$ μm for visible or near ir wavelength) to cut off propagation of the higher-order modes. In that case, the bandwidth is limited only by the material dispersion, or variation of refractive index in the core with wavelength, and bandwidths greater than 20 GHz km have been achieved [1.7]. Unfortunately, single-mode fibers are relatively expensive compared to the multimode type, and coupling and splicing problems are greatly aggravated by the small core diameter. An alternative approach is to use a graded index multimode fiber in which the refractive index of the core is graded from a maximum on axis to a minimum at the interface with the cladding. This grading of the index tends to equalize the effects of modal dispersion and it can be used to produce multimode fibers with bandwidths up to 3 GHz km [1.9].

In most applications, the large bandwidth of optical fibers will not be used to transmit a lone signal of that bandwidth, but rather to multiplex many signals of smaller bandwidth onto the same carrier light wave. This multiplexing capability, combined with the fact that the fiber diameter is typically hundreds of times smaller than that of a coaxial cable, means that the number of information channels per cross sectional area is on the order of 10^4 times larger when optical fibers are used. This is an important consideration when space is limited as in the case of aircraft, ships, or conduits under the streets of large cities.

A number of ways in which optical fibers offer improved performance over their metallic counterparts have been described. However, the replacement of copper wires or metallic rf waveguides by fibers also can be an effective means of reducing cost in many applications. The basic materials from which optical fibers are fabricated, glasses and plastics, are more abundant and less costly than copper. Thus, once development costs are recovered, fibers are inherently less expensive to produce than copper wires or cables. In addition, fibers have a diameter hundreds of times smaller than that of a coaxial cable, thus resulting in less volume and weight. As mentioned previously, the wide bandwidth of fibers permits the multiplexing of hundreds of information channels on each one, thus further reducing size and weight requirements. Savings in size and weight accomplished by using fibers instead of copper wires or cables can lead to a significant reduction of operating costs, as well as reducing the initial cost of system fabrication. For example, aircraft fuel consumption depends strongly on weight. A decrease of several hundred pounds in the total weight can mean thousands of dollars worth of fuel saved over the lifetime of the aircraft.

In discussing the various advantages of optical fibers, we already noted that large signal bandwidths on the order of ten gigahertz could be transmitted. However, it is not sufficient to have the capability of transmitting such signals if we do not also possess the ability to generate and process them. Electrical integrated circuits offer little hope of operating at frequencies above about 1 GHz, because elements linked together by wires or other forms of metallic interconnections inevitably have stray inductances and capacitances associated with them that limit frequency response. Thus, we are led to the concept of an optical integrated circuit in which information is transported by a beam of light.

1.1.2 Comparison of Optical Integrated Circuits with Electrical Integrated Circuits

The optical integrated circuit has a number of advantages when compared to either its counterpart, the electrical integrated circuit, or to conventional optical signal processing systems composed of relatively large discrete elements. The major advantages of the OIC are enumerated in Table 1.2.

The optical integrated circuit inherently has the same large characteristic bandwidth as the optical fiber because, in both cases, the carrier medium is a lightwave rather than an electrical current. Thus, the frequency limiting effects of capacitance and inductance can be avoided. The design and fabrication of a large scale OIC with a bandwidth to match than of an optical fiber, while feasible in principle, probably will require many years of technology development. However, practical applications of OIC's have already been accomplished (Chap. 17), and the future is promising.

It should be possible to multiplex hundreds of signals onto one optical waveguide channel by using the frequency division multiplexing scheme that has been diagrammed in Fig. 1.1. A four-channel DFB laser transmitter of this type has been produced by *Woolnough* et al. [1.10], incorporating DFB lasers along with JFET drive electronics on an InP substrate. The coupling of

Table 1.2. Comparative evaluation of optical integrated circuits

Advantage
Increased bandwidth
Expanded frequency (wavelength) division multiplexing
Low-loss couplers, including bus access types
Expanded multipole switching (number of poles, switching speed)
Smaller size, weight, lower power consumption
Batch fabrication economy
Improved reliability
Improved optical alignment, immunity to vibration

Major disadvantage
High cost of developing new fabrication technology

many optical signals into one waveguide can be conveniently and efficiently accomplished in the optical integrated circuit. Dual-channel directional couplers like those shown in Fig. 1.1 have been demonstrated to have coupling efficiency approaching 100% [1.11]. To implement the same type of bus access coupling to an optical fiber waveguide by using large discrete components such as the combiner-tap [1.12] would involve losses of at least 1 dB per insertion.

In addition to facilitating the coupling of many signals onto a waveguide, the optical integrated circuit also lends itself to convenient switching of signals from one waveguide to another. As shown diagrammatically in Fig. 1.4, this can be accomplished by using electro-optic switching. A deposited metal plate on top of, or between, the channels of a dual channel directional coupler can be used to control the transfer of optical power [1.13]. Electro-optic switches of this type are now commercially available from a number of different suppliers [1.14].

When compared to larger, discrete component optical systems, OIC's can be expected to have the same advantages that electrical integrated circuits enjoy over hand-wired discrete component circuits. These include smaller size, weight and lower power requirements, as well as improved reliability and batch fabrication economy. In addition, optical alignment and vibration sensitivity, which are difficult problems in discrete component optical systems, are conveniently controlled in the OIC.

Enumeration of the many advantages of optical integrated circuits and of optical fiber interconnections suggests that integrated optics may completely replace conventional electronics, and indeed some persons do hold that view. However, the prevalent opinion is that it is more likely that integrated-optic systems will substantially augment, but not entirely replace, electronic ones. The relatively high cost of developing the technology of OIC's may probably limit their application to situations where the superior performance obtainable

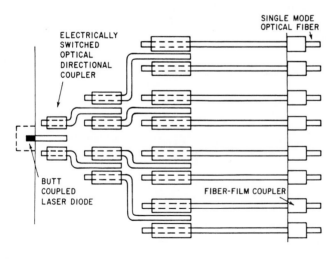

Fig. 1.4. Optical integrated circuit directional coupler tree with electro-optic switching

can justify the expense. Because optical fiber waveguides ultimately offer cost savings as well as performance improvements, they may very well replace wires and cables for most signal transmission. However, they cannot be used for low-frequency electrical energy transmission. Power transmission lines will most likely remain metallic. At present, integrated optics is still a rapidly developing field, and new applications are certain to emerge as the technology improves.

1.2 Substrate Materials for Optical Integrated Circuits

The choice of a substrate material on which to fabricate an optical integrated circuit depends most strongly on the function to be performed by the circuit. In most cases, the OIC may consist of a number of different optical devices such as sources, modulators, and detectors, and no one substrate material will be optimum for all of them. Thus, a compromise must be made. The first step is to decide whether a hybrid or a monolithic approach is preferred, as elaborated below.

1.2.1 Hybrid Versus Monolithic Approach

There are two basic forms of optical integrated circuits. One of these is the hybrid, in which two or more substrate materials are somehow bonded together to optimize performance for different devices. The other is the monolithic OIC, in which a single substrate material is used for all devices. Since most OIC's will require a source of light, monolithic circuits can only be fabricated in optically active materials, such as the semiconductors listed in Table 1.3. Passive materials like quartz or lithium niobate are also useful as substrate materials but generally an external light source, such as a semiconductor laser, must somehow be optically and mechanically coupled to the substrate. However, in recent years significant progress has been made in producing light emitters and amplifiers by incorporating erbium ions into passive substrate materials such as glasses and polymers, suggesting that it

Table 1.3. Substrate materials for optical integrated circuits

Passive (incapable of light generation)	*Active* (capable of light generation)
quartz	gallium arsenide
lithium niobate	gallium aluminum arsenide
lithium tantalate	gallium arsenide phosphide
tantalum pentoxide	gallium indium arsenide
niobium pentoxide	other III–V and II–VI direct bandgap
silicon	semiconductors
polymers	

may be practical to make monolithic OIC's in them. OIC's of this type are discussed in Sect. 1.2.4.

The major advantage of the hybrid approach is that the OIC can be fabricated using existing technology, piecing together devices which have been substantially optimized in a given material. For example, one of the earliest OIC's to perform a complex system function was the rf spectrum analyzer, which combined a commercially available GaAlAs diode laser and a silicon photodiode detector array with an acousto-optic modulator on a lithium niobate substrate [1.15, 16]. A hybrid butt-coupling approach was used to efficiently couple both the laser diode and the detector array to the LiNbO$_3$ substrate. In this case, the hybrid approach made possible the combining of already well developed technologies for GaAlAs heterojunction lasers [1.17], LiNbO$_3$ acousto-optic waveguide modulators [1.18] and Si photodiode arrays [1.19].

While the hybrid approach to OIC fabrication provides a convenient way for one to implement many desired functions, it has the disadvantage that the bonds holding the various elements of the circuit together are subject to misalignment, or even failure, because of vibration and thermal expansion. Also, the monolithic approach is ultimately cheaper if mass production of the circuit is desired, because automated batch processing can be used. This yields a low perunit cost after design and development costs are recovered. For these reasons, monolithic OIC's are likely to become the most common type in use once the technology has matured, even though the first commercially available OIC's have been produced in LiNbO$_3$ using the hybrid approach.

1.2.2 III–V and II–VI Ternary Systems

Most monolithic OIC's can be fabricated only in active substrates, in which light emitters can be formed. This essentially limits the choice of materials to semiconductors, such as those listed in Table 1.3. The III–V (or II–VI) ternary or quaternary compounds are particularly useful because the energy bandgap of the material can be changed over a wide range by altering the relative concentrations of elements. This feature is very important to the solution of one of the basic problems of monolithic OIC fabrication. Semiconductors characteristically emit light at a wavelength corresponding approximately to their bandgap energy. They also very strongly absorb light having a wavelength less than, or equal to, their bandgap wavelength. Thus, if a light emitter, waveguide, and detector are all fabricated in a single semiconductor substrate such as GaAs, light from the emitter will be excessively absorbed in the waveguide but not absorbed strongly enough in the detector. As will be explained in detail in following chapters, the composition of a ternary or quaternary material can be adjusted in the various component devices of the OIC to effectively eliminate these effects.

So far, most of the research in monolithic OIC's has used the gallium aluminum arsenide, $Ga_{(1-x)}Al_xAs$, system or the gallium indium arsenide phosphide $Ga_xIn_{(1-x)}As_{(1-y)}P_y$, system. The materials $Ga_{(1-x)}Al_xAs$ and $Ga_x In_{(1-x)}As_{(1-y)}P_y$, often designated by the abbreviated formulae GaAlAs and GaInAsP, have a number of properties that make them especially useful for OIC fabrication. The most important of these have been enumerated in Table 1.4. By changing the fractional atomic concentration of the constituents, the emitted wavelength can be varied from 0.65 μm (for AlAs) to 1.7 μm (for GaInAsP). GaAlAs and GaInAsP also have relatively large electro-optic and acousto-optic figures of merit, making them useful for optical switch and modulator fabrication. Widespread use of GaAlAs and GaInAsP has greatly reduced their cost as compared to that of other III–V or II–VI compounds. GaAs wafers suitable for use as OIC substrates, with diameters as large as three inches, can be purchased from a number of commercial suppliers.

The fact that GaAlAs technology is already well developed compared to that of other III-V ternaries make its use advantageous. It also has the unique property that the lattice constants of GaAs and AlAs are almost identical (5.646 and 5.369 Å, respectively [1.20]). Thus, layers of $Ga_{(1-x)}Al_xAs$ with greatly different Al concentrations can be epitaxially grown on top of each other with minimal interfacial lattice strain being introduced. This is particularly important in the fabrication of multilayered, heterojunction lasers, as will be discussed in Chap. 12. No other III–V or II–VI pair has lattice constants that are matched as well as those of GaAs and AlAs. As a result, interfacial strain is a major problem in the fabrication of multilayered devices in these materials. However, a strong desire to produce light sources emitting at wavelengths longer than those achievable with GaAlAs has led to the development of GaInAsP lattice matching technology capable of yielding highly efficient laser diodes emitting at wavelengths of 1.3 μm and 1.55 μm [1.21]. These wavelengths correspond to minimum values of absorption and dispersion in glass optical fibers used for long distance telecommunications.

Table 1.4. Properties of GaAs, GaAlAs and GaInAsP useful in optical integrated circuits

Transparency	0.6 μm to 12 μm
Emitted wavelength	0.65 μm to 1.7 μm
Lattice matching	Negligible lattice mismatch results in minimal strain
Switching	Large electro-optic and acousto-optic figures of merit
	$n_0^3 r_{41} \simeq 6\times10^{-11}\,\mathrm{m/V}$ $M = \dfrac{n_0^6 p^2}{\rho v_s^3} \simeq 10^{-13}\,\mathrm{s^3/kg}$
Technology	Epitaxy, doping, ohmic contacts, masking, etching all are well developed
Cost	Less than other III–V or II–VI materials

Work on GaAlAs and GaInAsP monolithic OIC's has already produced many of the desired optical devices in the required monolithic form. However, the level of integration so far has been limited to only a few devices per chip. Many of the monolithically integrated devices have performed as well as, or better than, their discrete counterparts. The relatively low levels of integration that have been achieved so far appear to stem from the personal preferences of the researchers involved rather than from any fundamental technical limitation. Some persons have preferred to work on heterojunction lasers, while others have concentrated on waveguides, or detectors. This specialization has been beneficial in the early years of the field, in that it has led to the invention and development of relatively sophisticated embodiments of many of the devices required in OIC's. Such devices as heterojunction, distributed feedback lasers, acousto-optic waveguide modulators, and electro-optically switched directional couplers are examples of well developed technology. In the recent years, some researchers have turned to the problem of actually integrating a number of devices on one substrate. This has led to monolithic OIC's, combining lasers, modulators or switches, and detectors [1.22–34]. Such a trend no doubt will continue in the future and should result in larger scale integration.

1.2.3 Hybrid OIC's in LiNbO$_3$

While much has been accomplished in the realm of monolithic OIC's, they have been slow to achieve commercial availability because of the very complex technology required for their fabrication. By contrast, a number of hybrid OIC's fabricated in LiNbO$_3$ are presently commercially available [1.14]. These are relatively simple, including an electro-optic modulator, a Mach-Zehnder interferometric modulator a 2×2 electro-optic switch and an optical amplifier. These OIC's take advantage of the wide wavelength range of transparency and large electro-optic coefficient of LiNbO$_3$, as well as some other beneficial properties of that material which are listed in Table 1.5.

Table 1.5. Properties of LiNbO$_3$ useful in optical integrated circuits

Transparency	0.2 μm to 12 μm low loss
Emitted wavelength	none
Switching	Large electro-optic and acousto-optic figures of merit
	$n_0^3 r_{33} \simeq 3 \times 10^{-10} \, \mathrm{m/V} \quad M = \dfrac{n_0^6 p^2}{\rho v_s^3} \simeq 7 \times 10^{-15} \, \mathrm{s^3/kg}$
Technology	Waveguide fabrication, masking, etching, polishing all are well developed
Cost	More than GaAs

In addition to the relatively simple $LiNbO_3$ OIC's that are commercially available, a number of relatively complex $LiNbO_3$ OIC's have been demonstrated in the laboratory. These OIC's, which include a fiber laser doppler velocimeter [1.35], an IO high frequency sampler [1.36] and an OIC for high-accuracy temperature measurement [1.37], will be discussed at length in Chap. 17. It is quite likely that relatively sophisticated $LiNbO_3$ OIC's of this nature will find their way to the commercial market within the next few years.

1.2.4 OIC's in Glasses and Polymers

Polymer integrated-optic devices and circuits are increasingly being considered for applications involving connectors, couplers and optical waveguide printed wire boards. Polymer-film layers are typically created by spinning or casting techniques on glass or other dielectric substrates [1.38]. Standard photolithographic techniques [1.39] are generally used to pattern the waveguide's lateral dimensions. However, direct writing with lasers or e-beams is also possible [1.40]. Polymers can be etched with solvents [1.41–42] or by reactive ion etching (RIE) [1.43–44]. Losses in etched waveguides typically range from 0.1 dB/cm at 850 nm wavelength to 0.3 dB/cm at 1300 nm. Loss of 0.3 dB/cm at 1300 nm wavelength has also been observed in single-mode polymer waveguides fabricated by injection molding of polymethymethacrylate (PMMA) substrates, with waveguide grooves (4–7 μm wide) filled with fully deuterated ethyleneglycoldimethacrylate (EGDMA) [1.45].

Polymer waveguides have been used in some fairly sophisticated optical signal transmission applications. For example, *Prakash* et al. [1.46] employed polyimide channel waveguides fabricated using direct laser writing as the interconnection media between 50 GHz modulated sources and high-speed photodetectors. They suggest that such interconnections could potentially be used in the implementation of sophisticated wafer-scale systems such as phased array radars.

The potential of polymer waveguides as interconnections between devices and/or circuit boards is also demonstrated by the eight-channel transmit-receive link fabricated by *Thomson* et al. [1.47]. This demonstration link involved a fiberglass epoxy (FR4) photonic-circuit board containing eight-channel transmitter and receiver packages, interconnected by DuPont Polyguide™ [1.38] waveguides, as shown in Fig. 1.5. The waveguide circuits include eight-channel arrays of straights cross-throughs, curves and self-aligning interconnects to multi-fiber ribbon at the edge of the board. The Polyguide waveguides were coupled to the TX and RX chips by means of 45° out-of-plane reflecting mirrors formed on the ends of the waveguides beneath the chips (see chap. 6 for more details of this coupling method.) To minimize optical losses in the curved and diagonal waveguides, the masks were made with 0.1 μm resolution. Data transmission at up to 1.25 Gbit/s per channel was achieved with this demonstration board with a bit error ratio less than 10^{-11}.

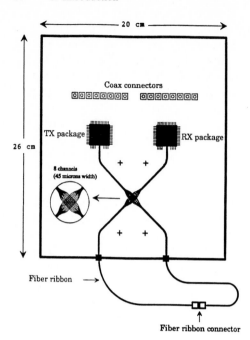

Fig. 1.5. Diagram of photonic circuit board 8-channel link demonstrator [1.47]

Adjacent channel crosstalk was −31dB, including electrical crosstalk on the receiver array chip and optical crosstalk.

Because of their low cost, there is no doubt that polymer and glass waveguides will play an important role in splitters, couplers and in interconnections between chips and boards. It would be extremely useful in regard to monolithic OIC's if light generation could also be accomplished effectively within glass or polymer layers. Some recent results are very encouraging. For example, *Oguma* et al. [1.48] have reported a tunable Er-doped glass Y-branched waveguide laser emitting at approximately 1.53 µm wavelength. The lasing wavelength was continuously tuned in the 1.532 to 1.534 µm range by controlling the electrical power to a thin film heater used as a phase shifter in one branch of the Y-branched waveguide laser. Also, an Er-doped silica-based planar waveguide amplifier has been produced by *Hattori* et al. [1.49] A maximum gain of 13.5 dB was achieved at 1.533 µm with this amplifier. The introduction of Er-doped lasers and amplifiers operating in the 1.5 µm optical-fiber telecommunications window is particularly important. Previously reported neodymium (Nd)-doped lasers and amplifiers in glass [1.50] and polymers [1.51] all have operated at 1.06 µm wavelength.

1.3 Organization of this Book

This book has been planned so as to serve both as a text for a course in integrated optics, and as a readable reference volume for scientists and engineers in industry. Hence, Chap. 1 has given an overview of the goals and basic principles of integrated optics and has compared this new approach to those alternative techniques now in use. The close relationship and interdependence between optical fiber waveguides and optical integrated circuits has been described. Thus, we are prepared to begin a detailed study of the component elements of an optical integrated circuit.

Chapters 2 through 5 are devoted to the fundamental element that literally ties the OIC together: the optical waveguide. Without effective, low-loss optical waveguides one cannot even contemplate an OIC. Hence, it is logical to begin the study of integrated optics by considering the waveguide. The theory of optical waveguides is discussed in Chaps. 2 and 3, beginning with the basic three-layer planar waveguide structure. A comparison is made between the geometric or "ray optic" approach and the electromagnetic field or "physical optic" approach. Theoretical derivations of the optical mode profiles and cutoff conditions for common waveguide geometries are compared with corresponding experimental results. Following the development of the theory of optical waveguides, techniques of waveguide fabrication are discussed in Chap. 4. Finally, in Chap. 5, optical losses in waveguides are described, and techniques used to measure those losses are discussed.

Once the design and fabrication of the optical waveguide has been considered, the next immediate question is that of efficiently coupling optical energy into or out of a waveguide. Chapter 6 describes a number of ways for accomplishing this aim. Since much of the laboratory research in OIC's involves the use of conventional (non-integrated) lasers, we consider prism and grating couplers, which are suitable for the beams of light emitted by such sources. Additionally, hybrid couplers like the transverse (end-butt) coupler and various fiber-to-waveguide couplers are described. Finally, in Chap. 7 fully integrated, monolithic waveguide-to-waveguide couplers like the dual-channel directional coupler are dealt with. This latter type of coupler is composed of two waveguiding channels spaced closely enough so that the evanescent "tails" of the optical modes overlap, thus producing coherent coupling of optical energy, in a manner analogous to that of the familiar slotted waveguide used as a coupler at microwave frequencies. As discussed in Chap. 8, the dual-channel directional coupler can also be used as an optical modulator or switch, because the strength of coupling between channels can be electrically controlled by adding properly designed electrodes.

The topic of optical beam modulation is very important to integrated optics because the wide bandwidths of optical fibers and OIC's can be utilized only if wide-band modulators are available to impress the signal information onto the optical beam. Hence, Chaps. 8 and 9 are devoted to detailed considerations of the various types of optical modulators that can be fabricated in a monolithic embodiment. These can be divided generally into two classes, electro-optic and acousto-optic, depending on whether the electrical signal is coupled to the optical beam by means of electro-optic change of the index of refraction, or by changes in index produced by acoustic waves. Of course, in the latter case, the electrical signal must first be coupled to the acoustic wave by an acoustic transducer.

While discrete (non-integrated) laser sources are adequate for laboratory research, and for some integrated optics applications, most of the benefits of an OIC technology cannot be obtained without using a monolithically integrated light source. The prime choice for such a source is the p-n junction laser diode, fabricated on a semiconductor substrate that also supports the rest of the optical integrated circuit. Because of the critical importance of this device to the field of integrated optics, and because of its complexity, Chaps. 10 through 13 have been dedicated to a thorough, systematic development of semiconductor laser diode theory and technology. Chapter 10 reviews the basic principles of light emission in semiconductors, explaining the phenomena of spontaneous and stimulated emission of photons. The important effect of the energy band structure on the emission process is discussed. In Chap. 11, attention is focused on specific types of semiconductor laser structures. The concept of optical field confinement is introduced and the development of the confined-field laser is traced up to the modern heterostructure diode laser, which will most likely become the standard light source in OIC's. Chapter 12 provides a detailed description of heterojunction, confined-field lasers. Finally, in Chap. 13, the distributed feedback (DFB) laser is discussed. This device features a diffraction grating that provides the optical feedback necessary for laser oscillation, thus eliminating the need for cleaved end-faces or other reflecting mirrors to form an optical cavity. Since efficient reflecting end surfaces are difficult to fabricate in a planar monolithic OIC, the DFB laser is an attractive alternative.

Chapter 14 may appear at first to be out of place, since it deals with the direct modulation of semiconductor lasers. However, the technique of directly modulating the light output of the laser diode by varying the input current is distinct from either electro-optic or acousto-optic modulation methods described in Chaps. 8 and 9 in that it cannot be understood without a knowledge of the characteristics of light generation in the semiconductor laser. Hence, presentation of the material in Chap. 14 has been postponed until after the explanation of the laser diode. Input current modulation is a very effective method for achieving the greater than 10 GHz modulation rates that are desired in OIC's. Hence, a thorough discussion has been given of the various techniques that have been used and the results that have been obtained to date.

Although the signal processing in an OIC is done in the optical domain, the output information is most often desired in the form of an electrical signal to interface with electronic systems. This requires and optical to electrical transducer, commonly called a photodetector. In Chap. 15 we describe various types of photodetectors that can be fabricated as elements of a monolithic OIC. The advantage of a waveguide detector as compared to the corresponding bulk device are enumerated. Also, techniques of locally modifying the effective bandgap of the substrate in the vicinity of the detector to increase conversion efficiency are described.

A recent development in the field of integrated optics has been the development of "quantum well' devices, in which a multilayer structure of very thin layers is used to create a "superlattice". This superlattice structure yields devices (e.g. lasers, detectors, modulators and switches) with improved operating characteristics, Chap. 16 is devoted to a review of these devices.

To conclude the study of integrated optics presented in this book, a number of applications of integrated optic devices and systems in consumer use or prototype testing at the present time are described in Chap. 17, and trends in the field are analysed.

Since integrated optics is a young field compared to that of electrical integrated circuits, it is difficult to envision exactly what course will be followed. However, it seems certain that many significant contributions to the technology of optical signal generation, transmission and processing will result.

2. Optical Waveguide Modes

The optical waveguide is the fundamental element that interconnects the various devices of an optical integrated circuit, just as a metallic strip does in an electrical integrated circuit. However, unlike electrical current that flows through a metal strip according to Ohm's law, optical waves travel in the waveguide in distinct optical modes. A mode, in this sense, is a spatial distribution of optical energy in one or more dimensions. In this chapter, the concept of optical modes in a waveguiding structure is discussed qualitatively, and key results of waveguide theory are presented with minimal proof to give the reader a general understanding of the nature of light propagation in an optical waveguide. Then, in Chap. 3, a mathematically sound development of waveguide theory is given.

2.1 Modes in a Planar Waveguide Structure

As shown in Fig. 2.1, a planar waveguide is characterized by parallel planar boundaries with respect to one (x) direction, but is infinite in extent in the lateral directions (z and y). Of course, because it is infinite in two dimensions, it cannot be a practical waveguide for optical integrated circuits, but it forms the basis for the analysis of practical waveguides of rectangular cross section. It has therefore been treated by a number of authors, including *McWhorter* [2.1], *McKenna* [2.2], *Tien* [2.3], *Marcuse* [2.4], *Taylor* and *Yariv* [2.5] and *Kogelnik* [2.6]. In Sect. 2.1.1 we follow the approach of *Taylor* and *Yariv* [2.5] to examine the possible modes in a planar waveguide, without fully solving the wave equation.

2.1.1 Theoretical Description of the Modes of a Three-Layer Planar Waveguide

To begin the discussion of optical modes, consider the simple three-layer planar waveguiding structure of Fig. 2.1. The layers are all assumed to be infinite in extent in the y and z directions, and layers 1 and 3 are also assumed to be semi-infinite in the x direction. Light waves are assumed to be propagating in the z direction. It has been stated previously that a mode is a spatial distribution of optical energy in one or more dimensions. An equivalent math-

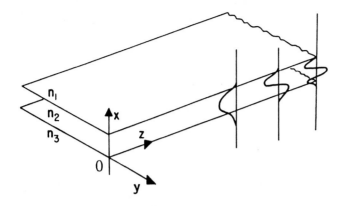

Fig. 2.1. Diagram of the basic three-layer planar waveguide structure. Three mode are shown, representing distributions of electric field in the x direction

ematical definition of a mode is that it is an electromagnetic field which is a solution of Maxwell's wave equation

$$\nabla^2 E(r, t) = [n^2(r)/c^2]\partial^2 E(r, t)/\partial t^2, \tag{2.1.1}$$

where E is the electric field vector, r is the radius vector, $n(r)$ is the index of refraction, and c is the speed of light in a vacuum. For monochromatic waves, the solutions of (2.1.1) have the form

$$E(r, t) = E(r)e^{i\omega t}, \tag{2.1.2}$$

where ω is the radian frequency. Substituting (2.1.2) into (2.1.1) we obtain

$$\nabla^2 E(r) + k^2 n^2(r)E(r) = 0, \tag{2.1.3}$$

where $k \equiv \omega/c$. If we assume, for convenience, a uniform plane wave propagating in the z direction, i.e., $E(r) = E(x, y)\exp(-i\beta z)$, β being a propagation constant, then (2.1.3) becomes

$$\partial^2 E(x,y)/\partial x^2 + \partial^2 E(x,y)/\partial y^2 + [k^2 n^2(r) - \beta^2]E(x,y) = 0. \tag{2.1.4}$$

Since the waveguide is assumed infinite in the y direction, by writing (2.1.4) separately for the three regions in x, we get

$$
\begin{aligned}
&\text{Region 1} \quad \partial^2 E(x,y)/\partial x^2 + (k^2 n_1^2 - \beta^2)E(x,y) = 0 \\
&\text{Region 2} \quad \partial^2 E(x,y)/\partial x^2 + (k^2 n_2^2 - \beta^2)E(x,y) = 0 \\
&\text{Region 3} \quad \partial^2 E(x,y)/\partial x^2 + (k^2 n_3^2 - \beta^2)E(x,y) = 0,
\end{aligned}
\tag{2.1.5}
$$

where $E(x, y)$ is one of the Cartesian components of $E(x, y)$. The solutions of (2.1.5) are either sinusoidal or exponential functions of x in each of the regions, depending on whether $(k^2 n_i^2 - \beta^2)$, $i = 1, 2, 3$, is greater than or less than zero. Of course, $E(x, y)$ and $\partial E(x, y)/\partial x$ must be continuous at the interface between layers. Hence the possible modes are limited to those shown in Fig. 2.2.

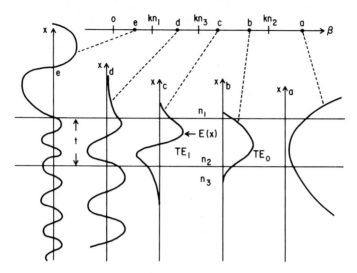

Fig. 2.2. Diagram of the possible modes in a planar waveguide [2.5]

Consider how the mode shape changes as a function of β, for the case of constant frequency ω and $n_2 > n_3 > n_1$. This relative ordering of the indices is quite a common case, corresponding, for example, to a waveguiding layer of index n_2 formed on a substrate with smaller index n_3, surrounded by air of index n_1. As we will see in Chap. 3, it is a necessary condition for waveguiding in Layer 2 that n_2 be greater than both n_1 and n_3. When $\beta > kn_2$, the function $E(x)$ must be exponential in all three regions and only the mode shape shown as (a) in Fig. 2.2 could satisfy the boundary conditions of $E(x)$ and $\partial E(x)/\partial x$ being continuous at the interfaces. This mode is not physically realizable because the field increases unboundedly in Layers 1 and 3, implying infinite energy. Modes (b) and (c) are well confined guided modes, generally referred to as the zeroth order and first order transverse electric modes, TE_0 and TE_1 [2.7]. For values of β between kn_2 and kn_3 such modes can be supported. If β is greater than kn_1, but less than kn_3, a mode like that in (d) will result. This type of mode, which is confined at the air interface but sinusoidally varying at the substrate, is often called a *substrate radiation* mode. It can be supported by the waveguide structure, but because it is continually losing energy from the waveguiding Region 2 to the substrate Region 3 as it propagates, it tends to be damped out over a short distance. Hence it is not very useful in signal trans-mission, but, in fact, it may be very useful in coupler applications such as the tapered coupler. This type of coupler will be discussed in Chap. 6. If β is less than kn_1, the solution for $E(x)$ is oscillatory in all three regions of the waveguide structure. These modes are not guided modes because the energy is free to spread out of the waveguiding Region 2. They are generally referred to as the air radiation modes of the waveguide structure. Of course, radiation is also occurring at the substrate interface.

2.1.2 Cutoff Conditions

We shall see in Chap. 3, when (2.1.1) is formally solved, subject to appropriate boundary conditions at the interface, that β can have any value when it is less than kn_3, but only discrete values of β are allowed in the range between kn_3 and kn_2. These discrete values of β correspond to the various modes TE_j, $j = 0, 1, 2, \ldots$ (or TM_k, $k = 0, 1, 2, \ldots$). The number of modes that can be supported depends on the thickness t of the waveguiding layer and on ω, n_1, n_2 and n_3. For given t, n_1, n_2, and n_3 there is a cutoff frequency ω_c below which waveguiding cannot occur. This ω_c corresponds to a long wavelength cutoff λ_c.

Since wavelength is often a fixed parameter in a given application, the cutoff problem is frequently stated by asking the question, "for a given wavelength, what indices of refraction must be chosen in the three layers to permit waveguiding of a given mode?" For the special case of the so-called asymmetric waveguide, in which n_1 is very much less than n_3, it can be shown (Chap. 3) that the required indices of refraction are related by

$$\Delta n = n_2 - n_3 \perp (2m+1)^2 \lambda_0^2 / (32 n_2 t^2), \tag{2.1.6}$$

where the mode number $m = 0, 1, 2, \ldots$, and λ_0 is the vacuum wavelength. The change in index of refraction required for waveguiding of the lower-order modes is surprisingly small. For example, in a gallium arsenide waveguide with n_2 equal to 3.6 [2.8] and with t on the order of λ_0, (2.1.6) predicts that a Δn on the order of only 10^{-2} is sufficient to support waveguiding of the TE_0 mode.

Because only a small change in index is needed, a great many different methods of waveguide fabrication have proven effective in a variety of substrate materials [2.9]. The more important of these have been listed in Table 2.1 so that the reader will be familiar with the names of the techniques when they are mentioned in the following discussion of experimental observations of waveguide performance. A thorough explanation of the methods of waveguide fabrication is given in Chap. 4.

Table 2.1. Methods of fabricating waveguides for optical integrated circuits

1) Deposited thin films (glass, nitrides, oxides, organic polymers)
2) Photoresist films
3) Ion bombarded glass
4) Diffused dopant atoms
5) Heteroepitaxial layer growth
6) Electro-optic effect
7) Metal film stripline
8) Ion migration
9) Reduced carrier concentration in a semiconductor
   a) epitaxial layer growth
   b) diffusion counterdoping
   c) ion implantation counterdoping or compensation

2.1.3 Experimental Observation of Waveguide Modes

Since the waveguides in optical integrated circuits are typically only a few micrometers thick, observation of the optical mode profile across a given dimension cannot be accomplished without a relatively elaborate experimental set-up, featuring at least $1000 \times$ magnification. One such system [2.10], which works particularly well for semiconductor waveguides, is shown in Fig. 2.3. The sample, with its waveguide at the top surface, is fixed atop an x-y-z micropositioner. Microscope objective lenses, used for input beam coupling and output image magnification, are also mounted on micropositioners to facilitate the critical alignment that is required. The light source is a gas laser, emitting a wavelength to which the waveguide is transparent. For example, a helium-neon laser operating at 1.15 μm is good for GaAs, GaAlAs and GaP waveguides, while one emitting at 6328 Å can be used for GaP but not for GaAlAs or GaAs. For visual observation of the waveguide mode, the output face of the waveguide can be imaged onto either a white screen or an image converter (IC) screen depending on whether visible wavelength or infrared (ir) light is used. The lowest order mode ($m = 0$) appears as a single band of light, while higher order modes have a correspondingly increased number of bands, as shown in Fig. 2.4. The light image appears as a band rather than a spot because it is confined by the waveguide only in the x direction. Since the waveguide is much wider than it is thick the laser beam is essentially free to diverge in the y direction.

To obtain a quantitative display of the mode profile, i.e. optical power density vs. distance across the face of the waveguide, a rotating mirror is used to scan the image of the waveguide face across a photodetector that is masked to a narrow slit input. The electrical signal from the detector is then fed to the

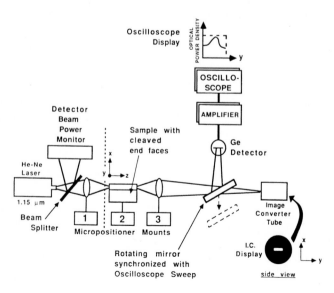

Fig. 2.3. Diagram of an experimental setup than can be used to measure optical mode shapes [2.10]

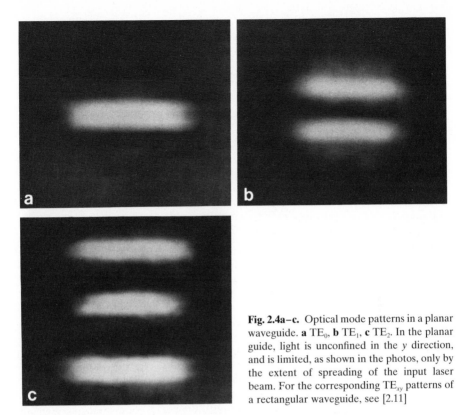

Fig. 2.4a–c. Optical mode patterns in a planar waveguide. **a** TE$_0$, **b** TE$_1$, **c** TE$_2$. In the planar guide, light is unconfined in the y direction, and is limited, as shown in the photos, only by the extent of spreading of the input laser beam. For the corresponding TE$_{xy}$ patterns of a rectangular waveguide, see [2.11]

vertical scale of an oscilloscope on which the horizontal sweep has been synchronized with the mirror scan rate. The result is in the form of graphic displays of the mode shape, like those shown in Fig. 2.5. Note that the modes have the theoretically predicted sinusoidal-exponential shape, by remembering that what is observed is optical power density, or intensity, which is proportional to E^2. Details of the mode shape, like the rate of exponential decay (or extinction) of the evanescent "tail" extending across the waveguide-substrate and waveguide-air interfaces, depend strongly on the values of Δn at the interface. As can be seen in Fig. 2.5, the extinction is much sharper at the waveguide-air interface where $\Delta n \cong 2.5$ than at the waveguide-substrate plane where $\Delta n \cong 0.01$–0.1.

A system like that shown in Fig. 2.3 is particularly useful for analysis of mode shapes in semiconductor waveguides, which generally support only one or two modes because of the relatively small Δn at the waveguide-substrate interface. Generally, the position of the focused input laser beam can be moved toward the center of the waveguide to selectively pump the zeroth order mode, or toward either the air or substrate interface to select the first order mode. It becomes very difficult to visually resolve the light bands in the case of higher-order, multimode waveguides because of spatial overlapping, even though the modes may be electromagnetically distinct and non-coupled

TE$_0$ and TE$_1$ MODE PROFILES

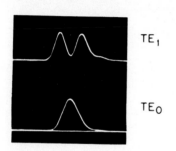

AIR ◄━┤GUIDE├━► SUB.

Fig. 2.5. Optical mode shapes are measured using the apparatus of Fig. 2.3. The waveguide in this case was formed by proton implantation into a gallium arsenide substrate to produce a 5 μm thick carrier-compensated layer [2.12]

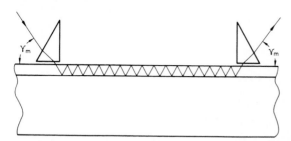

Fig. 2.6. The prism coupler used as a device for modal analysis

one to another. Waveguides produced by depositing thin films of oxides, nitrides or glasses onto glass or semiconductor substrates usually are multimode, supporting 3 or more modes, because of the larger waveguide-substrate Δn [2.13–15]. For waveguides of this type, a different experimental technique, employing prism coupling, is most often used to analyze the modes.

The prism coupler will be discussed in detail in Chap. 6. At this point it suffices to say that the prism coupler has the property that it selectively couples light into (or out of) a particular mode, depending on the angle of incidence (or emergence). The mode-selective property of the prism coupler, which is illustrated in Fig. 2.6, results from the fact that light in each mode within a waveguide propagates at a different velocity, and continuous phase-matching is required for coupling. The particular angle of incidence required to couple light into a given mode or the angle of emergence of light coupled out of a given mode can both be accurately calculated from theory, as will be seen in Chap. 6. The prism coupler can thus be used to analyze the modes of a waveguide. This can be done in two ways.

In one approach, the angle of incidence of a collimated, monochromatic laser beam on an input coupler prism is varied and the angles for which a propagating optical mode is introduced into the waveguide are noted. The propagation of optical energy in the waveguide can be observed by merely placing a photodetector at the output end of the waveguide. One can then determine which modes the waveguide is capable of supporting by calculating from the angle of incidence data.

An alternative method uses the prism as an output coupler. In this case, monochromatic light is introduced into the waveguide in a manner so as to excite all of the waveguide modes. For example, a diverging laser beam, either from a semiconductor laser, or from a gas laser beam passed through a lens to produce divergence, is focused onto the input face of the waveguide. Since the light is not collimated, but rather enters the waveguide at a variety of angles, some energy is introduced into all of the waveguide modes for which the waveguide is above cutoff at the particular wavelength used. If a prism is then used as an output coupler, light from each mode emerges from the prism at a different angle. Again, the particular modes involved can be determined by calculation from the emergence angle data. Since the thickness of the waveguide is much less than its width, the emerging light from each mode appears as a band, producing a series of so-called "m" lines [2.16] as shown in Fig. 2.7, corresponding to the particular mode number.

When the prism coupler is used to analyze the modes of a waveguide, the actual mode shape, or profile, cannot be determined in the same way as that of the scanning mirror approach of Fig. 2.3. However, the prism coupler method lets one determine how many modes can be supported by a multimode waveguide, and, as will be seen in Chap. 6, the phase velocity (hence the

Fig. 2.7. Photograph of "m" lines produced by prism coupling of light out of a planar waveguide. (Photo courtesy of U.S. Army ARRADCOM, Dover, NJ)

effective index of refraction) for each mode can be calculated from incidence and emergence angle data.

2.2 The Ray-Optic Approach to Optical Mode Theory

In Sect. 2.1, we considered the propagation of light in a waveguide as an electromagnetic field which mathematically represented a solution of Maxwell's wave equation, subject to certain boundary conditions at the interfaces between planes of different indices of refraction. Plane waves propagating along the z direction, supported one or more optical modes. The light propagating in each mode traveled in the z direction with a different phase velocity, which is characteristic of that mode. This description of wave propagation is generally called the physical-optic approach. An alternative method, the so-called ray-optic approach [2.6, 17–19], is also possible but provides a less complete description. In this latter formulation, the light propagating in the z direction is considered to be composed of plane waves moving in zig-zag paths in the x-z plane undergoing total internal reflection at the interfaces bounding the waveguide. The plane waves comprising each mode travel with the same phase velocity. However, the angle of reflection in the zig-zag path is different for each mode, making the z component of the phase velocity different. The plane waves are generally represented by rays drawn normal to the planes of constant phase as shown in Fig. 2.8, which explains the name *ray-optic*.

2.2.1 Ray Patterns in the Three-Layer Planar Waveguide

The ray patterns shown in Fig. 2.8 correspond to two modes, say the TE_0 and TE_1, propagating in a three layer waveguide with $n_2 > n_3 > n_1$. The electric (E) and magnetic (H) fields of these plane waves traveling along zig-zag paths would add vectorially to give the E and H distributions of the waves comprising the same two modes, propagating in the z direction, that were described by the physical-optic model of Sect. 2.1. Both the ray-optic and physical-optic formulations can be used to represent either TE waves, with components E_y, H_z, and H_x, or TM waves, with components H_y, E_z and E_x.

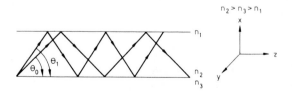

Fig. 2.8. Optical ray pattern within a multimode planar waveguide

The correlation between the physical-optic and ray-optic approaches can be seen by referring back to (2.1.5). The solution to this equation in the waveguiding Region 2 has the form [2.5]:

$$E_y(x, z) \propto \sin(hx + \gamma), \tag{2.1.7}$$

where a TE mode has been assumed, and where h and γ are dependent on the particular waveguide structure. Substituting (2.1.7) into (2.1.5) for Region 2, one obtains the condition

$$\beta^2 + h^2 = k^2 n_2^2. \tag{2.1.8}$$

Remembering that $k \equiv \omega/c$, it can be seen that β, h and kn_2 are all propagation constants, having units of (length)$^{-1}$. A mode with a z direction propagation constant β_m and an x direction propagation constant h can thus be represented by a plane wave travlling at an angle $\theta_m = \tan^{-1}(h/\beta_m)$ with respect to the z direction, having a propagation constant kn_2, as diagrammed in Fig. 2.9. Since the frequency is constant, $kn_2 \equiv (\omega/c)n_2$ is also constant, while θ_m, β_m and h are all parameters associated with the mth mode, with different values for different modes.

To explain the waveguiding of light in a planar three-layer guide like that of Fig. 2.8 by the ray-optic method, one needs only Snell's law of refraction, coupled with the phenomenon of total internal reflection. For a thorough discussion of these basic concepts of optics see, for example, *Condon* [2.20] or *Billings* [2.21]. Consider a ray of light propagating within a three-layer waveguide structure as shown in Fig. 2.10. The light rays of Fig. 2.10a, b and c correspond to a radiation mode, a substrate mode, and a guided mode, respectively. The angles of incidence and refraction, φ_i, with $i = 1, 2, 3$, are measured with respect to the normals to the interface planes, as is common practice in optics. From Snell's law

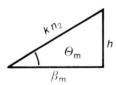

Fig. 2.9. Geometric (vectorial) relationship between the propagation constants of an optical waveguide

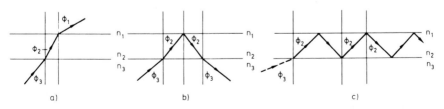

Fig. 2.10a–c. Optical ray patterns for **a** air radiation modes; **b** substrate radiation modes; **c** guided mode. In each case a portion of the incident light is reflected back into layer 3; however, that ray has been omitted from the diagrams

$$\sin \varphi_1/\sin \varphi_2 = n_2/n_1 \qquad (2.1.9)$$

and

$$\sin \varphi_2/\sin \varphi_3 = n_3/n_2. \qquad (2.1.10)$$

Beginning with very small angles of incidence, φ_3, near zero, and gradually increasing φ_3, we find the following behavior. When φ_3 is small, the light ray passes freely through both interfaces, suffering only refraction, as in Fig. 2.10a. This case corresponds to the radiation modes discussed in Sect. 2.1. As φ_3 is increased beyond the point at which φ_2 exceeds the critical angle for total internal reflection at the $n_2 - n_1$ interface, the light wave becomes partially confined as shown in Fig. 2.10b, corresponding to a substrate radiation mode. The condition for total internal reflection at the $n_2 - n_1$ interface is given by [2.21]

$$\varphi_2 \gtrsim \sin^{-1}(n_1/n_2). \qquad (2.1.11)$$

or, combining (2.1.11) and (2.1.10),

$$\varphi_3 \gtrsim \sin^{-1}(n_1/n_3). \qquad (2.1.12)$$

As φ_3 is further increased beyond the point at which φ_2 also exceeds the critical angle for total internal reflection at the $n_2 - n_3$ interface, the lightwave becomes totally confined, as shown in Fig. 2.10c, corresponding to a guided mode. In this case, the critical angle is given by

$$\varphi_2 \gtrsim \sin^{-1}(n_3/n_2), \qquad (2.1.13)$$

or, combining (2.1.13) and (2.1.10),

$$\varphi_3 \gtrsim \sin^{-1}(1) = 90°. \qquad (2.1.14)$$

The conditions given by (2.1.11) and (2.1.13) for determining what type of modes can be supported by a particular waveguide as a function of φ_2 are exactly equivalent to the conditions given by (2.1.5) as a function of β. For example, (2.1.5) indicates that only radiation modes result for β less than kn_1. Referring to Fig. 2.9, note that,

$$\sin \varphi_2 \gtrsim \beta/kn_2. \qquad (2.1.15)$$

Thus, if $\beta \lesssim kn_1$,

$$\sin \varphi_2 \lesssim kn_1/kn_2 = n_1/n_2, \qquad (2.1.16)$$

which is the same condition given by (2.1.11). Similarly, if β is greater than kn_1 but less than kn_3, (2.1.5) indicates that substrate radiation modes will be supported. Only when $\beta \gtrsim kn_3$, can confined waveguide modes occur. From Fig. 2.9, if $\beta \gtrsim kn_3$,

$$\sin \varphi_2 = \beta/kn_2 \gtrsim kn_3/kn_2 \gtrsim n_3/n_2. \qquad (2.1.17)$$

Equation (2.1.17), obtained from physical-optic theory, is merely a repeat of (2.1.13) that resulted from the ray-optic approach. Finally, if β is greater than kn_2,

$$\sin\varphi_2 = \beta/kn_2 \gtrless 1. \tag{2.1.18}$$

Equation (2.1.18) is, of course, a physically unrealizable equality, corresponding to the physically unrealizable "a" type of modes of Fig. 2.2. Thus an equivalence has been demonstrated between the ray-optic and physical-optic approaches in regard to the determination of mode type.

2.2.2 The Discrete Nature of the Propagation Constant β

The correspondence between the ray-optic and physical optic formalisms extends beyond merely determining what type modes can be supported. It has been mentioned previously, and will be demonstrated mathematically in Chap. 3, that the solution of Maxwell's equation subject to the appropriate boundary conditions requires that only certain discrete values of β are allowed. Thus, there are only a limited number of guided modes that can exist when β is in the range

$$kn_3 \leq \beta \leq kn_2. \tag{2.1.19}$$

This limitation on β can be visualized quite conveniently using the ray-optic approach. The plane wavefronts that are normal to the zig-zag rays of Fig. 2.8 are assumed to be infinite, or at least larger than the cross section of the waveguide that is intercepted; otherwise they would not fit the definition of a plane wave, which requires a constant phase over the plane. Thus, there is much overlapping of the waves as they travel in the zig-zag path. To avoid decay of optical energy due to destructive interference as the waves travel through the guide, the total phase change for a point on a wavefront that travels from the $n_2 - n_3$ interface to the $n_2 - n_1$ interface and back again must be a multiple of 2π. This leads to the condition,

$$2kn_2t\sin\theta_m - 2\varphi_{23} - 2\varphi_{21} = 2m\pi, \tag{2.1.20}$$

where t is the thickness of the waveguiding Region 2, θ_m is the angle of reflection with respect to the z direction, as shown in Fig. 2.8, m is the mode number, and φ_{23} and φ_{21} are the phase changes suffered upon total internal reflection at the interfaces. The phases $-2\varphi_{23}$ and $-2\varphi_{21}$ represent the Goos-Hänchen shifts [2.22, 23]. These phase shifts can be interpreted as penetration of the zig-zag ray (for a certain depth δ) into the confining layers 1 and 3 before it is reflected [Ref. 2.6, pp. 25-29].

The values of φ_{23} and φ_{21} can be calculated from [2.24]:

$$\tan\varphi_{23} = (n_2^2\sin^2\varphi_2 - n_3^2)^{1/2} /(n^2\cos\varphi_2)$$
$$\tan\varphi_{21} = (n_2^2\sin^2\varphi_2 - n_1^2)^{1/2} /(n^2\cos\varphi_2) \tag{2.1.21}$$

for TE waves, and

$$\tan\varphi_{23} = n_2^2(n_2^2\sin^2\varphi_2 - n_3^2)^{1/2} /(n_3^2 n_2\cos\varphi_2)$$
$$\tan\varphi_{21} = n_2^2(n_2^2\sin^2\varphi_2 - n_1^2)^{1/2} /(n_1^2 n_2\cos\varphi_2) \tag{2.1.22}$$

for TM waves.

It can be seen that substitution of either (2.1.21) or (2.1.22) into (2.1.20) results in a transcendental equation in only one variable, θ_m, or φ_m, where

$$\varphi_m = \frac{\pi}{2} - \theta_m. \tag{2.1.23}$$

For a given m, the parameters n_1, n_2, n_3 and t, φ_m (or θ_m) can be calculated. Thus a discrete set of reflection angles φ_m are obtained corresponding to the various modes. However, valid solutions do not exist for all values of m. There is a cutoff condition on allowed values of m for each set of n_1, n_2, n_3 and t, corresponding to the point at which φ_m becomes less than the critical angle for total internal reflection at either the $n_2 - n_3$ or the $n_2 - n_1$ interface, as discussed in Sect. 2.2.1.

For each allowed mode, there is a corresponding propagation constant β_m given by

$$\beta_m = kn_2\sin\varphi_m = kn_2\cos\theta_m. \tag{2.1.24}$$

The velocity of the light parallel to the waveguide is then given by

$$v = c(k/\beta), \tag{2.1.25}$$

and one can define an effective index of refraction for the guide as

$$n_{\text{eff}} = c/v = \beta/k. \tag{2.1.26}$$

Chapter 2 has described the optical modes that can exist in a three-layer planar waveguide. We have seen that the modes can be described either by a physical-optic method, based on a solution of Maxwell's wave equation, or by a ray-optic method, relying on geometrical ray tracing principles of classical optics. In Chap. 3, the mathematical model underlying the mode theory will be developed in greater detail.

Problems

2.1   We wish to fabricate a planar waveguide in GaAs for light of wavelength $\lambda_0 = 1.15\,\mu\text{m}$ that will operate in the single (fundamental) mode. If we assume a planar waveguide like that of Fig. 2.1 with the condition $n_2 - n_1 \gg n_2 - n_3$, what range of values can $n_2 - n_3$ have if $n_2 = 3.4$ and the thickness of the waveguiding layer $t = 3\,\mu\text{m}$?

2.2   Repeat Problem 2.1 for the case $\lambda_0 = 1.06\,\mu\text{m}$, all other parameters remaining unchanged.

2.3   Repeat Problems 2.1 and 2.2 for a waveguide of thickness $t = 6\,\mu\text{m}$.

2.4   In a planar waveguide like that of Fig. 2.8 with $n_2 = 2.0$, $n_3 = 1.6$, and $n_1 = 1$, what is the angle of propagation of the lowest order mode (θ_0) when cutoff occurs? Is this a maximum or a minimum angle for θ_0?

2.5   Sketch the three lowest order modes in a planar waveguide like that of Fig. 2.8 with $n_1 = n_3 < n_2$.

2.6   A mode is propagating in a planar waveguide as shown with $\beta_m = 0.8 kn_2$. How many reflections at the $n_1 - n_2$ interface does the ray experience in traveling a distance of 1 cm in the z direction?

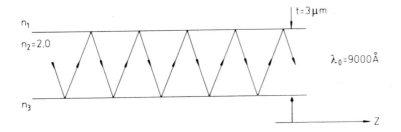

2.7   Show that the Goos-Hänchen phase shift goes to zero as the cutoff angle is approached for a waveguided optical mode.

2.8   Calculate the Goos-Hänchen shifts for a TE mode guided with $\beta = 1.85 k$ in a guide like that of Fig. 2.8, with $n_1 = 1.0$, $n_2 = 2.0$, $n_3 = 1.7$.

2.9   Show by drawing the vectorial relationship between the propagation constants (as in Fig. 2.9) How β, kn_2 and h change in relative magnitude and angle as one goes from the lowest-order mode in a waveguide progressively to higher-order modes.

3. Theory of Optical Waveguides

Chapter 2 has reviewed the key results of waveguide theory, particularly with respect to the various optical modes that can exist in the waveguide. A comparison has been made between the physical-optic approach and the ray-optic approach in describing light propagation in a waveguide. In this chapter, the electromagnetic wave theory of the physical-optic approach is developed in detail. Emphasis is placed on the two basic waveguide geometries that are used most often in optical integrated circuits, the planar waveguide and the rectangular waveguide.

3.1 Planar Waveguides

As was mentioned previously, the planar waveguide has a fundamental geometry that has been considered by many authors [2.1–6] as the basis for more sophisticated waveguide structure. In Sect. 3.1.1 we will follow, for the most part, the development of *Taylor* and *Yariv* [2.5].

3.1.1 The Basic Three-Layer Planar Waveguide

Consider the basic three-layer waveguide structure shown in Fig. 3.1. The light confining layers, with indices of refraction n_1 and n_3, are assumed to extend to infinity in the $+x$ and $-x$ directions, respectively. The major significance of this assumption is that there are no reflections in the x direction to be concerned with, except for those occurring at the $n_1 - n_2$ and $n_2 - n_3$ interfaces. For the case of TE plane waves traveling in the z direction, with propagation constant β, Maxwell's wave equation (2.1.1) reduces to

$$\nabla^2 E_y = \frac{n_i^2}{c^2} \frac{\partial^2 E_y}{\partial t^2} \quad i = 1, 2, 3, \ldots \tag{3.1.1}$$

with solutions of the form

$$E_y(x, z, t) = \mathcal{E}_y(x) e^{i(\omega t - \beta z)} \tag{3.1.2}$$

The subscript i in (3.1.1), of course, corresponds to a particular one of the three layers of the waveguide structure. For TE waves, it will be recalled that E_x and E_z are zero. Note also in (3.1.2) that $\mathcal{E}_y$ has no y or z dependence because the

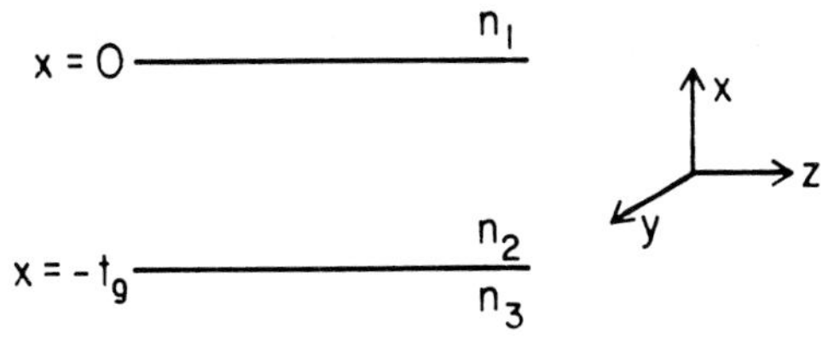

$x = 0$ ——————— n_1

$x = -t_g$ ——————— n_2 / n_3

Fig. 3.1. Basic three-layer planar waveguide structure

planar layers are assumed to be infinite in these directions, precluding the possibility of reflections and resultant standing waves.

The transverse function $\mathscr{E}_y(x)$ has the general form

$$\mathscr{E}_y(x) \begin{cases} A \ \exp(-qx) & 0 \leq x \leq \infty \\ B \ \cos(hx) + C \sin(hx) & -t_g \leq x \leq 0 \\ D \ \exp[p(x+t_g)] & -\infty \leq x \leq -t_g \end{cases} \tag{3.1.3}$$

where A, B, C, D, q, h, and p are all constants that can be determined by matching the boundary conditions, which requires the continuity [3.1] of $\mathscr{E}_y$ and $\mathscr{H}_z = (i/\omega\mu) \, \partial\mathscr{E}_y/\partial x$. Since the permeability μ and frequency ω are assumed to be constant, the second condition translates into a requirement that $\partial\mathscr{E}_y/\partial x$ be continuous. The constants A, B, C and D can thus be determined by making $\mathscr{E}_y$ and $\partial\mathscr{E}_y/\partial x$ continuous at the boundary between Region 1 and Region 2 ($x = 0$), and $\mathscr{E}_y$ continuous at $x = -t_g$. The procedure provides three equations in four unknowns, so that the solution for $\mathscr{E}_y$ can be expressed in terms of a single constant C'

$$\mathscr{E}_y = \begin{cases} C' \exp(-qx), & (0 \leq x \leq \infty); \\ C'\left[\cos(hx) - (q/h)\sin(hx)\right], & (-t_g \leq x \leq 0); \\ C'\left[\cos(ht_g) + (q/h)\sin(ht_g)\right]\exp[p(x+t_g)], & (-\infty \leq x \leq -t_g). \end{cases} \tag{3.1.4}$$

To determine q, h, and p, substitute (3.1.4) into (3.1.2), using the resulting expression for $E_y(x, z, t)$ in (3.1.1) for each of the three regions obtaining

$$q = (\beta^2 - n_1^2 k^2)^{1/2},$$
$$h = (n_2^2 k^2 - \beta^2)^{1/2},$$
$$p = (\beta^2 - n_3^2 k^2)^{1/2}, \tag{3.1.5}$$
$$k \equiv \omega/c.$$

Note in (3.1.5) that q, h and p are all given in terms of the single unknown β, which is the propagation constant in the z direction. By making $\partial E_y/\partial x$ continuous at $x = -t_g$, as required, a condition on β is derived. Taking $\partial E_y/\partial x$ from (3.1.4) and making it continuous at $x = -t_g$ yields the condition

$$-h\sin(-ht_g) - h(q/h)\cos(-ht_g),$$
$$= p\left[\cos(ht_g) + (q/h)\sin(ht_g)\right] \tag{3.1.6}$$

or, after simplification,

$$\tan(ht_g) = \frac{p+q}{h(1-pq/h^2)}. \tag{3.1.7}$$

The transcendental equation (3.1.7), in conjunction with (3.1.5), can be solved either graphically, by plotting right and left hand sides as a function of β and noting the intersection points, or numerically on a computer. Regardless of the method of solution, the result is a set of discrete allowed values of β, corresponding to the allowed modes. For each β_m, the corresponding values of q_m, h_m and p_m can be determined from (3.1.5).

The one remaining unknown constant C' in (3.1.4) is arbitrary. However, it is convenient to normalize so that $\mathscr{E}_y(x)$ represents a power flow of one Watt per unit width in the y direction. Thus, a mode for which $E_y = A\mathscr{E}_y(x)$ has a power flow of $|A|^2$ W/m. In this case, the normalization condition is [3.2]

$$-\frac{1}{2}\int_{-\infty}^{\infty} E_y H_x^* dx = \frac{\beta_m}{2\omega\mu}\int_{-\infty}^{\infty}\left[\mathscr{E}_y^{(m)}(x)\right]^2 dx = 1. \tag{3.1.8}$$

Substituting (3.1.4) into (3.1.8) yields

$$C_m' = 2h_m\left[\frac{\omega\mu}{|\beta_m|(t_g + 1/q_m + 1/p_m)(h_m^2 + q_m^2)}\right]^{1/2}. \tag{3.1.9}$$

For orthogonal modes

$$\int_{-\infty}^{\infty}\mathscr{E}_y^{(l)}\mathscr{E}_y^{(m)} dx = \frac{2\omega\mu}{\beta_m}\delta_{l.m}. \tag{3.1.10}$$

For the case of TM modes, the development exactly parallels that which has just been performed for the TE case, except that the non-zero components are H_y, E_x, and E_z rather than E_y, H_x and H_z. The resulting field components are

$$H_y(x, z, t) = \mathscr{H}_y(x)e^{i(\omega t - \beta z)}, \tag{3.1.11}$$

$$E_x(x, z, t) = \frac{i}{\omega\varepsilon}\frac{\partial H_y}{\partial z} = \frac{\beta}{\omega\varepsilon}\mathscr{H}_y(x)e^{i(\omega t - \beta z)}, \tag{3.1.12}$$

$$E_z(x, z, t) = -\frac{i}{\omega\varepsilon}\frac{\partial H_y}{\partial x}. \tag{3.1.13}$$

The transverse magnetic component $\mathcal{H}_y(x)$ is given by

$$\mathcal{H}_y(x) = \begin{cases} -C'\left[\dfrac{h}{q}\cos(ht_g) + \sin(ht_g)\right]\exp\left[p(x+t_g)\right], (-\infty \leq x \leq -t_g); \\[2ex] C'\left[-\dfrac{h}{q}\cos(hx) + \sin(hx)\right], \quad (-t_g \leq x \leq 0); \\[2ex] -C'\dfrac{h}{q}\exp\left[-qx\right], \qquad\qquad (0 \leq x \leq \infty). \end{cases} \qquad (3.1.14)$$

where h, q and p are again defined by (3.1.5), and where

$$\bar{q} = \frac{n_2^2}{n_1^2}q. \qquad (3.1.15)$$

When boundary conditions are matched in a manner that is analogous to the TE case, it is found that only those values of β are allowed for which

$$\tan(ht_g) = \frac{h(\bar{p}+\bar{q})}{h^2 - \bar{p}\bar{q}}, \qquad (3.1.16)$$

where

$$\bar{p} = \frac{n_2^2}{n_3^2}p. \qquad (3.1.17)$$

The constant C' in (3.1.14) can be normalized so that the field represented by (3.1.11–14) carries one Watt per unit width in the y direction, leading to [2.5]

$$C'_m = 2\sqrt{\frac{\omega\varepsilon_0}{\beta_m t'_g}}, \qquad (3.1.18)$$

where

$$t'_g \equiv \frac{\bar{q}^2 + h^2}{\bar{q}^2}\left(\frac{t_g}{n_2^2} + \frac{q^2+h^2}{\bar{q}^2+h^2}\frac{1}{n_1^2 q} + \frac{p^2+h^2}{\bar{p}^2+h^2}\frac{1}{n_3^2 p}\right). \qquad (3.1.19)$$

3.1.2 The Symmetric Waveguide

A special case of the basic three-layer planar waveguide that is of particular interest occurs when n_1 equals n_3. Such symmetric waveguides are frequently used in optical integrated circuits, for example, when a waveguiding layer with index n_2 is bounded on both surfaces by identical layers with somewhat lesser index n_1. Multilayer GaAlAs OIC's often utilize this type of waveguide. The equations developed in Sect. 3.1.1 apply to this type of waveguide, but a major simplification is possible in the determination of which modes may be supported. In many cases, it is not required to know the β's for the various modes. The only question being whether the waveguide is capable of guiding a particular mode or not.

A closed-form expression for the cutoff condition for TE modes can be derived in this case by referring to (2.1.5) and noting that, at cutoff (the point at which the field becomes oscillatory in Regions 1 and 3), the magnitude of β is given by

$$\beta = kn_1 = kn_3. \tag{3.1.20}$$

Substituting (3.1.20) into (3.1.5) we find that

$$\left. \begin{array}{l} p = q = 0 \quad \text{and} \\ h = k(n_2^2 - n_1^2)^{1/2} = k(n_2^2 - n_3^2)^{1/2} \end{array} \right\}. \tag{3.1.21}$$

Substituting (3.1.21) into (3.1.7) yields the condition

$$\tan(ht_g) = 0 , \tag{3.1.22}$$

or

$$ht_g = m_s\pi, \quad m_s = 0, 1, 2, 3, \dots . \tag{3.1.23}$$

Combining (3.1.21) and (3.1.23) yields

$$k(n_2^2 - n_1^2)^{1/2} t_g = m_s\pi. \tag{3.1.24}$$

Thus, for waveguiding of a given mode to occur, one must have

$$\Delta n = (n_2 - n_1) > \frac{m_s^2 \lambda_0^2}{4t_g^2(n_2 + n_1)}, \quad m_s = 0, 1, 2, 3,\dots \tag{3.1.25}$$

where $k \equiv \omega/c = 2\pi/\lambda_0$ has been used. The cutoff condition given in (3.1.25) determines which modes can be supported by a waveguide with a given Δn and ratio of λ_0/t_g. It is interesting to note that the lowest-order mode ($m_s = 0$) of the symmetric waveguide is unusual in that it does not exhibit a cutoff as all other modes do. In principle, any wavelength could be guided in this mode even with an incrementally small Δn. However, for small Δn and/or large λ_0/t_g confinement would be poor, with relatively large evanescent tails of the mode extending into the substrate.

If $n_2 \cong n_1$, the cutoff condition (3.1.25) becomes

$$\Delta n = (n_2 - n_1) > \frac{m_s^2 \lambda_0^2}{8t_g^2 n_2}, \quad m_s = 0, 1, 2, 3,\dots \tag{3.1.26}$$

or, if $n_2 \gg n_1$, it is given by

$$\Delta n = (n_2 - n_1) > \frac{m_s^2 \lambda_0^2}{4t_g^2 n_2}, \quad m_s = 0, 1, 2, 3,\dots . \tag{3.1.27}$$

3.1.3 The Asymmetric Waveguide

Another important special case of the three-layer planar waveguide is the asymmetric waveguide, in which $n_3 \gg n_1$. Of course, n_2 must still be greater than n_3 if waveguiding is to occur. The asymmetric waveguide is often found, for example, in optical integrated circuits in which a thin film waveguide is deposited or otherwise formed on a substrate of somewhat smaller index, while the top surface of the waveguiding layer is either left open to the air or, perhaps, coated with a metal layer electrode. It is possible to derive for the case of the asymmetric guide an approximate closed form expression for the cutoff condition by using a geometrical argument comparing it to the symmetric guide [3.3].

Consider an asymmetric waveguide, as shown in Fig. 3.2, that has a thickness t_g equal to half the thickness of a corresponding symmetric waveguide. The two lowest order TE modes of the symmetric guide ($m_s = 0, 1$) and of the asymmetric guide ($m_a = 0, 1$) are both shown in the figure. Note that, for well-confined modes, the lower half of the $m_s = 1$ mode of the symmetric guide corresponds closely to the $m_a = 0$ mode of the asymmetric guide of half thickness. This fact can be used as a mathematical device to permit one to obtain a closed form expression for the cutoff condition in the case of the asymmetric waveguide.

Solving (3.1.7) for the case of the symmetric waveguide with thickness equal to $2t_g$ in the same fashion as in Sect. 3.1.2 yields the condition

$$\Delta n = n_2 - n_3 > \frac{m_s^2 \lambda_0^2}{4(n_2 + n_3)(2t_g)^2} \qquad m_s = 0, 1, 2, 3, \ldots. \tag{3.1.28}$$

However, the asymmetric waveguide supports only modes corresponding

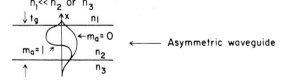

	Sym. $2t_g$	Asym. t_g
$m_{s,a} =$	1	0
	3	1
	5	2
	7	3

Mode correspondence ⟶

Fig. 3.2. Diagram of the modes in symmetric and asymmetric planar waveguides

to the odd modes of a symmetric guide of twice its thickness. Hence, the cutoff condition for the asymmetric guide is given by

$$\Delta n = n_2 - n_3 > \frac{m_a^2 \lambda_0^2}{16(n_2 + n_3)t_g^2},$$ (3.1.29)

where m_a are the elements of the subset consisting of odd values of m_s. This can be conveniently expressed by

$$m_a = (2m+1), \quad m = 0, 1, 2, 3, \ldots.$$ (3.1.30)

Assuming that $n_2 \cong n_3$ (3.1.29) becomes

$$\Delta n = n_2 - n_3 > \frac{(2m+1)^2 \lambda_0^2}{32 n_2 t_g^2}, \quad m_s = 0, 1, 2, 3, \ldots.$$ (3.1.31)

While the cutoff conditions (3.1.31) and (3.1.25) are valid only for the special cases defined, they offer a convenient means to estimate how many modes can be supported by a particular waveguide. To answer this question in the general case, or to determine the β_m for the various modes, one would have to solve the transcendental equation (3.1.7).

Although (3.1.31) has been derived for the case of TE waves, it can be shown [3.4] that it also holds for TM waves as long as $n_2 \cong n_3$. Thus, the asymmetric waveguide is seen to have a possible cutoff for all modes, unlike the symmetric waveguide for which the TE_0 mode cannot be cut off. This makes the asymmetric waveguide particularly useful as an optical switch, as will be discussed in Chap. 8.

3.2 Rectangular Waveguides

The planar waveguides discussed in the previous section are useful in many integrated optic applications in spite of the fact that they provide confinement of the optical fields in only one dimension. Even relatively complex optical integrated circuits, such as the rf spectrum analyzer of *Mergerian* and *Malarkey* [3.5], can be fabricated using planar waveguides. However, other applications require optical confinement in two dimensions. Use of a "stripe" geometry waveguide of rectangular cross-section can yield a laser with reduced threshold current and single mode oscillation [3.6], or an electro-optic modulator with reduced drive power requirement [3.7]. Sometimes two-dimensional confinement is required merely to guide light from one point on the surface of an OIC to another, to interconnect two circuit elements in a manner analogous to that of the metallic stripes used in an electrical integrated circuit.

3.2.1 Channel Waveguides

The basic rectangular waveguide structure consists of a waveguide region of index n_1 surrounded on all sides by a confining medium of lesser index n_2, as shown in Fig. 3.3. Such waveguides are often called *channel* guides, *strip* guides, or *3-dimensional* guides. It is not necessary that the index in the confining media be the same in all regions. A number of different materials, all with indices less than n_1, may be used to surround the guide. However, in that case, the modes in the waveguide will not be exactly symmetric. The exact solution of the wave equation for this general case is extremely complicated, and has not been obtained yet. However, *Goell* [3.8] has used cylindrical space harmonics to analyze guides with aspect ratios (width to height) between one

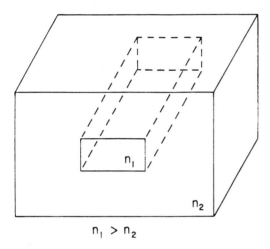

$n_1 > n_2$

Fig. 3.3. Basic rectangular dielectric waveguide structure

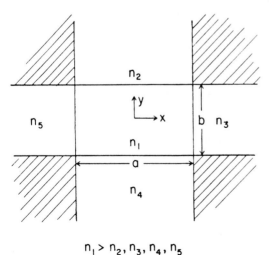

$n_1 > n_2, n_3, n_4, n_5$

Fig. 3.4. Cross-sectional view of a rectangular dielectric waveguide bounded by regions of smaller index of refraction

and two, and *Schlosser* and *Unger* [3.9] have analyzed guides with large aspect ratios by using rectangular harmonics and numerical methods.

Marcatili [3.10] has derived an approximate solution to the rectangular channel waveguide problem, by analyzing the structure shown in Fig. 3.4, which is still fairly general. The key assumption made in Marcatili's analysis is that the modes are well guided, i.e., well above cutoff, so that the field decays exponentially in Regions 2, 3, 4, and 5, with most of the power being confined to Region 1. The magnitudes of the fields in the shaded corner regions of Fig. 3.4 are small enough to be neglected. Hence, Maxwell's equations can be solved by assuming relatively simple sinusoidal and exponential field distributions, and by matching boundary conditions only along the four sides of Region 1. The waveguide is found to support a discrete number of guided modes that can be grouped into two families, E_{pq}^x and E_{pq}^y, where the mode numbers p and q correspond to the number of peaks in the field distribution in the x and y directions, respectively. The transverse field components of the E_{pq}^x modes are E_x and H_y, while those of the E_{pq}^y modes are E_y and H_x. The E_{11}^y (fundamental) mode is sketched in Fig. 3.5. Note that the shape of the mode is characterized by extinction coefficients η_2, ξ_3, η_4, and ξ_5 in the regions where it is exponential, and by propagation constants k_x and k_y in Region 1.

Quantitative expressions for k_x, k_y, η and ξ can be determined for the E_{pq}^y modes as follows. The field components in the five regions shown in Fig. 3.4 (designated by $v = 1, 2, 3, 4, 5$) have the form:

$$H_{xv} = \exp(-ik_z z + i\omega t) \begin{cases} M_1 \cos(k_x x + \alpha)\cos(k_y y + \beta) & \text{for} \quad v = 1 \\ M_2 \cos(k_x x + \alpha)\exp(-ik_{y2}y) & \text{for} \quad v = 2 \\ M_3 \cos(k_y y + \beta)\exp(-ik_{x3}x) & \text{for} \quad v = 3 \\ M_4 \cos(k_x x + \alpha)\exp(ik_{y4}y) & \text{for} \quad v = 4 \\ M_5 \cos(k_y y + \beta)\exp(ik_{x5}x) & \text{for} \quad v = 5. \end{cases} \tag{3.2.1}$$

$$H_{yv} = 0, \tag{3.2.2}$$

$$H_{zv} = -\frac{i}{k_z} \frac{\partial^2 H_{xv}}{\partial x \partial y}, \tag{3.2.3}$$

$$E_{xv} = -\frac{1}{\omega \varepsilon_0 n_v^2 k_z} \frac{\partial^2 H_{xv}}{\partial x \partial y}, \tag{3.2.4}$$

$$E_{yv} = -\frac{k^2 n_v^2 - k_{yv}^2}{\omega \varepsilon_0 n_v^2 k_z} H_{xv}, \tag{3.2.5}$$

$$E_{zv} = -\frac{i}{\omega \varepsilon_0 n_v^2} \frac{\partial H_{xv}}{\partial y}, \tag{3.2.6}$$

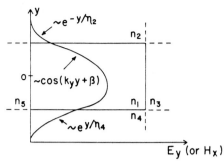

Fig. 3.5. Sketch of a typical E^y_{11} mode

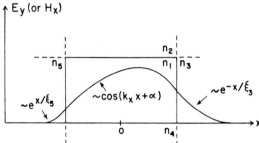

where M_v is an amplitude constant, ω is the angular frequency and ε_0 is the permittivity of free space. The phase constants, α and β locate the field maxima and minima in Region 1, and k_{xv} and k_{yv} ($v = 1, 2, 3, 4, 5$) are the transverse propagation constants along the x and y directions in the various media. Matching the boundary conditions requires the assumption that

$$k_{x1} = k_{x2} = k_{x4} = k_x, \tag{3.2.7}$$

and

$$k_{y1} = k_{y3} = k_{y5} = k_y . \tag{3.2.8}$$

Also, it can be shown that

$$k_z = (k_1^2 - k_x^2 - k_y^2)^{1/2}, \tag{3.2.9}$$

where

$$k_1 = kn_1 = \frac{2\pi}{\lambda_0} n_1, \tag{3.2.10}$$

is the propagation constant of a plane wave with free-space wavelength λ_0 in a medium of refractive index n_1. Assuming that n_1 is only slightly larger than the other n_v, as is usually the case in an OIC, leads to the condition

$$k_x \quad \text{and} \quad k_y \ll k_z . \tag{3.2.11}$$

Note that (3.2.11) corresponds, in ray-optics terminology, to a grazing incidence of the ray at the surfaces of the waveguiding Region 1. Calculations show that the two significant components for the E_{pq}^y modes are H_x and E_y.

Matching field components at the boundaries of Region 1 yields the transcendental equations

$$k_x a = p\pi - \tan^{-1} k_x \xi_3 - \tan^{-1} k_x \xi_5, \qquad (3.2.12)$$

and

$$k_y b = q\pi - \tan^{-1} \frac{n_2^2}{n_1^2} k_y \eta_2 - \tan^{-1} \frac{n_4^2}{n_1^2} k_y \eta_4, \qquad (3.2.13)$$

where the $\tan^{-1}$ functions are to be taken in the first quadrant, and where

$$\xi_{\substack{3\\5}} = \frac{1}{\left|k_{x\substack{3\\5}}\right|} = \frac{1}{\left[\left(\dfrac{\pi}{A_{\substack{3\\5}}}\right)^2 - k_x^2\right]}, \qquad (3.2.14)$$

$$\eta_{\substack{2\\4}} = \frac{1}{\left|k_{y\substack{2\\4}}\right|} = \frac{1}{\left[\left(\dfrac{\pi}{A_{\substack{2\\4}}}\right)^2 - k_y^2\right]^{1/2}}, \qquad (3.2.15)$$

and

$$A_v = \frac{\pi}{(k_1^2 - k_v^2)^{1/2}} = \frac{\lambda_0}{2(n_1^2 - n_v^2)^{1/2}}, \quad v = 2, 3, 4, 5. \qquad (3.2.16)$$

The transcendental equations (3.2.12) and (3.2.13) cannot be solved exactly in closed form. However, one can assume for well confined modes that most of the power is in Region 1. Hence,

$$\left(\frac{k_x A_{\substack{3\\5}}}{\pi}\right)^2 \ll 1 \quad \text{and} \quad \left(\frac{k_y A_{\substack{2\\4}}}{\pi}\right)^2 \ll 1. \qquad (3.2.17)$$

Using the assumptions of (3.2.17), approximate solutions of (3.2.12) and (3.2.13) for k_x and k_y can be obtained by expanding the $\tan^{-1}$ functions in a power series, keeping only the first two terms. Thus,

$$k_x = \frac{p\pi}{a}\left(1 + \frac{A_3 + A_5}{\pi a}\right)^{-1}, \qquad (3.2.18)$$

$$k_y = \frac{q\pi}{b}\left(1 + \frac{n_2^2 A_2 + n_4^2 A_4}{\pi n_1^2 b}\right)^{-1}. \qquad (3.2.19)$$

Substituting (3.2.18) and (3.2.19) in (3.2.9), (3.2.14) and (3.2.15) expressions for k_z, ξ_3, ξ_5, η_2 and η_4 can be obtained.

$$k_2 = \left[k_1^2 - \left(\frac{\pi p}{a} \right)^2 \left(1 + \frac{A_3 + A_5}{\pi a} \right)^{-2} - \left(\frac{\pi q}{b} \right)^2 \left(1 + \frac{n_2^2 A_2 + n_4^2 A_4}{\pi n_1^2 b} \right)^{-2} \right]^{1/2} \cdot \quad (3.2.20)$$

$$\xi_3 \atop 5} = \frac{A_3 \atop 5}{\pi} \left[1 - \left(\frac{p A_3 \atop 5}{a} \frac{1}{1 + \dfrac{A_3 + A_5}{\pi a}} \right)^2 \right]^{-1/2} \cdot \quad (3.2.21)$$

$$\eta_2 \atop 4} = \frac{A_2 \atop 4}{\pi} \left[1 - \left(\frac{q A_2 \atop 4}{b} \frac{1}{1 + \dfrac{n_2^2 A_2 + n_4^2 A_4}{\pi n_1^2 b}} \right)^2 \right]^{-1/2} \cdot \quad (3.2.22)$$

The E_{pq}^y modes are polarized such that E_y is the only significant component of electric field; E_x and E_z are negligibly small. It can be shown for the case of the E_{pq}^x modes that E_x is the only significant electric field component, with E_y and E_z being negligible. To develop relationships for the E_{pq}^x modes corresponding to those that have been derived from the E_{pq}^y modes, one can merely change E to H, μ_0 to $-\varepsilon_0$, and vice-versa in the various equations. As long as the assumption is made that n_1 is only slightly larger than the indices of the surrounding media, i.e.,

$$\frac{1}{n_1}(n_1 - n_v) \ll 1, \quad (3.2.23)$$

then k_z, ξ_3, and η_2 are still given by (3.2.20), (3.2.21), and (3.2.22), respectively, for the E_{pq}^y modes just as for the E_{pq}^y modes.

Marcatili's analysis of the rectangular three-dimensional waveguide is very useful in designing such structures, even though it features an approximate solution to Maxwell's equations. It must be remembered that the theory assumes well confined modes. When waveguide dimensions a and b are small enough, compared to the wavelength, the theory becomes inaccurate for that mode [3.10].

3.2.2 Strip-Loaded Waveguides

It is possible to make a three-dimensional waveguide, in which there is confinement in both the x and y dimensions, without actually surrounding the waveguide with materials of lesser index. This is done by forming a strip of

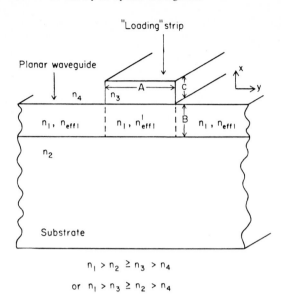

Fig. 3.6. Diagram of dielectric strip-loaded waveguide. Effective index of refraction, as well as the bulk index, are indicated in the wave-guiding layer. $n'_{\text{effl}} > n_{\text{effl}}$

dielectric material of lesser index, n_3, on top of a planar waveguide, with index n_1, as shown in Fig. 3.6. Such a structure is usually called either a *strip-loaded* waveguide, or an *optical stripline*. The presence of the loading strip on top of the waveguiding layer makes the effective index in the region beneath it, n'_{effl} larger than the effective index, n_{effl} in the adjacent regions. Thus there can be confinement in the y direction as well as in the x direction. The physical nature of this phenomenon can be visualized best by using the ray optic approach. Consider a particular mode propagating in the z direction but consisting of a plane wave following the usual zig-zag path in the waveguiding layer (Fig. 2.10). Since n_3 is larger than n_4, the wave penetrates slightly more at the $n_1 - n_3$ interface than it would at the $n_1 - n_4$ interface; thus, the effective height of the waveguide is greater under the loading strip than it is in the regions on either side. This means that the zig-zag path of the plane wave would be slightly longer under the loading strip, leading to the result that

$$n'_{\text{effl}} = \frac{\beta'}{k} > n_{\text{effl}} = \frac{\beta}{k}. \tag{3.2.24}$$

Furuta et al. [3.11] have used the effective index of refraction method to analyze a strip-loaded guide like that of Fig. 3.6, and have shown that its waveguiding properties are equivalent to those of a dielectric waveguide like that shown in Fig. 3.7, where the equivalent index in the side confining layers is given by

$$n_{\text{eq}} = (n_1^2 - n'^2_{\text{effl}} + n^2_{\text{effl}})^{1/2}. \tag{3.2.25}$$

The propagation constants of the rectangular waveguide of Fig. 3.7 can be determined by using Marcatili's method, described in Sect. 3.2.1. The effectiveness of this approach has been demonstrated [3.11] by comparing theoreti-

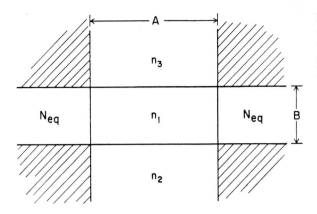

Fig. 3.7. Cross-sectional view of rectangular dielectric waveguide equivalent to the strip-loaded waveguide of Fig. 3.6

cal predictions with experimental observations of waveguiding of 6328 Å light in strip-loaded waveguides. In that case, the waveguides were thin (approx. 0.6 μm) glass films ($n_1 = 1.712$), on glass substrates with $n_2 = 1.662$. The index of the loading strip was 1.592 and its cross-sectional dimensions were 0.7 μm thick × 14 μm wide. *Ramaswamy* [3.12] has also used the effective index method to analyze strip-loaded guides and found good agreement with experimental results.

A strip-loaded waveguide can also be made with the index of the loading strip, n_3, equal to that of the guide, n_1. That type of waveguide is usually called a *ridge*, or *rib*, guide. *Kogelnik* [3.13] has applied the effective index method to that type of guide to show that propagation is characterized by a phase constant

$$\beta = kN, \tag{3.2.26}$$

where N is given by

$$N^2 = n_{\text{eff1}}^2 + b(n_{\text{eff1}}'^2 - n_{\text{eff1}}^2). \tag{3.2.27}$$

In this case, n'_{eff1} is the effective index in a guide having a height equal to the sum of the thicknesses of the waveguiding layer and the loading strip. The parameter b is a normalized guide index, given by

$$b = (N^2 - n_2^2)/(n_1^2 - n_2^2). \tag{3.2.28}$$

Metallic loading strips can also be used to produce optical striplines. In that case, two metal loading strips are placed on the surface of the waveguiding layer, on either side of the region in which confinement is desired, as shown in Fig. 3.8. Since penetration of the guided waves is deeper at the $n_1 - n_4$ interface than at the $n_1 - n_3$ interface, the desired confinement in the y dimension is obtained just as in the case of the dielectric strip-loaded waveguide. Metal strip-loaded waveguides are particularly useful in applications such as electro-optic modulators, where surface metal electrodes are desired, since

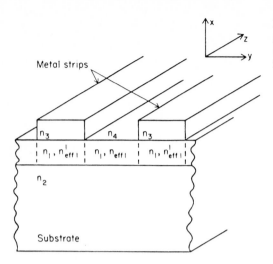

Fig. 3.8. Metal strip-loaded waveguide. Effective indices of refraction, as well as the bulk index, are indicated in the waveguiding layer. $n'_{\text{eff1}} < n_{\text{eff1}}$

these metal stripes can perform the additional function of defining the waveguide [3.14].

In principle, it is expected that strip-loaded waveguides should have less optical loss than rectangular channel dielectric waveguides, because scattering due to side-well roughness is reduced. Experimental results seem to support that hypothesis. *Blum* et al. [3.15] have reported a loss coefficient of $1\,\text{cm}^{-1}$ in GaAs strip-loaded guides, while *Reinhart* et al. [3.16] have measured waveguide loss of less than $2\,\text{cm}^{-1}$ in GaAs-GaAlAs rib waveguides. The greatest concern regarding the usefulness of strip-loaded waveguides is that the small effective index difference in the y dimension, produced by the loading effect, will be insufficient to limit radiation loss occurring from bends in the guide. However, 90° bends with radii as small as 2.5 mm have been made [3.11] in glass strip-loaded guides with no observation of excessive radiation loss.

Chapter 3 has described basic theoretical models that can be used to calculate the propagation characteristics of the waveguide types that are commonly used in optical integrated circuits. Additional information on this topic is available, for example, in the books by *Unger* [3.17] and *Tamir* [3.18], and in a variety of journal articles [3.19–3.41], which have been listed by title in the references for this chapter to facilitate referral. In the next chapter, methods of fabricating these waveguides will be described.

Problems

3.1 For a basic three-layer planar waveguide such as that shown in Fig. 3.1, show that the phase constant h in Region 2, and the extinction coefficients q and p in Region 1 and 3, respectively, are given by

$$q = (\beta^2 - n_1^2 k^2)^{1/2}$$

$$p = (\beta^2 - n_3^2 k^2)^{1/2}$$

$$h = (n_2^2 k^2 - \beta^2)^{1/2},$$

where β is the propagation constant in the z direction and $k \equiv \omega/c$.

3.2   Show that the change in index of refraction Δn required for waveguiding of the mth order mode in an asymmetric waveguide is given by

$$\Delta n = n_2 - n_3 > \frac{(2m+1)^2 \lambda_0^2}{32 n_2 t_g^2}, \quad m = 0, 1, 2, 3,$$

where all terms are as defined in Sect. 3.1.3.

3.3   Sketch the cross-sectional electric field distributions of the E_{12}^y and E_{22}^y modes in a rectangular waveguide.

3.4   Sketch the cross-sectional optical power distributions of the E_{33}^y and E_{23}^y modes in a rectangular waveguide.

3.5   A *symmetric* planar waveguide has a waveguiding layer with thickness = $3\,\mu m$ and index $n_2 = 1.5$. The waveguide is excited with a He-Ne laser ($\lambda_0 = 0.6328\,\mu m$). Assume that the surrounding medium is air (i.e., $n_1 = n_3 = 1$)
a) How many TE modes can the waveguide support?
b) Suppose that instead of the surrounding medium being air it is dielectric so that $n_1 = n_3 = 1.48$, how many TE modes can now propagate?
c) In the case of the waveguide of part (b), how thin can the waveguiding layer be made and still allow the lowest order möde to propagate?

3.6   A planar dielectric waveguide is fabricated by sputtering a thin glass film ($n_2 = 1.62$) on top of a glass substrate ($n_3 = 1.346$). The light source is a He-Ne laser operating at wavelength $\lambda_0 = 6328\,\text{Å}$.
a) What is the maximum thickness of the waveguiding layer for the guide to propagate only one mode?
b) If the propagation constant in the direction of propagation is $\beta = 1.53 \times 10^5$ cm^{-1}, what is the bounce angle θ_m of the mode? (Make a sketch and label the angle for clarity.)

3.7   For an asymmetric planar waveguide like that shown in Fig. 3.1 with $n_1 = 1.0$, $n_2 = 1.65$, $n_3 = 1.52$, $t_g = 1.18\,\mu m$ and $\lambda_0 = 0.63\,\mu m$, find the number of allowed TE modes.

4. Waveguide Fabrication Techniques

In Chap. 3, the theoretical considerations relevant to various types of waveguides were discussed. In every case, waveguiding depended on the difference in the index of refraction between the waveguiding region and the surrounding media. A great many techniques have been devised for producing that required index difference. Each method has particular advantages and disadvantages, and no single method can be said to be clearly superior. The choice of a specific technique of waveguide fabrication depends on the desired application, and on the facilities available. In this Chapter, various methods of waveguide fabrication are reviewed, and their inherent features are discussed.

4.1 Deposited Thin Films

One of the earliest-used, and most effective, methods of waveguide fabrication is the deposition of thin films of dielectric material. In recent years glass and polymer waveguides have greatly increased in importance because they are inexpensive compared to those made of semiconductors or lithium niobate. Millions of inexpensive waveguide devices are needed to fully implement the international-communications network, sometimes called the "information superhighway". In this section, the word deposition is broadly defined to include methods of liquid-source deposition, such as spinning or dipping, as well as vacuum-vapor deposition and sputtering.

4.1.1 Sputtered Dielectric Films

Thermally stimulated vacuum-evaporation, which is the standard method of producing thin films for conventional applications, such as anti-reflection coatings, is seldom used for waveguide fabrication, since it produces films with relatively high loss at visible wavelengths (10 dB/cm). This high loss is due to inclusion of contaminant atoms that act as absorption and scattering centers [4.1]. Instead, sputter deposition of molecules from a solid source is used. Sputtering is the process whereby atoms or molecules are removed from the surface of a source (target) material, in vacuo, by bombardment with ions having energies in excess of about 30 eV to about 2 keV. The atoms (or molecules) removed from the surface of the target are allowed to deposit on the surface of a substrate, forming a thin film layer. The sputtered thin film is slowly built up by an accumulation of individual particles arriving at the surface with significant kinetic energy. This process produces a very uniform layer because the deposited atoms are kinetically distributed over the surface. Contaminant atoms are mostly excluded because the process is carried out at a relatively low temperature compared to that required for vacuum-vapor-deposition, and the target material can be highly purified before use. As a result, good quality optical films can be produced, with losses on the order of 1 dB/cm [4.2].

One method of sputtering [4.3] a thin film layer is by plasma discharge, as diagrammed in Fig. 4.1. The target and substrate are placed in a vacuum system and a gas is let in at a pressure of $(2 \text{ to } 20) \times 10^{-3}$ torr. A high voltage bias is applied between anode and cathode so that a plasma discharge is established. Ions generated in the plasma discharge are accelerated toward the cathode. They strike the target, thus transferring their momentum to atoms near the surface of the target material, which are thereby sputtered off, and then deposited on the substrate.

In order for the sputtered atoms to adhere well to the substrate so as to produce a uniform, low-loss layer, the substrate must be thoroughly cleaned in an appropriate solvent or etchant prior to being placed in the vacuum system. Substrate cleaning is a basic step in any thin-film deposition process, and methods vary, depending on the substrate and film materials. A review of substrate cleaning procedures has been given by *Zernike* [4.4].

The position of the substrate in Fig. 4.1 is merely representative of one of the many different geometric configurations that can be used, since atoms sputtered from the target tend to deposit on every exposed surface within the vacuum system. For best uniformity of the deposited layer, the distance from

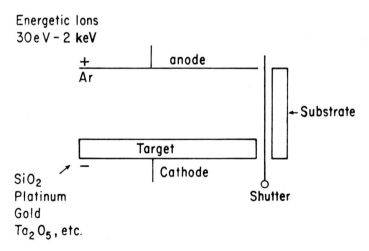

Fig. 4.1. Plasma discharge system for sputter deposition of thin films

substrate to target should be large compared to the substrate dimensions. However, increased substrate-target separation reduces the deposition rate, so that a longer time is required to produce a layer of a given thickness. Often the substrate is attached to the anode because it provides a parallel geometry that is conducive to layer uniformity. However, electrical isolation must be provided if a separate substrate bias is desired, and substrate heating due to incident electrons may be significant. The purpose of the shutter, shown in Fig. 4.1, is to shield the substrate from deposition during the first minute or so after the plasma discharge is activated, since many adsorbed contaminant atoms are released during that period.

Usually one of the inert gases, argon, neon, or krypton, is used in the plasma discharge chamber to avoid contamination of the deposited layer with active atoms. However, in some cases it is advantageous to employ reactive sputtering, in which atoms sputtered from the target react with the bombarding ions to form an oxide or nitride deposited film. For example, a silicon target can be used in the presence of ammonia gas to form silicon nitride according to the reaction

$$3Si + 4NH_3 \rightarrow Si_3N_4 + 6H_2. \tag{4.1.1}$$

The bias voltage applied between anode and cathode may be either dc, or rf with a dc component. rf sputtering usually produces better quality films and is not subject to charge build-up problems in low conductivity targets but, of course, a more complicated power supply is needed. Detailed descriptions of both rf and dc sputtering are available elsewhere [4.5, 6].

Plasma discharge sputtering is also useful for depositing metal films, as well as dielectric layers. Thus, metallic electrical contacts and field plates can be deposited on top of, or adjacent to, dielectric waveguides, as required.

As an alternative to generating the bombarding ions in a plasma discharge, one can use a collimated beam of ions produced by a localized source, or ion gun, as shown in Fig. 4.2. The advantages of ion beam sputtering, as compared to the plasma discharge method, are that deposition can be done in a high vacuum $<10^{-6}$ torr, and that the focused ion beam strikes only the target. No contaminant atoms are sputtered off of the walls of the chamber. Since the focused ion beam is often smaller than the target, the beam is usually

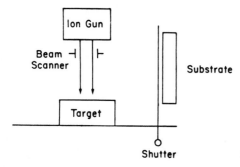

Fig. 4.2. Ion beam sputter deposition system

electrically scanned in a raster pattern over the target to insure uniform sputtering.

Sputter deposition, whether plasma discharge or ion beam, can be used to fabricate a wide variety of dielectric waveguides. To cite just a few examples: *Zernike* has produced Corning 7059 glass waveguides on both glass and KDP substrates [4.4], while *Agard* has sputtered niobium pentoxide waveguides on quartz substrates [4.7], and *Hinkernell* has deposited ZnO on SiO_2 [4.8]. Reactive sputtering has been used to produce waveguides of Ta_2O_5 [Ref. 4.4, p. 211] and Nb_2O_5 [4.9]. Tantalum pentoxide and niobium pentoxide waveguides can also be produced by sputter deposition of Ta [4.10] and Nb [4.11], followed by thermal oxidation.

4.1.2 Deposition from Solutions

Many materials can be used to form waveguides by applying a solution which dries to form a dielectric film and by spinning the substrate on a photoresist spinner to spread the layer evenly, by dipping the substrate into the solution and slowly withdrawing it, or by casting or injection-molding techniques. Table 4.1 lists some common waveguide materials that can be applied in this fashion, along with the appropriate solvent in each case. The advantage of these waveguides is that they are inexpensive and can be applied without any sophisticated equipment. However, material purity is often low compared to that obtainable in sputtered films, and uniformity is relatively poor. Nevertheless, some surprisingly good results have been obtained. *Ulrich* and *Weber* [4.13] have reported epoxy films with losses of only 0.3 dB/cm, at 6328 Å wavelength, made by a modified dipping technique in which a horizontal substrate was first coated with the liquid and then turned vertically to allow the excess epoxy to drain. *Neyer et al.* [4.18] have used injection molding of polymethylmethacrylate (PMMA) filled with ethyleneglycoldimethacrylate (EGDMA) to form waveguides with 0.3 dB/cm loss at 1300 nm wavelength.

Polyimide channel waveguides fabricated by direct laser writing have been used as the optical interconnect medium between 50 GHZ modulated sources and high-speed photodetectors [4.19]. Such interconnects could potentially be *utilized* in the implementation of sophisticated water-scale systems such as phased array radars.

Table 4.1. Waveguide materials applied by spinning and dipping

Material	Solvent
Photoresist [4.12]	Acetone
Epoxy [4.13]	Proprietary Compounds
Polymethylmethacrylate [4.14]	Chloroform, Toluene
Polyurethane [4.13, 15]	Xylene
Polyimide [4.16. 17]	

4.1.3 Organosilicon Films

Tien et al. [4.20] have produced thin film waveguides by rf discharge polymerization of organic chemical monomers. Films prepared from vinyltrimethylsilane (VTMS) and hexamethyldisiloxane (HMDS) monomers were deposited on glass substrates. The VTMS films had an index of 1.531 at 6328 Å, about 1% larger than the index of the substrate of ordinary glass (1.512). The corresponding index of HMDS films was 1.488; thus, a substrate of Corning 744 Pyrex glass, which has an index of 1.4704, was used. Deposition was accomplished by introducing monomers and argon through separate leak valves into a vacuum system which had been evacuated to a pressure of less than 2×10^{-6} torr. A discharge was then initiated by applying a 200 W, 13.56 MHz rf voltage between an anode and a cathode on which the substrate rested. VTMS films grew at a rate of about 2000 Å/min, while the rate for HMDS films was 1000 Å/min, for metered pressures of 0.3 torr (monomer) and 0.1 torr (Ar). The resulting polymer films were smooth, pinhole free, and transparent from 4000 Å to 7500 Å. Optical losses were exceptionally low (<0.004 dB/cm). In addition, it was found that the index of the films could be varied either by mixing of the two monomers prior to deposition, or by chemical treatment after deposition.

4.2 Substitutional Dopant Atoms

While thin-film deposition techniques have proved very effective for producing waveguides in glasses and other amorphous materials, they generally cannot be used to produce layers of crystalline materials, such as semiconductors, or ferroelectrics like $LiNbO_3$ and $LiTaO_3$. This is true because the elements of these compound materials usually do not deposit congruently; the relative concentrations of different elements are not maintained as atoms are transferred from source to substrate. Even when conditions can be established so that congruent deposition occurs, the grown layer will generally not be single crystal and epitaxial (of the same crystal structure as the substrate on which it is grown). To avoid this problem, many different approaches have been developed to produce waveguides in, or on, crystalline materials without seriously disrupting the lattice structure. A number of these methods involve the introduction of dopant atoms which substitutionally replace some of the lattice atoms, bringing about an increase in the index of refraction.

4.2.1 Diffused Dopants

Since waveguides for visible and near-infrared wavelength usually have thicknesses of only a few micrometers, diffusion of dopant atoms from the surface of the substrate is a viable approach to waveguide fabrication. Standard diffusion techniques are used [4.21], in which the substrate is placed in a furnace, at typically 700° to 1000°C, in the presence of a source of dopant atoms. This source may be a flowing gas, liquid, or solid surface film. For example, diffusion of atoms from titanium or tantalum metal surface layers is often used to produce waveguides in $LiNbO_3$ and $LiTaO_3$ [4.22–24], while *Chen* [4.25] has used an organo-metallic layer, applied from a solution, as the diffusion source for waveguide formation in $LiNbO_3$. Metal diffused waveguides in $LiNbO_3$ and $LiTaO_3$ can be produced with losses that are smaller than 1 dB/cm.

The most commonly used method for producing waveguides in $LiNbO_3$ is to deposit a layer of titanium on the surface by rf sputtering or e-beam evaporation. Then the substrate is placed in a diffusion furnace at about 1000°C and heated in a flowing gas atmosphere containing oxygen. For example, argon which has been passed through an H_2O bubbler is often used. As the Ti coated $LiNbO_3$ substrate is moved into the furnace, oxidation of the deposited Ti begins when its temperature reaches about 500°C. By the time the substrate has reached the desired diffusion temperature, set somewhere in the range of 1000–1100°C, oxidation is complete and the Ti metal has been replaced by a layer of polycrystalline TiO_2. This TiO_2 subsequently reacts with the $LiNbO_3$ to form a complex compound at the surface which acts as the diffusion source of Ti atoms. The extent of Ti diffusion depends on the diffusion coefficient and on time and temperature. It follows the usual Fick's laws for diffusion from a limit source of atoms yielding a Gaussian distribution of diffused Ti atoms as a function of depth from the surface. The advantage of this multi-staged diffusion process, as compared to diffusion directly from a metallic Ti layer, is that the out-diffusion of Li atoms from the $LiNbO_3$ is prevented. Li out-diffusion would otherwise result in changes in the index of refraction to a much greater depth than that desired for single mode waveguides. With the oxidation process, the change in index occurs only over the depth of the diffused Ti profile, which is accurately controllable.

Metal diffusion can also be used to produce waveguides in semiconductors [4.26]. Generally, a p-type dopant is diffused into a n-type substrate, or vice versa, so that a p-n junction is formed, thus providing electrical isolation as well as optical waveguiding. For example, *Martin* and *Hall* [4.27] have used diffusion of Cd and Se into ZnS and Cd into ZnSe to make both planar and channel waveguides. Losses in 10μm wide, 3μm deep channel waveguides formed by diffusion of Cd into ZnSe were measured to be less than 3 dB/cm.

Diffusion of substitutional dopant atoms to produce waveguides is not limited to semiconductors; it can also be done in glass and polymers. For example, *Zolatov* et al. [4.28] have studied waveguides fabricated by diffusing silver into glass. The observed change in index of refraction resulting from the presence of the diffused Ag atoms was $\Delta n \cong 0.073$ at wavelengths from 6328 Å to 5461 Å. Because diffusion must be performed below the relatively low temperatures at which glasses melt, it is a slow process. Hence, the alternative methods of ion exchange and migration are often used to introduce the required dopant atoms.

Many polymers have excellent dopant-diffusion capabilities, as described by *Booth* [4.29]. Solvent bath dopants can be selectively diffused into guide regions to increase the index, or low moleoular weight dopants or monomers can be thermally out-diffused to create guide regions.

4.2.2 Ion Exchange and Migration

A typical ion migration and exchange process is diagrammed in Fig. 4.3. The substrate material is a sodium-doped glass. When an electric field is applied as shown, and the glass substrate is heated to about 300 °C, Na^+ ions migrate toward the cathode. The surface of the glass is submerged in molten thallium nitrate. Some of the Na^+ ions are exchanged for Tl^+, resulting in the formation of a higher index layer at the surface.

Izawa and *Nakagome* [4.30] have used an ion exchange and migration method to form a buried waveguide slightly beneath the surface. Their setup is similar to that of Fig. 4.3 except that a borosilicate glass substrate is used, at a temperature of 530 °C. A two stage process is employed. In the first stage, the salts surrounding the substrate (anode) are thallium nitrate, sodium nitrate and potassium nitrate, so that Na^+ and K^+ ions in the glass are exchanged with Tl^+ ions, forming a higher-index layer at the surface. In the second stage, only sodium nitrate and potassium nitrate are present at the anode. Hence, the Na^+ and K^+ ions diffuse back into the surface, while the Tl^+ ions move deeper into the substrate. The resulting structure has a higher-index, Tl^+ doped layer located at some depth below the surface, covered on top and bottom by the normal borosilicate glass. Thus, a *buried* waveguide has been formed. Ion exchange can also be used to produce waveguides in $LiNbO_3$. In that case, protons are exchanged for lithium ions to produce the desired increase in index of refraction. The source of protons, benzoic acid, is heated above its melting point (122 °C) to a temperature of about 235 °C. The masked lithium niobate substrate is then placed in the melt for a period of one hour or less. This treatment is sufficient to produce a step-like index change of (extraordinary) $\Delta n_e = 0.12$ and (ordinary) $\Delta n_o = -0.04$ [4.31]. Design procedures and software tools for designing ion exchange waveguides have been published [4.32].

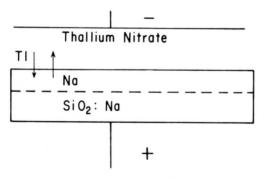

Fig. 4.3. Ion migration system for waveguide fabrication

4.2.3 Ion Implantation

Substitutional dopant atoms can also be introduced by ion implantation. In the ion implantation doping process, ions of the desired dopant are generated, then accelerated through, typically, 20–300 keV before impinging on the substrate. Of course, the process must be carried out in a vacuum. A basic system for implantation doping is shown in Fig. 4.4. Detailed descriptions of ion implantation doping techniques are available elsewhere [4.33]; hence, they will not be repeated here. However, it should be noted that the basic elements of the implantation system are an ion source, accelerating electrodes, electrostatic or magnetic deflection elements combined with a slit to form an ion (mass) separator, and a beam deflector to scan the collimated ion beam in a raster pattern over the substrate. The depth of penetration of the implanted ions depends on their mass, energy, and on the substrate material and its orientation. Tabulated data are available for most ion/substrate combinations that are of interest [4.34]. Knowing the penetration characteristics and the implanted ion dose per cm^2, which can be determined very accurately by measuring beam current density and implant time, one can calculate the profile of implanted ion concentration with depth.

In most cases, ion implantation must be followed by annealing at elevated temperatures to remove implantation-caused lattice damage and to allow the implanted dopant atoms to move into substitutional sites in the lattice. After annealing, waveguides made by substitutional dopant implantation have much the same characteristics as diffused waveguides. However, the implantation process allows greater control of the dopant concentration profile, since ion energy and dose can be varied to produce either flat or other desired distributions. Diffusion always yields either a Gaussian or a complementary error function distribution of dopant atoms, depending on the type of source used. An example of a waveguide formed by implantation of a substitutional dopant is the strip-loaded waveguide of *Leonberger* et al. [4.35], in which Be$^+$ implantation was used to produce a p$^+$ loading stripe on top of an n$^-$ epitaxial planar waveguide. Implantation of substitutional dopants has also been used by *Rao* et al. [4.36] to produce waveguides in CdTe, GaP, and ZnTe.

Aside from the substitutional dopant effect, ion implantation can also be used to produce waveguides in which the increase in index of refraction is the result of lattice disorder produced by non-substitutional implanted ions. An example of this type of waveguide has been produced by *Standley* et al. [4.37]. They found that implantation of fused quartz substrates with a variety of ions ranging from helium to bismuth produced a layer of increased index. Ion energy was in the range of 32–200 keV. The best results were obtained using lithium atoms, for which the empirical relation between ion dose and index of refraction at 6328 Å was given by

$$n = n_0 + 2.1 \times 10^{-21} C, \qquad\qquad (4.2.1)$$

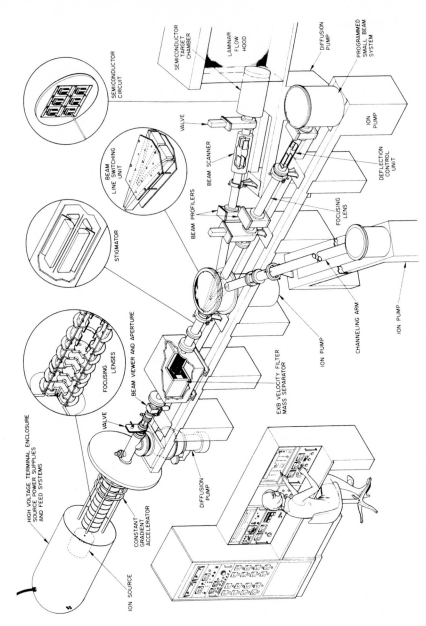

SEMICONDUCTOR
CIRCUIT

SEMICONDUCTOR
TARGET CHAMBER

LAMINAR
FLOW
HOOD

DIFFUSION
PUMP

PROGRAMMED
SMALL BEAM
SYSTEM

VALVE

BEAM
LINE SWITCHING
UNIT

BEAM SCANNER

ION
PUMP

DEFLECTION
CONTROL
UNIT

BEAM PROFILERS

STIGMATOR

FOCUSING
LENS

ION PUMP

CHANNELING ARM

ION PUMP

FOCUSING
LENSES

BEAM VIEWER AND APERTURE

EXB VELOCITY FILTER
MASS SEPARATOR

HIGH VOLTAGE TERMINAL ENCLOSURE
SOURCE POWER SUPPLIES
AND FEED SYSTEMS

VALVE

DIFFUSION
PUMP

CONSTANT
GRADIENT
ACCELERATOR

ION SOURCE

Fig. 4.4. Drawing of a 300-kV ion-implantation system with three arms for separate applications of the ion beam (Diagram courtesy of Hughes Research Laboratories, Malibu, CA)

where n_0 is the preimplantation index and C is the ion concentration per cm^3. Optical losses less than 0.2 dB/cm were achieved after annealing for 1 hr at 300 °C.

A third type of optical waveguide can be produced by ion implantation in certain semiconductors that are subject to defect center trapping. Proton bombardment can be used to generate lattice damage, which results in the formation of compensating centers, to produce a region of very low carrier concentration in a substrate with relatively large carrier concentration. The index of refraction is slightly larger in the low carrier concentration region, because free carriers normally reduce the index. The implanted protons do not by themselves cause a significant increase in index; the index difference results from carrier trapping which lowers the carrier concentration. This type of waveguide, which was first demonstrated in GaAs [4.38, 39], is discussed in detail in Sect. 4.3, along with other types of carrier-concentration-reduction waveguides.

4.3 Carrier-Concentration-Reduction Waveguides

In a semiconductor, any free carriers that are present contribute negatively to the index of refraction; i.e., they reduce the index below the value that it would have in a completely carrier-depleted sample of the material. Thus, if the carriers are somehow removed from a region, that region will have a larger index than the surrounding media and could function as a waveguide.

4.3.1 Basic Properties of Carrier-Concentration-Reduction Waveguides

A quantitative expression for the reduction in index of refraction produced by a given concentration per cm^3 of free carriers, N, can be developed by analogy with the change in index produced in a dielectric by a plasma of charged particles [4.40]. In that case

$$\Delta\left(\frac{\varepsilon}{\varepsilon_0}\right) = 2n\Delta n = -\frac{\omega_p^2}{\omega^2} = -\frac{Ne^2}{\varepsilon_0 m^* \omega^2}, \tag{4.3.1}$$

where ω_p is the plasma frequency, m^* is the effective mass of the carriers, and the other parameters are as defined previously. If we define n_0 as the index of refraction of the semiconductor when free carriers are absent, the index with N carriers per cm^3 present is

$$n = n_0 - \frac{Ne^2}{2n\varepsilon_0 m^* \omega^2}. \tag{4.3.2}$$

Thus, the difference in index between a waveguiding layer with N_2 carriers per cm^3 and a confining layer with N_3 carriers per cm^3 is

$$\Delta n = n_2 - n_3 = \frac{(N_3 - N_2)e^2}{2n_2\varepsilon_0 m^* \omega^2},$$

(4.3.3)

where $N_3 > N_2$, and where it has been assumed that $n_2 \cong n_3$.

Some interesting features of carrier-concentration-reduction waveguides are evident in (4.3.3). First, it can be seen that once N_2 has been reduced by a factor of about 10 below the value of N_3, further reduction does not substantially change the index difference Δn. The significance of this fact is that proton bombarded waveguides of the carrier-concentration-reduction type have a much more abrupt change in index at the interface between the waveguide and the confining layer than might be expected from the distribution profile of bombardment-created damage. This effect is illustrated in Fig. 4.5.

A second interesting feature of carrier-concentration-reduction waveguides can be seen by considering an asymmetric waveguide like that of Fig. 3.2, with $n_1 \ll n_2, n_3$. In that case, the carrier concentration difference required to permit waveguiding of the lowest order ($m = 0$) mode is found by combining (4.3.3) and (3.1.31), which yields

$$N_3 - N_2 \cong \frac{\varepsilon_0 m^* \omega^2 \lambda_0^2}{16 e^2 t_g^2}.$$

(4.3.4)

Since $\omega = 2\pi v$ and $v\lambda_0 = c$, (4.3.4) can be written

$$N_3 - N_2 \cong \frac{\pi^2 c^2 m^* \varepsilon_0}{4 e^2 t_g^2}.$$

(4.3.5)

Notice that the cutoff condition of (4.3.5) is independent of wavelength! Thus, in the case of a carrier-concentration-reduction type waveguide, no greater difference in concentration is required to guide light at, say, 10 μm wavelength than that required for 1 μm light. Of course, the shape of the mode, particularly the size of the evanescent tail extending into Region 3, would be different,

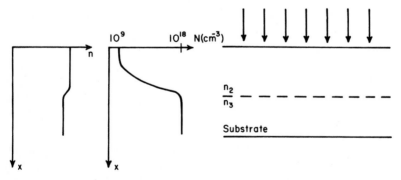

Fig. 4.5. Diagram of proton bombardment method of waveguide fabrication, showing the resultant carrier concentration (N) and index of refraction (n) profiles with depth

which would result in different losses at the two wavelengths. The reason for the unusual lack of dependence of the cutoff condition on wavelength is that both the required Δn and the Δn produced by a given value of $N_3 - N_2$ are proportional to λ_0^2, as can be seen from (4.3.4) and (3.1.31).

4.3.2 Carrier Removal by Proton Bombardment

Carrier-concentration-reduction waveguides have been produced by proton bombardment in both GaAs and GaP [4.39, 41, 42]. In general, the optical losses are greater than 200 dB/cm after bombardment with a proton dose that is greater than 10^{14} cm^{-2}. However, after annealing at temperatures below 500 °C, losses can be reduced below 3 dB/cm [4.38]. Typical anneal curves for samples of GaAs implanted with 300 keV protons are shown in Fig. 4.6. After annealing to remove excess optical absorption associated with the damage centers, the remaining loss is mostly due to free carrier absorption in the tail of the mode that extends into the substrate (see Fig. 2.5 for a representation of modes in proton bombarded waveguide). Substitution of typical values of m^* for GaAs and GaP, respectively, into (4.3.5) indicates that, for $t_g = 3\,\mu m$, substrate carrier concentrations of $N_3 = 6 \times 10^{17}$ cm^{-3} and $N_3 = 1.5 \times 10^{18}$ cm^{-3} are required to guide even the lowest-order mode. Hence, free carrier absorption in the mode substrate tail can be a serious problem. Increasing the ratio t_g/λ_0 reduces the relative size of the mode tail, mitigating the effect of related free carrier absorption. By careful control of ion dose and/or anneal conditions it is

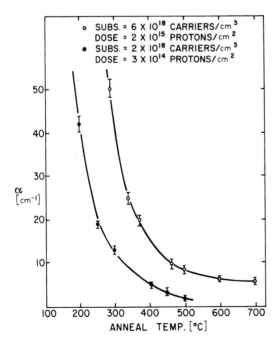

Fig. 4.6. Loss in proton bombarded waveguides. (Data are given in terms of the exponential loss coefficient α, defined by $I(z) = I_o \exp(-\alpha z)$, where I_o is the optical intensity at the starting point, $z = 0$)

possible to produce proton bombarded waveguides with losses less than 2 dB/cm for use at wavelengths ranging from 1.06 μm to 10.6 μm [4.39].

In addition to GaAs and GaP, ZnTe [4.43] and ZnSe [4.44] are materials in which proton-bombardment generated carrier compensation can be used to produce waveguides. The method is effective in both n- and p-type substrates, suggesting that deep level traps are involved rather than shallow acceptors or donors. Proton bombarded waveguides are useful over the normal transparency range of the substrate material. Thus, for example, waveguides for 6328 Å wavelength can be made in GaP, but not in GaAs. Waveguides for 1.06 μm, 1.15 μm and 10.6 μm have been made in both GaAs and GaP.

4.4 Epitaxial Growth

In a monolithic optical integrated circuit formed on a semiconductor substrate, epitaxial growth is the most versatile method of fabricating wave guides. This is true because the chemical composition of the epitaxially grown layer can be varied to adjust both the index of refraction and the wavelength range of transparency of the waveguide [4.45].

4.4.1 Basic Properties of Epitaxially Grown Waveguides

In monolithic semiconductor OIC's, there exists a basic problem of wavelength incompatibility that must be overcome. A semiconductor has a characteristic optical emission wavelength, corresponding to its bandgap energy, which is approximately the same as its absorption-edge wavelength. Thus, if a light emitting diode or laser is fabricated in a semiconductor substrate, it will emit light that is strongly absorbed in a waveguide formed in the same substrate material. Also, this light will not be efficiently detected by a detector formed in that substrate because its wavelength will correspond to the *tail* of the absorption-edge. In order to produce an operable OIC, the effective bandgap energies for absorption and emission must be altered in the various elements of the circuit so that

$$E_g \text{ waveguide} > E_g \text{ emitter} > E_g \text{ detector.} \tag{4.4.1}$$

Epitaxial growth of a ternary (or quaternary) material offers a convenient means of producing this required change of bandgap energy, as well as a change in index of refraction. In general, changes in bandgap on the order of tenths of an eV can be produced by changing the atomic composition of the grown layer by as little as 10%. Corresponding changes in index of refraction occur attendantly. With careful design, one can fabricate waveguides that have low loss at the emitted wavelength, along with efficient detectors. The epitaxial growth technique can best be illustrated by describing its application to $Ga_{(1-x)} Al_x As$, which is the most commonly used material for monolithic OIC fabrication.

4.4.2 Ga$_{(1-x)}$Al$_x$ As Epitaxially Grown Waveguides

The growth of Ga$_{(1-x)}$Al$_x$As epitaxial layers on a GaAs substrate for OIC fabrication is usually carried out by liquid phase epitaxy (LPE), in a tube furnace, at temperatures in the range from 700–900 °C. A more-or-less standard approach has been developed in which a number of liquid melts, of relatively small volume (~0.1 cm³), are contained in a movable graphite *slidebar* assembly that allows the substrate to be transported sequentially from one melt to the next so as to grow a multi-layer structure. Details of this approach have been reviewed by *Garmire* [4.46].

Epitaxial growth of multi-layered Ga$_{(1-x)}$Al$_x$As structures with different Al concentration in each layer, as shown in Fig. 4.7, can be used to create waveguides with transparency for light of wavelengths longer than the bandgap (absorption-edge) wavelength, which varies depending on Al concentration, as shown in Fig. 4.8.

The curves in Fig. 4.8 were generated by shifting the experimentally measured absorption edge of GaAs [4.47, 48] by an amount corresponding to a change in bandgap calculated from [4.49]

$$E_g(x) = 1.439 + 1.042x + 0.468x^2, \tag{4.4.2}$$

where x is the atomic fraction of Al in the Ga$_{(1-x)}$Al$_x$As. Al concentrations >35% are not often used because above that level the bandgap is indirect, resulting in deleterious effects which are discussed in Chap. 10. Thus, the Al concentration x can be selected so that the waveguide is transparent at the desired wavelength.

Optical losses as low as 1 dB/cm are obtainable. Once the Al concentration in the waveguiding layer has been selected, the next step in the waveguide design is to select the Al concentration y in the confining layer.

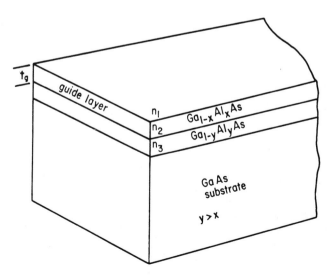

Fig. 4.7. Ga$_{(1-x)}$Al$_x$As planar waveguide

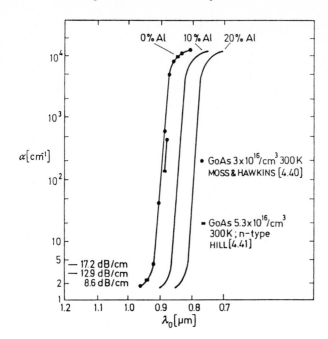

Fig. 4.8. Interband absorption as a function of wavelength and Al concentration

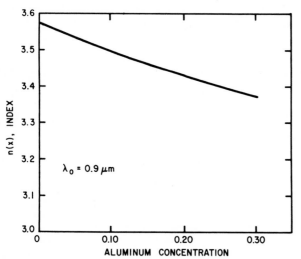

Fig. 4.9. Index of refraction of $Ga_{(1-x)}Al_xAs$ as a function of Al concentration

Increasing the Al concentration in $Ga_{(1-x)}Al_xAs$ causes the index of refraction to decrease. The dependence is shown in Fig. 4.9, which is a plot of the empirically determined Sellmeier equation [4.50, 51]

$$n^2 = A(x) + \frac{B}{\lambda_0^2 - C(x)} - D(x)\lambda_0^2,$$

(4.4.3)

where x is the atomic fraction of Al atoms in the $Ga_{(1-x)}Al_xAs$, and where A, B, C, D are functions of x as given in Table 4.2. Because of the slight nonlinearity of the n vs. x curve, the index difference between two layers with unequal Al concentrations depends not only on the difference in concentration, but also in its absolute level. This effect is illustrated in Fig. 4.10 for two different Al concentrations in the waveguiding layer, 0 and 20%, respectively.

For the case of an asymmetric $Ga_{(1-x)}Al_xAs$ waveguide, as shown in Fig. 4.7, the cutoff condition (3.1.31) can be used in conjunction with (4.4.3) to calculate what Al concentration is required to produce waveguiding of a given mode. Figure 4.11 shows the results of such a calculation for the case of

Table 4.2. Sellmeier equation coefficients (for λ_0 in μm)

Material	A	B	C	D
GaAs	10.906	0.97501	0.27969	0.002467
$Ga_{1-x}Al_xAs$	$10.906 - 2.92x$	0.97501	$(0.52886 - 0.735x)^2$ $x \leq 0.36$ $(0.30386 - 0.105x)^2$ $x \geq 0.36$	$[(0.002467) \cdot (1.41x + 1)]$

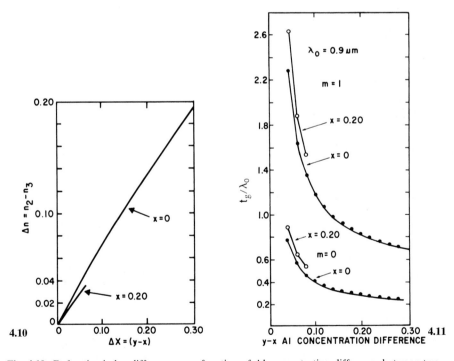

Fig. 4.10. Refractive index difference as a function of Al concentration difference between two layers of $Ga_{1-x}Al_xAs$ for two values of absolute Al concentration

Fig. 4.11. Guide thickness to wavelength ratio as a function of Al concentration between layer and substrate

waveguiding of the two lowest order modes of 0.9 µm wavelength light in $Ga_{(1-x)}Al_xAs$ waveguides of various thicknesses and Al concentrations [4.52]. Note the sharp increase in Al concentration difference $(y - x)$ required for thickness/wavelength ratios less than about 0.8, particularly in the case of the $m = 1$ mode. However, for $t_g/\lambda_0 \gtrsim 1.0$, a concentration difference of only 10% is sufficient to waveguide both the $m = 0$ and the $m = 1$ mode.

$Ga_{(1-x)}Al_xAs$ multilayer waveguides, produced by epitaxial growth, have been used in a variety of optical devices, of both discrete and integrated form. *Alferov* et al. [4.53] have used this type of waveguide to make one of the earliest confined field lasers, a topic that will be discussed in greater detail in Chap. 12. $Ga_{(1-x)}Al_xAs$ epitaxial waveguides have been used to make modulators and beam deflectors for ir light of 10.6 µm wavelength [4.54], and they also have been utilized in optical integrated circuits for 1 µm wavelength [4.55, 56].

To conclude this section on $Ga_{(1-x)}Al_xAs$ waveguides grown by liquid phase epitaxy, it should be noted that alternatives to the limited-melt, slidebar growth method are available. For example, *Craford* and *Groves* [4.57] have grown $Ga_{(1-x)}Al_xAs$ layers by vapor phase epitaxy (VPE), and *Kamath* [4.58] has used a vertical dipping *infinite melt* technique of LPE. The major advantage of both of these methods is that they permit the use of large area substrates. At the present time, slidebar LPE growth predominates because of the ease of fabricating multilayer structures and the accurate control of Al concentration that it provides. However, as demand for large monolithic OIC's increases in the future, interest in the alternative growth methods may increase as well.

4.4.3 Epitaxial Waveguides in Other III-V and II-VI Materials

Most of the work done on epitaxially grown OIC's has been done in $Ga_{(1-x)}Al_xAs$ because of the extraordinary close match of the lattice constants of GaAs and AlAs, 5.64 Å and 5.66 Å, respectively [4.59]. However, other III-V and II-VI materials are of interest. Waveguides have been grown in $Ga_{(1-x)}In_xAs$ [4.60] and in $CdS_xSe_{(1-x)}$ [4.61]. $Ga_{(1-x)}In_xAs$ is particularly interesting because the addition of In to GaAs shifts the bandgap, and hence the absorption-edge, to longer wavelengths. This suggests the possibility of $Ga_{(1-x)}In_xAs$ OIC's in which the operating wavelengths of emitters and detectors could be made longer by the addition of indium, while more pure GaAs could be used for waveguides to minimize absorption. Such circuits would be particularly useful in conjunction with fiber-optic interconnections, since the minimum absorption of glass fibers occurs for about 1.2 µm wavelength [4.59] Unfortunately, the lattice constants of GaAs and InAs do not match very well, being 6.06 Å for InAs as opposed to 5.64 Å for GaAs [Ref. 4.46, p. 257]. This mismatch results in serious lattice strain at the interfaces between layers of different In concentration, thereby increasing optical scattering and introducing phase distortion into the waveguided modes. This problem can be miti-

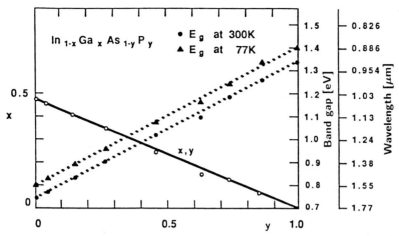

Fig. 4.12. Measured band gaps of lattice matched $In_{(1-x)}Ga_xAs_{(1-y)}P_y$ alloys as a function of alloy composition. [4.62]

gated by using layers of the quaternary compound $Ga_{(1-x)}In_xAs_yP_{(1-y)}$, which provides an additional degree of freedom in simultaneously optimizing both wavelength and lattice matching. The measured bandgaps of lattice matched $In_{(1-x)}Ga_xAs_{(1-y)}P_y$ alloys as a function of alloy composition are shown in Fig. 4.12 [4.62]. The dotted curves give the bandgap as a function of phosphorous concentration, while the solid curve shows the concentrations of Ga(x) and and P(y) required for the lattice matched condition. To use these curves for light emitter design, one projects horizontally from the bandgap scale at the desired wavelength to the dotted curve corresponding to the desired operating temperature. The intersection determines the P concentration (y). (The corresponding As concentration is given by $1 - y$.) The concentration of Ga(x) required for lattice matching is then read from the solid x, y curve at the point corresponding to the desired P concentration (y), using the values of x shown on the left vertical scale. The corresponding In concentration is given by $1 - x$. Because of lattice matching and bandgap control obtainable, InGaAsP heterojunction lasers have become prevalent in many applications today, operating at 1300 and 1550 nm wavelengths. InGaAsP lasers have also been made to operate at 2550 nm [4.63]. The integration of such lasors with In Ga AsP waveguides (and photodetectors) provides the basis for many optical integrated circuits.

4.4.4 Molecular Beam Epitaxy

In addition to the liquid phase epitaxy described in the preceding paragraphs, multi-layered structures of semiconductors can also be grown by the technique of molecular beam epitaxy (MBE). In MBE, the constituent atoms are delivered to the substrate surface during growth by beams of accelerated molecules (or atoms). Since neither atoms nor molecules are charged particles they can

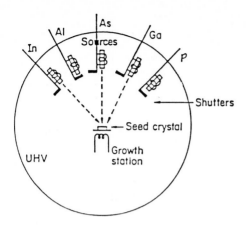

Fig. 4.13. A schematic of an MBE growth system

only be accelerated by thermal excitation. A schematic diagram of an MBE growth chamber is shown in Fig. 4.13.

Sources for the growth of either GaAlAs or GaInAsP are shown. These sources consist of small, resistively heated crucibles that have an elongated shape with an opening at one end to produce a directed beam of thermally accelerated atoms. Each source has a remotely controlled shutter so that the beam can be turned on or off as desired at different points in the growth cycle. The entire assembly is enclosed in an ultra-high-vacuum chamber (UHV) so that the atoms are not scattered by atmospheric molecules. For a thorough discussion of the techniques of molecular beam epitaxy the reader is referred to the books by *Chernov* [4.64] and *Ploog* and *Graf* [4.65].

The key advantage of MBE as compared to conventional LPE growth is that it provides superb control over purity, doping and thickness. To obtain the best control over purity, MBE systems generally consist of three separate chambers, a loading chamber, a growth chamber and an analysis chamber. This arrangement also increases the wafer throughput of the system because growth can be continuing on one substrate wafer while another is being loaded and prepared for growth. Remote control transfer mechanisms are used to move the wafer from one chamber to another [4.66]. In a carefully designed MBE system control of doping concentration and layer thickness with uniformity of ±1% over a 2-inch diameter wafer can be achieved [4.67]. Layers of thickness less than 100 Å can be controllably produced by MBE, permitting one to fabricate multilayered superlattices for use in quantum well devices [4.68]. This topic is discussed more thoroughly in Chap. 16.

4.4.5 Metal-organic Chemical Vapor Deposition

Another technique of epitaxial growth that has been applied to opto-electronic device fabrication is metal-organic chemical vapor deposition (MOCVD), sometimes called *metal – organic vapor phase epitaxy* (MOVPE), [4.69–72]. In the case of MOCVD, the constituent atoms are delivered to the substrate as a gaseous flow within a growth reactor furnace. Metal-organic gases such as PH_2, AsH_2, TEGa, TEIn and TEAl are used. The substrate temperature during growth is typically 750 °C, a relatively low temperature compared to those used for LPE. The growth reactor can be a simple tube furnace such as those used for dopant diffusion. However, very complex (and expensive) control valves, filters and exhaust vents must be used to regulate the gaseous flow. These are required for the sake of safety because the metal-organic gases are extremely toxic and the carrier gas (H_2) is explosive. Despite this inconvenience, MOCVD is growing in popularity as a method for fabricating opto-electronic devices because, like MBE, it offers precise control of dopant concentration and layer thickness. Quantum well structures can be grown by MOCVD. In addition, MOCVD can be used with large area substrate wafers. This latter property makes MOCVD very attractive as a method for large-volume production of lasers, LEDs and other opto-electronic devices [4.72].

4.5 Electro-Optic Waveguides

GaAs and $Ga_{(1-x)}Al_xAs$ exhibit a strong electro-optic effect, in which the presence of an electric field produces a change in the index of refraction. Thus, if a metal field plate is formed over a GaAs substrate, so as to produce a Schottky barrier contact, as shown in Fig. 4.14, and a voltage is applied so as to reverse bias the Schottky barrier, the electric field in the depletion layer can cause a sufficient change in the index of refraction to produce a waveguiding layer, with index n_2 larger than that of the substrate, n_1. Fortunately, it is very easy to form a Schottky barrier on GaAs or $Ga_{(1-x)}Al_xAs$. Almost any metal, except silver, will form a Schottky barrier rather than an ohmic contact, when deposited on n-type material without any heat treatment. When a reverse-bias voltage is applied to the Schottky barrier, a depletion layer forms, as in p-n junction. The index of refraction is increased in this layer by two different mechanisms.

First, the carrier depletion causes an increase as discussed in Sect. 4.3. Secondly, the presence of the electric field causes a further increase. For a detailed description of this linear electro-optic effect see, for example, *Yariv*

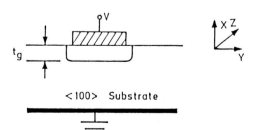

Fig. 4.14. Diagram of a basic electro-optic waveguide

[4.73]. The change in index of refraction for the particular orientation shown in Fig. 4.14, and for TE waves, is given by

$$\Delta n = n^3 r_{41} \frac{V}{2t_g}, \qquad (4.5.1)$$

where n is the index of refraction with no field present, V is the applied voltage, t_g is the thickness of the depletion layer, and r_{41} is the element of the electro-optic tensor [4.73] appropriate to the chosen orientations of crystal and electric field. The electro-optic effect is non-isotropic, and other orientations will not necessarily exhibit the same change in index. For example, in the case of a crystal oriented as in Fig. 4.14 but for TM waves (with the E vector polarized in the x direction), the change in index is zero. Thus, both the orientations of the crystal substrate and the polarization of the waves to be guided must be carefully considered in the design of an electro-optic waveguide.

The thickness of the waveguide, i.e. the depletion layer thickness, depends on the substrate carrier concentration, as well as on the applied voltage. Assuming a reasonable substrate carrier concentration of 10^{16} cm^{-3}, application of $V = 100$ V, i.e., the maximum allowed if avalanche breakdown is to be avoided, would produce a depletion layer thickness $t_g = 3.6\,\mu$m. This result is based on an abrupt junction calculation [4.74]. In that case, the resulting change in index, from (4.5.1), would be $\Delta n = 8.3 \times 10^{-4}$, where we have used $n = 3.4$ and $n^3 r_{41} = 6 \times 10^{-11}$ m/V for GaAs [4.52]. From (3.1.31), it can be seen that the lowest order mode would be guided in this case, but just barely. ($\Delta n = 5.7 \times 10^{-4}$ is required.)

From the foregoing example, it is obvious that the constraints placed on waveguide design by the necessity of avoiding avalanche breakdown are serious. Unless one has available unusually lightly doped substrate material (carrier concentration $<10^{16}$ cm^{-3}), only the lowest order mode can be guided. This problem can be alleviated by growing a relatively thick ($>10\,\mu$m) epitaxial layer of very lightly doped GaAs on the substrate before depositing the Schottky contact. Such layers can be grown with carrier concentration as low as 10^{14} cm^{-3}. In this case, care must be taken to use a low enough substrate carrier-concentration so that the epilayer does not form a carrier-concentration-reduction type waveguide.

The greatest advantage of the electro-optic waveguide, as compared to the others which have been described, is that it is electrically switchable. Thus, it can be used in switches and modulators, as is described in more detail in Chap. 8. Varying the applied voltage changes not only the index of refraction in the waveguiding region, but also the thickness of the waveguide, thus bringing the guide either above or below cutoff for a particular mode, as desired.

The field plate Schottky contact can be in the form of a narrow stripe, either straight or curved. In that case, a rectangular channel waveguide is formed underneath the stripe when the voltage is increased above the cutoff threshold level. Some additional methods of fabricating channel, or stripe, waveguides are described in the next section.

4.6 Methods for Fabricating Channel Waveguides

The starting point for the fabrication of many types of channel waveguides is a substrate on which a planar waveguide has been formed by any of the methods previously described. The lateral dimensions of a number of different channel waveguides can then simultaneously be determined on the surface of the wafer by using standard photolithographic techniques, such as are used in the fabrication of electrical integrated circuits.

4.6.1 Ridged Waveguides Formed by Etching

The usual method of fabricating a ridged waveguide is to coat the planar waveguide sample with photoresist, expose the resist to uv light or x-rays through a contact printing mask that defines the waveguide shape, then develop the resist to form a pattern on the surface of the sample, as shown in Fig. 4.15 [4.75]. The photoresist [4.76] is an adequate mask for either wet-chemical [4.77] or ion-beam-sputter etching [4.78]. Ion-beam-sputter etching, or *micromachining* as it is sometimes called, produces smoother edges, particularly on curves, but it also causes some lattice damage which must be removed by annealing if minimum optical losses are desired. A photograph of a ridged channel waveguide formed in GaAs by ion-beam micromachining is shown in Fig. 4.16 [4.78]. Photoresist makes a surprisingly good mask against bombarding ions because the mass of its organic molecules is so much less than the mass of Ar^+ or Kr^+ that there is relatively little energy transfer, as compared to that transferred in an Ar-GaAs interaction. For example, the difference in sputter-

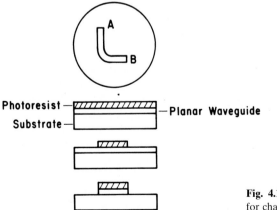

Fig. 4.15. Photoresist masking technique for channel waveguide fabrication

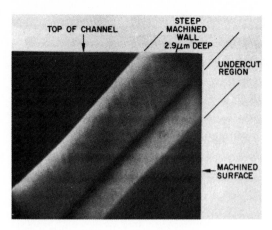

Fig. 4.16. Ion beam etched channel waveguide. (Photo courtesy of Hughes Research Laboratories, Malibu, CA)

ing yields under 2 kV Ar⁺ ion bombardment is such that a 1 μm thick photoresist mask allows the removal of 5 μm of GaAs in unmasked regions. Residual lattice damage after ion machining of GaAs can be removed by annealing the sample for about a half hour at 250 °C.

Wet chemical etching can also be used in conjunction with a photoresist mask to define channel waveguide structures. Chemical etching produces no lattice damage, but it is very difficult to control the etch depth and profile. Most etchants are preferential with regard to crystal orientation, thus leading to ragged edges on curved sections of waveguides. However, preferential chemical etch effects can sometimes be used advantageously, as is done to define the rectangular mesas of semiconductor lasers [4.79].

An interesting method that appears to combine the advantages of ion-beam machining with those of wet chemical etching is an ion-bombardment-enhanced etching technique applied to LiNbO$_3$ by *Kawabe* et al. [4.80]. They produced 2 μm wide ridged waveguides in a Ti-diffused planar waveguiding layer by using a poly (methylmethacrylate) resist mask and electron beam lithography. The etching was accomplished by first bombarding the sample with a dose of 3×10^{15} cm^{-2} 60 keV Ar⁺ at room temperature. Because of the relatively large energy of the Ar⁺ ions, much lattice damage was generated but little sputtering of surface atoms occurred. The bombardment was followed by wet etching in diluted HF, which preferentially removes the damaged layer [4.81]. For 60 keV Ar⁺, the depth removed is 700 Å, but the bombardment per etch can be repeated a number of times to produce deeper etching. Ion-bombardment-enhanced etching combines more accurate pattern definition of ion-beam micro-machining with the damage-free crystal surface that is obtained with wet chemical etching.

4.6.2 Strip-Loaded Waveguides

Instead of etching the lateral walls of a channel waveguide into a previously formed planar waveguide structure, a loading strip of dielectric can be depos-

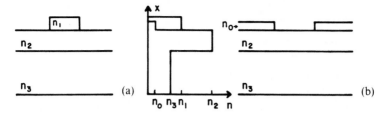

Fig. 4.17 a, b. Index of refraction profile in **a** dielectric strip loaded waveguide and **b** metal strip loaded waveguide

ited on the top surface to provide lateral confinement, as described in Sect. 3.2.2. The shape of this strip can be defined by the same photolithographic process that is used to make ridged waveguides, as described in Sect. 4.6.1. Again, either ion beam or wet chemical etching can be used in conjunction with the photoresist masking. *Blum* et al. [4.81] have made optical striplines of this type in GaAs by using a heavily doped GaAs loading strip, with index n_1 slightly smaller than that in the waveguiding layer n_2, as shown in Fig. 4.17. They used a wet chemical etch of NaOH and H_2O_2, with photoresist masking. Also shown in Fig. 4.17 is a metal strip-loaded waveguide, with index n_0 in the metal. This type of waveguide can also be fabricated by photoresist masking and either ion-beam or chemical etching.

4.6.3 Masked Ion Implantation or Diffusion

Not all techniques for fabricating channel waveguides begin with a planar waveguide structure. Instead, channel waveguides can be formed by either diffusion [4.82] or ion implantation [4.83, 84] of suitable dopant atoms directly into the substrate, but through a mask. Such waveguides are often called *buried* channel waveguides since they lie beneath the surface. Photoresist is not an effective mask in this case, because it cannot withstand the high temperature required for diffusion, nor does it have enough mass to block incident high-energy ions. Usually, deposited oxides, such as SiO_2 or Al_2O_3, are used as diffusion masks, while noble metals such as Au or Pt are used as implantation masks. Photoresist is used, however, to define the pattern on the masking oxide or metal layer.

When waveguides are directly implanted or diffused into the substrate, the optical quality of the substrate material is critically important. It must have low optical absorption losses, as well as a smooth surface to avoid scattering. One of the most significant advantages of buried channel waveguide production by implantation or diffusion through a mask is that it is a planar process; the surface of the OIC is not disrupted by ridges or valleys, so that optical coupling into and out of the OIC is facilitated, and problems of surface contamination by dust or moisture are minimized. Another advantage of this process is that it is less elaborate than those described in previous sections, because there is no need to fabricate a planar waveguide structure over the whole OIC chip before defining the strip waveguides. In order to define masked patterns with dimensions on the order of 1 μm, electron beam lithography can be used [4.85, 86]. Also higher definition etched patterns can be obtained by using reactive ion etching [4.87, 88].

Both ion-implantation and ion-etching processes are usually done in conjunction with masks to define the lateral geometry of the waveguide. However state-of-the-art focused ion-beam systems are capable of forming beam diameters on the order of 50 nm, with a current deusity of $1 \, A/cm^2$. Using these systems, unique processing techniques such as maskless ion implantation and etching have been developed [4.89]

Problems

4.1   We wish to fabricate a planar waveguide for light of wavelength $\lambda_o = 1.15$ μm that will operate in the single (fundamental) mode. If we use the proton-bombardment, carrier-concentration-reduction method to form a 3 μm thick waveguide in GaAs, what are the minimum and maximum allowable carrier concentrations in the substrate? (Calculate for the two cases of p-type or n-type substrate material if that will result in different answers.)

4.2   If light of $\lambda_o = 1.06$ μm is used, what are the answers to Problem 4.1?

4.3   Compare the results of Problems 4.1 and 4.2 with those of Problems 2.1 and 2.2, note the unique wavelength dependence of the characteristics of the carrier-concentration-reduction type waveguide.

4.4   We wish to fabricate a planar waveguide for light of $\lambda_o = 0.9$ μm in $Ga_{(1-x)}Al_xAs$. It will be a double layer structure on a GaAs substrate. The top (waveguiding) layer will be 3.0 μm thick with the composition $Ga_{0.9}Al_{0.1}As$. The lower (confining) layer will be 10 μm thick and have the composition $Ga_{0.17}Al_{0.83}As$. How many modes will this structure be capable of waveguiding?

4.5   What is the answer to Problem 4.4 if the wavelength is $\lambda_0 = 1.15$ μm?

4.6   A planar asymmetric waveguide is fabricated by depositing a 2.0 μm thick layer of Ta_2O_5 ($n = 2.09$) onto a quartz substrate ($n = 1.5$).
a) How many modes can this waveguide support for light of 6328 Å (vacuum wavelength)?
b) Approximately what angle does the ray representing the highest-order mode make with the surface of the waveguide?

4.7   A three-layer planar waveguide structure consists of a 5 μm thick waveguiding layer of GaAs with electron concentration $N = 1 \times 10^{14} \, cm^{-3}$, covered on top with a confining layer of $Ga_{0.75}Al_{0.25}As$ with $N = 1 \times 10^{18}$ cm^{-3}, and on bottom with a confining layer of GaAs with $N = 1 \times 10^{18}$ cm^{-3}. If this waveguide is used to guide light of $\lambda_0 = 1.06$ μm, how many modes can it support?

4.8 For an asymmetric planar waveguide like that of Fig. 3.1 with $n_1 = 1.0$, $n_2 = 1.65$, $n_3 = 1.52$, $t_g = 1.18$ μm, and $\lambda_0 = 0.63$ μm, find the number of allowed TE modes.

4.9 If the waveguide of Prob. 4.4 were to be modified so as to support one additional mode, would it be practical to produce the required change is n by using the electro-optic effect? (Assume a uniform electric field to be applied over the thickness of the waveguide in the direction normal to the surface, with $r_{41} = 1.4 \times 10^{-12}$ m/V).

5. Losses in Optical Waveguides

Chapters 2 and 3 have explained cutoff conditions in waveguides and described the various optical modes which can be supported. Following the question as to which modes propagate, the next most important characteristic of a waveguide is the attenuation, or loss, that a light wave experiences as it travels through the guide. This loss is generally attributable to three different mechanisms: scattering, absorption and radiation. Scattering loss usually predominates in glass or dielectric waveguides, while absorption loss is most important in semiconductors and other crystalline materials. Radiation losses become significant when waveguides are bent through a curve.

Thus far we have used either light waves or their associated rays to represent the optical fields. However, loss mechanisms can often be described more advantageously by using the quantum mechanical description, in which the optical field is viewed as a flux of particle-like quantized units of electromagnetic energy, or photons [5.1]. Photons can be either scattered, absorbed or radiated as the optical beam progresses through the waveguide, thus reducing the total power transmitted. When photons are absorbed, they are annihilated by giving up their energy to either the atoms or subatomic particles (usually electrons) of the absorbing material. In contrast, when photons are scattered or radiated, they maintain their identity by changing only their direction of travel, and sometimes their energy (as in Raman scattering). Nevertheless, scattered or radiated photons are removed from the optical beam, thus constituting a loss as far as the total transmitted energy is concerned.

5.1 Scattering Losses

There are two types of scattering loss in an optical waveguide: volume scattering and surface scattering. Volume scattering is caused by imperfections, such as voids, contaminant atoms and crystalline defects, within the volume of the waveguide. The loss per unit length due to volume scattering is proportional to the number of imperfections (scattering centers) per unit length. Also, the volume scattering loss depends very strongly on the relative size of the imperfections, as compared to the wavelength of light in the material. In all but the crudest of waveguides, volume imperfections are so small compared to wavelength, and so few in number, that the volume scattering loss is negligible compared to surface scattering loss.

5.1.1 Surface Scattering Loss

Surface scattering loss can be significant even for relatively smooth surfaces, particularly in the case of higher-order modes, because the propagating waves interact strongly with the surfaces of the waveguide. This effect can be visualized best by considering the ray-optic description of the guided wave shown in Fig. 5.1. A wave traveling in the guide experiences many bounces. In a length L, the number of reflections from each surface is given by

$$N_R = \frac{L}{2t_g \cot \theta_m}. \tag{5.1.1}$$

As was demonstrated in Problem 2.6, for the case of a Ta_2O_5 waveguide with $t_g = 3$ μm, $n_2 = 2.0$, and $\beta_m = 0.8 \, kn_2$, light of wavelength $\lambda_0 = 9000$ Å undergoes 1250 reflections from each surface for each cm traveled. Scattering loss then occurs at each reflection. Since θ_m is larger for the higher-order modes, they experience greater loss because of surface scattering.

To quantitatively describe the magnitude of optical loss, the exponential attenuation coefficient is generally used. In that case, the intensity (power per area) at any point along the length of the waveguide is given by

$$I(z) = I_0 e^{-\alpha z}, \tag{5.1.2}$$

where I_0 is the initial intensity at $z = 0$. It can be shown (Problem 5.2) that the loss in dB/cm is related to α by

$$\mathscr{L}\left[\frac{dB}{cm}\right] = 4.3\alpha[\text{cm}^{-1}]. \tag{5.1.3}$$

Tien [5.2] has derived an expression for scattering loss due to surface roughness, based on the Rayleigh criterion. He has shown that loss to be

$$\alpha_s = A^2 \left(\frac{1}{2} \frac{\cos^3 \theta'_m}{\sin \theta'_m}\right) \left(\frac{1}{t_g + (1/p) + (1/q)}\right), \tag{5.1.4}$$

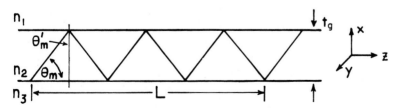

Fig. 5.1. Diagram of a ray optic approach to determination of scattering loss

where θ'_m is as shown in Fig. 5.1, p and q are the extinction coefficients in the confining layers (Sect. 3.1.1), and A is given by

$$A = \frac{4\pi}{\lambda_2}(\sigma_{12}^2 + \sigma_{23}^2)^{1/2},$$ (5.1.5)

where λ_2 is the wavelength in the guiding layer, and σ_{12}^2 and σ_{23}^2 are the variances of surface roughness. It will be recalled that the statistical variance of a variable x is given by

$$\sigma^2 = S[x^2] - S^2[x],$$ (5.1.6)

where $S[x]$ is the mean value of x and

$$S[x^2] = \int_{-\infty}^{\infty} x^2 f(x)dx,$$ (5.1.7)

where $f(x)$ is the probability density function.

The expression given in (5.1.4) is based on the Rayleigh criterion, which says: if the incident beam at the surface has power P_i, then the specularly reflected beam has power

$$P_r = P_i \exp\left[-\left(\frac{4\pi\sigma}{\lambda_2}\cos\theta'_m\right)^2\right].$$ (5.1.8)

This Rayleigh criterion holds only for long correlation lengths, but that is a reasonably good assumption in most cases.

Note in (5.1.4) and (5.1.5) that the attenuation coefficient α is basically proportional to the square of the ratio of the roughness to the wavelength in the material, as represented by A^2. This ratio is then weighted by a factor varying inversely with the waveguide thickness plus the terms $1/p$ and $1/q$, which are related to the penetrating *tails* of the mode. Obviously, well-confined modes may be more affected by surface scattering than modes with large evanescent tails. If $1/p$ and $1/q$ are large compared to t_g, scattering will be reduced. Physically, the penetration of the wave at the interface makes it less sensitive to surface roughness, tending to average out the effect of variations. The factor $\cos^3 \theta'_m/\sin \theta'_m$ in (5.1.4) accounts for the greater loss for higher-order modes (with small θ'_m) because of more reflections at the surface per unit length traveled in the direction of propagation. Experimentally measured [5.3] values of α_s for a Ta_2O_5 waveguide, and light of 6328 Å wavelength, correlate very well with those calculated using (5.1.4), indicating $\alpha_s = 0.3 \, cm^{-1}$ for the $m = 0$ mode, as compared to $\alpha_s = 2.8$ for the $m = 3$ mode.

Although Tien's theoretical model for scattering loss [5.2] is only an approximation, it provides a convenient closed-form expression for α_s. A more sophisticated theory of surface scattering in slab waveguides has been developed by *Marcuse* as part of a larger effort to calculate total waveguide losses [5.4]. Marcuse's theory treats surface scattering as a form of radiation loss in which irregularities in the surfaces of the waveguide couple energy from propagating modes into radiation modes (and into other propagating modes as well). With proper approximations, and in the limit of long correlation length, the results of Marcuse's theory correlate well with those predicted by (5.1.4).

The far-field radiation patterns of light scattered by surface irregularities have been studied by *Suematsu* and *Furuya* [5.5], and by *Miyanaga* et al. [5.6]. In general, the substrate scattering in highly directional, having many very narrow lobes at specific angles, while the air scattering occurs in a single broad lobe; this air lobe is peaked at an angle with respect to the direction of propagation which is dependent on the correlation length. *Gottlieb* et al. [5.7] have experimentally observed the out-of-plane scattering of 7059 glass waveguides on substrates of thermally oxidized silicon, and have found the results to be in agreement with the theoretical predictions of *Suematsu* and *Furuya*.

Surface scattering is generally the dominant loss in dielectric film waveguides, such as glasses and oxides, contributing about 0.5–5 dB/cm to the losses of the lowest-order mode, and more for higher-order modes [5.8, 9]. This loss is consistent with surface variations of about 0.1 µm, which are typically observed in deposited thin film waveguides. In semiconductor waveguides, thickness variations can usually be held to approximately 0.01 µm, and also, absorption losses are much larger, so that surface scattering is not as important.

5.2 Absorption Losses

Absorption losses in amorphous thin films and in crystalline ferroelectric materials, such as $LiTaO_3$ or $LiNbO_3$, are generally negligibly small compared to scattering loss, unless contaminant atoms are present [5.10, 11]. However, in semiconductors, significant loss occurs because of both interband or *band edge absorption* and free carrier absorption.

5.2.1 Interband Absorption

Photons with energy greater than the bandgap energy are strongly absorbed in semiconductors by giving up their energy to raise electrons from the valence band to the conduction band. This effect is generally very strong, resulting in absorption coefficients that are larger than 10^4 cm^{-1} in direct bandgap semiconductors. See, for example, the absorption curve for GaAs shown in Fig. 4.8. To avoid interband absorption, one must use a wavelength that is significantly longer than the absorption edge wavelength of the waveguide material. Chapter 4 discussed the technique of adding additional elements to a binary compound to control the bandgap. Such as adding Al to GaAs to form $Ga_{(1-x)}Al_xAs$ or adding P and In to form $In_{(1-x)}Ga_xAs_{(1-y)}P_y$. This method is generally effective in III–V and II–VI compounds, as long as care is taken in the proper choice of the additional elements so as to obtain an acceptable lattice match between epitaxial layers. Proper concentrations of the elements must be present to sufficiently shift the absorption edge so that the operating wavelength lies beyond the *tail* of the absorption curve. This effect is illustrated for the case of $Ga_{(1-x)}Al_xAs$ in Fig. 5.2, where experimentally determined absorption data are presented for wavelengths slightly longer than the absorption edge. A multiple abscissa scale is used in Fig. 5.2 to illustrate the shift of the absorption edge to shorter wavelengths as the Al concentration is increased. It can be seen, for example, that an Al concentration of $x - 30\%$ results in a reduction of absorption loss to 3 dB/cm for a wavelength of 9000 Å. With no Al, the loss would be about $50\,cm^{-1}$ or 215 dB/cm. The level of absorption loss that can be tolerated, of course, depends on the particular application. However, since most OIC's are sized on the order of a centimeter, 3 dB/cm should be acceptable in most cases. Table 5.1 shows the Al concentration required to reduce interband absorption to less than either 8.6 dB/cm or 3 dB/cm, as desired, for light from three different semiconductor laser sources that have been used in OIC's [5.13].

Interband absorption can also be avoided by using a hybrid approach, which employs a laser source emitting a wavelength significantly longer than the absorption edge of the waveguide material. For example, CdTe and

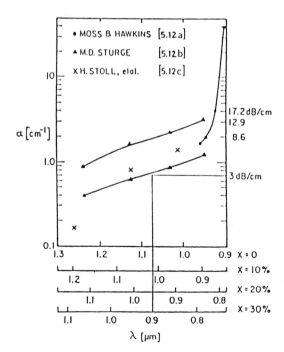

Fig. 5.2. Absorption in the long wavelength tail of the $Ga_{(1-x)}Al_xAs$ bandedge. Abscissa scales are shown for four different concentrations of aluminum (in atomic percent)

Table 5.1. Absolute aluminum concentration requirements

Source wavelength	Required aluminum concentration in the guide	
	$\alpha = 2\,\mathrm{cm}^{-1}$ (8.6 dB/cm)	$\alpha = 0.7\,\mathrm{cm}^{-1}$ (3 dB/cm)
0.85 µm GaAlAs	17%	40%
0.90 µm GaAs	7%	32%
0.95 to 1.0 µm Si:GaAs	0%	20%

GaAs waveguides have been used for CO_2 laser emission at 10.6 µm [5.14, 15], and GaP waveguides have been used with He-Ne laser light at 6328 Å [5.16].

Regardless of which approach is used to avoid interband absorption in semiconductor waveguides, additional steps must be taken to eliminate free carrier absorption if practical waveguides are to be realized.

5.2.2 Free Carrier Absorption

Free carrier absorption, sometimes called *intra*band absorption, is that which occurs when a photon gives up its energy to an electron already in the conduction band, or to a hole in the valence band, thus raising it to higher energy. Usually free carrier absorption is taken to include absorption in which electrons are raised out of shallow donor states near the conduction band edge, or holes are excited into the valence band from shallow acceptor states near the valence band edge.

An expression for the absorption coefficient α_{fc} due to free carrier absorption can be derived from classical elecromagnetic theory. It is worthwhile to review that derivation because it gives insight into the nature of free carrier absorption and also provides an attendant derivation of the change in index of refraction that is produced by the presence of free carriers.

The motion of an electron in the presence of an applied field $E_0 \exp(i\omega t)$ must satisfy the differential equation [5.17]

$$m * \frac{d^2 x}{dt^2} + m * g \frac{dx}{dt} = -eE_o \mathrm{e}^{i\omega t}, \tag{5.2.1}$$

where g is a damping coefficient and x is the displacement. The first term in (5.2.1) is the familiar force term (mass × acceleration); the second term represents a linear damping of the electron motion by interaction with the lattice, and the term to the right of the equals sign is the applied force. The steady state solution of (5.2.1) is

$$x = \frac{(eE_0)/m*}{\omega^2 - i\omega g} e^{i\omega t}. \tag{5.2.2}$$

The dielectric constant of a material is given, in general, by

$$K = \frac{\varepsilon}{\varepsilon_0} = 1 + \frac{\overline{P}}{\varepsilon_0 \overline{E}}, \tag{5.2.3}$$

where $\overline{P}$ is the polarization. In the presence of free carriers

$$\overline{P} = \overline{P}_0 + \overline{P}_1, \tag{5.2.4}$$

where $\overline{P}_0$ is the component present without carriers, i.e. the polarization of the dielectric, and $\overline{P}_1$ is the additional polarization due to the shift of the electron cloud in the field. Thus,

$$K = \frac{\varepsilon}{\varepsilon_0} = 1 + \frac{\overline{P}_0}{\varepsilon_0 \overline{E}} + \frac{\overline{P}_1}{\varepsilon_0 \overline{E}} \tag{5.2.5}$$

or

$$K = n_0^2 + \frac{\overline{P}_1}{\varepsilon_0 \overline{E}}, \tag{5.2.6}$$

where n_0 is the index of refraction of the material without carriers present. Assuming an isotropic material, in which $\overline{P}$ and $\overline{E}$ are in the same direction,

$$\overline{P}_1 = - Ne\overline{x}, \tag{5.2.7}$$

where N is the free carrier concentration per cm^3 and x is the displacement already given in (5.2.2). Substituting (5.2.7) and (5.2.2) into (5.2.6), we get

$$K = n_0^2 - \frac{(Ne^2)/(m*\varepsilon_0)}{\omega^2 - i\omega g}. \tag{5.2.8}$$

Separating the real and imaginary parts of K, we find

$$K_r = n_0^2 - \frac{(Ne^2)/(m*\varepsilon_0)}{\omega^2 + g^2}, \tag{5.2.9}$$

and

$$K_i = \frac{(Ne^2 g)/(m*\omega\varepsilon_0)}{\omega^2 + g^2}. \tag{5.2.10}$$

The damping coefficient g can be evaluated from the known steady-state solution of (5.2.1). Since, at steady state, $d^2x/dt^2 = 0$,

$$m * g \frac{dx}{dt} = eE. \tag{5.2.11}$$

From the definition of mobility, μ, we use

$$\frac{dx}{dt} = \mu E, \tag{5.2.12}$$

which yields

$$g = \frac{e}{\mu m *}. \tag{5.2.13}$$

Consider the magnitude of g relative to ω. For the typical case of n-type GaAs, $\mu \cong 2000 \text{ cm}^2/\text{V s}$ and $m* = 0.08 \, m_0$. Thus, $g = 1.09 \times 10^6 \text{ s}^{-1}$. Since $\omega \sim 10^{15} \text{ s}^{-1}$ at optical frequencies, g can most certainly be neglected in the denominators of (5.2.9) and 5.2.10). Making that approximation, and also substituting for g from (5.2.13), we get

$$K_r = n_0^2 - \frac{Ne^2}{m * \varepsilon_0 \omega^2} \tag{5.2.14}$$

and

$$K_i = \frac{Ne^3}{(m*)^2 \varepsilon_0 \omega^3 \mu}. \tag{5.2.15}$$

The exponential loss coefficient α is related to the imaginary part of the dielectric constant by

$$\alpha = \frac{kK_i}{n}, \tag{5.2.16}$$

where n is the index of refraction and k is the magnitude of the wavevector. Hence, for the case of free carrier absorption,

$$\alpha_{fc} = \frac{kK_i}{n} = \frac{Ne^3}{(m*)^2 n\varepsilon_0 \omega^2 \mu c}, \tag{5.2.17}$$

where $k \equiv \omega/c$ has been used. Since $c = v\lambda_0$ and $\omega = 2\pi v$, (5.2.17) can be rewritten as

$$\alpha_{fc} = \frac{Ne^3 \lambda_0^2}{4\pi^2 n(m*)^2 \mu\varepsilon_0 c^3}. \tag{5.2.18}$$

For the typical case of 1.15 μm light guided in n-type GaAs, with $n = 3.4$, $m* = 0.08 \, m_0$, and $\mu = 2000 \text{ cm}^2/\text{V s}$, (5.2.18) becomes

$$\alpha_{fc}[\text{cm}^{-1}] \cong 1 \times 10^{-18} N[\text{cm}^{-3}].$$ (5.2.19)

Thus, free carrier absorption in heavily doped ($N > 10^{18} \text{cm}^{-3}$) GaAs can be expected to produce losses of the order of 1–10 cm^{-1}. The major loss due to free carrier absorption occurs in the evanescent tail of optical modes propagating on heavily doped substrates or confining layers (Sect. 4.3.2.). However, by properly choosing the ratio of guide thickness to wavelength one can minimize these losses, as shown by the data of *Mentzer* et al. [5.18] given in Fig. 5.3.

Before leaving this theoretical development, note in (5.2.14) that the change in index of refraction resulting from the presence of the carriers is $-(Ne^2)/(m^*\varepsilon_0\omega^2)$, as was given previously, without proof, in (4.3.1).

The classical expression for α_{fc} given in (5.2.18) exhibits a λ_0^2 dependence. However, that rarely is exactly observed in practical situations. It must be remembered that the model on which (5.2.18) is based assumes a constant,

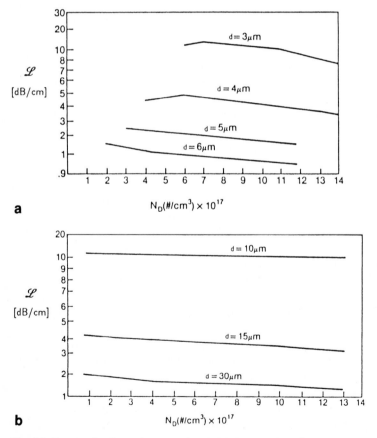

Fig. 5.3. Propagation losses in proton bombarded waveguides (d = waveguide thickness, N_D = substrate doping concentration). **a** $\lambda_0 = 1.3$ μm propagation losses versus substrate doping. **b** $\lambda_0 = 10.6$ μm propagation losses versus substrate doping

wavelength independent, damping coefficient g. In actuality, the damping that occurs because of interaction with the lattice is a varying quantity, which depends on whether acoustic phonons, optical phonons, or ionized impurities are involved. In general, all three will be involved to some extent and the resultant free carrier absorption coefficient can best be represented by [5.19]

$$\alpha_{fc} = A\lambda_0^{1.5} + B\lambda_0^{2.5} + C\lambda_0^{3.5}, \tag{5.2.20}$$

where A, B and C are constants giving the relative proportions due to acoustic phonons, optical phonons, and ionized impurities, respectively. Free carrier absorption in n-type compound semiconductors have been studied by *Fan* [5.20], and an effective wavelength dependence has been determined for a number of different materials.

In general, (5.2.18) will give a reasonably accurate estimate of free carrier loss in the visible and near-ir wavelength range, even though it is only an approximation.

5.3 Radiation Losses

Optical energy can be lost from waveguided modes by radiation, in which case photons are emitted into the media surrounding the waveguide and are no longer guided. Radiation can occur from planar waveguides as well as from channel waveguides.

5.3.1 Radiation Loss from Planar and Straight Channel Waveguides

Radiation losses from either planar or straight channel waveguides are generally negligible for well confined modes that are far from cutoff. However, at cutoff, all of the energy is transferred to the substrate radiation modes, as discussed in Chap. 2. Since the higher-order modes of a waveguide are always either beyond cutoff or are, at least, closer to cutoff than the lower-order modes, radiation loss is greater for higher-order modes. In an ideal waveguide the modes are orthogonal, so that no energy will be coupled from the lower-order modes to the higher-order modes. However, waveguide irregularities and inhomogeneities can cause mode conversion, so that energy is coupled from lower-order to higher-order modes [5.21]. In that case, even though a particular mode may be well confined, it may suffer energy loss through coupling to higher-order modes with subsequent radiation. This problem is not usually encountered in typical waveguides of reasonably good quality, and radiation losses can generally be neglected compared to scattering and absorption losses. The one important exception is the case of curved channel waveguides.

5.3.2 Radiation Loss from Curved Channel Waveguides

Because of distortions of the optical field that occurs when guided waves travel
through a bend in a channel waveguide, radiation loss can be greatly increased.
In fact, the minimum allowable radius of curvature of a waveguide is generally
limited by radiation losses rather than by fabrication tolerances. Since
waveguide bends are a necessary part of all but the simplest OIC's the radiation losses from a curved waveguide must be considered in circuit design.

A convenient way of analyzing radiation loss is the velocity approach
developed by *Marcatili* and *Miller* [5.22]. The tangential phase velocity of
waves in a curved waveguide must be proportional to the distance from the
center of curvature because otherwise the phase front would not be preserved.
To appreciate this, consider the case of a waveguide mode assumed to be
propagating in a circular bend of radius R, with a propagation constant β_z, as
shown in Fig. 5.4. There is a certain radius $(R + X_r)$ beyond which the phase
velocity would have to exceed the velocity of unguided light (in the confining
medium, with index n_1), in order to preserve the phase front. Since $d\theta/dt$ must
be the same for all waves along the phase front, two resultant equalities are
that

$$(R + X_r)\frac{d\theta}{dt} = \frac{\omega}{\beta_0},$$

(5.3.1)

and

$$R\frac{d\theta}{dt} = \frac{\omega}{\beta_z},$$

(5.3.2)

where β_0 is the propagation constant of unguided light in Medium 1, and β_z is
the propagation constant in the waveguide at radius R.

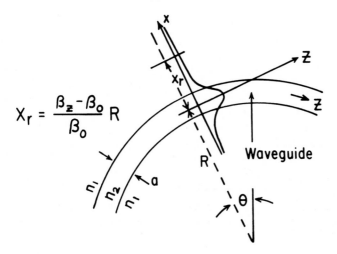

$$X_r = \frac{\beta_z - \beta_0}{\beta_0} R$$

Fig. 5.4. Diagram
illustrating the velocity
approach to the
determination of
radiation loss

Combining (5.3.1) and (5.3.2) leads to

$$X_r = \frac{\beta_z - \beta_0}{\beta_0} R. \tag{5.3.3}$$

The radiation process can be visualized as follows. Photons of the optical mode located at radii greater than $R + X_r$ cannot travel fast enough to keep up with the rest of the mode. As a result, they split away and are radiated into Medium 1. The question which arises is then, "How far must the photons travel before they can be considered as having been removed from the guided mode?". An estimate of this length can be made by analogy to the emission of photons from an abruptly terminated waveguide, as shown in Fig. 5.5. *Miller* [5.23] has shown that light emitted into a medium from an abruptly terminated waveguide remains collimated to within a waveguide thickness over a length Z_c given by

$$Z_c = \frac{a}{\varphi} = \frac{a^2}{2\lambda_1}, \tag{5.3.4}$$

where α and ϕ are the near field beam width and far-field angle, as shown in Fig. 5.5, and where λ_1 is the wavelength in the medium surrounding the waveguide. The derivation of (5.3.4) is based on the fundamental relation from diffraction theory

$$\sin\frac{\varphi}{2} = \frac{\lambda_1}{a}, \tag{5.3.5}$$

which assumes a sinusoidal distribution of fields in the aperture and requires $\alpha > \lambda_1$.

The exponential attenuation coefficient is related to the power lost per unit length traveled in the guide by (see Problem 5.1)

$$\alpha = \frac{-1}{P(z)} \frac{dP(z)}{dz}, \tag{5.3.6}$$

where $P(z)$ is the power transmitted. Thus, if we define P_1 as the power in the tail of the mode beyond X_r (i.e., the power to be lost by radiation within a

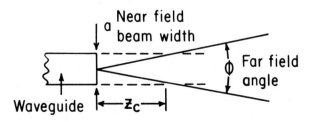

Fig. 5.5. Spread of light emitted from a truncated waveguide

length Z_c), and P_t as the total power carried by the waveguide, the attenuation coefficient is given by

$$\alpha \cong \frac{1}{P_t} \frac{P_l}{Z_c}, \tag{5.3.7}$$

The distance Z_c can be conveniently determined from (5.3.4), but P_l must be calculated by integration of the power contained in the optical mode for radii grater than $(R + X_r)$.

If it is assumed that the fields have the form

$$E(x) = \sqrt{C_0} \cos(hx) \quad \text{for} \quad -\frac{a}{2} \le x \le \frac{a}{2}, \tag{5.3.8}$$

and

$$E(x) = \sqrt{C_0} \cos\left(\frac{ha}{2}\right) \exp\left[-\left(\frac{|x|-(a/2)}{q}\right)\right] \quad \text{for} \quad |x| \ge \frac{a}{2}, \tag{5.3.9}$$

then

$$P_l = \int_{X_r}^{\infty} E^2(x)dx = C_0 \frac{q}{2} \cos^2\left(\frac{ha}{2}\right) \exp\left[-\frac{2}{q}\left(X_r - \frac{a}{2}\right)\right] \tag{5.3.10}$$

and

$$P_t = \int_{-\infty}^{\infty} E^2(x)dx = C_0 \left[\frac{a}{2} + \frac{1}{2h}\sin(ha) + q\cos^2\left(\frac{ha}{2}\right)\right]. \tag{5.3.11}$$

Substituting (5.3.10) and (5.3.11) into (5.3.7) yields

$$\alpha = \frac{\frac{q}{2}\cos^2\left(\frac{ha}{2}\right)\exp\left(-\frac{2}{q}\frac{\beta_z - \beta_0}{\beta_0}R\right)\left[2\lambda_1\exp\left(\frac{a}{q}\right)\right]}{\left[\frac{a}{2} + \frac{1}{2h}\sin(ha) + q\cos^2\left(\frac{ha}{2}\right)\right]a^2} \tag{5.3.12}$$

While the expression for α in (5.3.12) appears quite complex, close scrutiny reveals that it has the relatively simple form,

$$\alpha = C_1 \exp(-C_2 R), \tag{5.3.13}$$

where C_1 and C_2 are constants that depend on the dimensions of the waveguide, and on the shape of the optical mode. The key feature of (5.3.13) is that the radiation loss coefficient depends exponentially on the radius of curvature. The minimum radius of curvature allowable for radiation loss

Table 5.2. Waveguide radiation loss data [5.24]

Case	Index of refraction		Width a	C_1	C_2	R
	Waveguide	surrounding	[μm]	[dB/cm]	[cm^{-1}]	for $\alpha = 0.1$ dB/cm
1	1.5	1.00	0.198	2.23×10^5	3.47×10^4	4.21 μm
2	1.5	1.485	1.04	9.03×10^3	1.46×10^2	0.78 mm
3	1.5	1.4985	1.18	4.69×10^2	0.814	10.4 cm

smaller than 0.1 dB/cm has been calculated by *Goell* [5.24a] for several typical dielectric waveguides. Experimental data [5.24b] also indicate that such low levels of loss are achievable. However, as can be seen in Table 5.2, the radiation loss can be significant, particularly when the difference in index of refraction between the waveguide and the surrounding medium is very small.

5.4 Measurement of Waveguide Losses

The fundamental method of determining waveguide loss is to introduce a known optical power into one end of the waveguide and measure the power emerging from the other end. However, there are many problems and inaccuracies inherent in using such a basic approach. For example, coupling losses at the input and output are generally not known, and can largely obscure the true waveguide loss per cm. Also, if the waveguide is of the multimode type, losses attributable to individual modes cannot be separately determined. A number of different methods of loss measurement have been devised in order to circumvent problems such as these. The proper choice of a measurement technique depends on what type of waveguide is being used, on what type of loss predominates, and on the magnitude of the loss being measured.

5.4.1 End-Fire Coupling to Waveguides of Different Length

One of the simplest, and also most accurate, methods of measuring waveguide loss is to focus light of the desired wavelength directly onto a polished or cleaved input face of a waveguide as shown in Fig. 5.6, and then measure the total power transmitted. Such direct coupling is often referred to as *end-fire* coupling. The measurement is repeated for a relatively large number of waveguide samples that have different lengths, but are otherwise identical. Often this series of measurements is performed by beginning with a relatively long waveguide sample, then repetitively shortening the sample by cleaving, or cutting and polishing. Care must be taken before each measurement to align the laser beam and the sample for optimum coupling, by maximizing the observed output power. When this is done, the resulting loss data fall in a

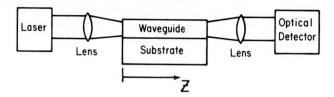

Fig. 5.6. Experimental set-up for measurement of waveguide attenuation employing end-fire coupling

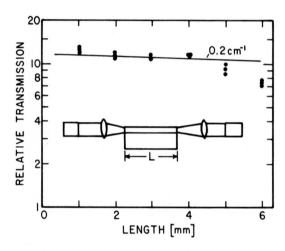

Fig. 5.7. Typical attenuation data. (These data were obtained using the set-up of Fig. 5.6 to measure loss in a Ti diffused waveguide in LiNbO$_3$)

straight line when plotted on semilog paper, as shown in Fig. 5.7. The loss coefficient can be determined from the slope of the transmission versus length curve, or equivalently, from the relation

$$\alpha = \frac{\ln(P_1/P_2)}{Z_2 - Z_1}, \quad \text{for} \quad Z_2 > Z_1, \tag{5.4.1}$$

where P_1 and P_2 are the transmitted power for waveguides of two different lengths Z_1 and Z_2.

The extent of the scatter of the data points is a measure of the consistency of sample input/output coupling loss, which depends on face preparation, and on sample alignment. If there has been significant change in these parameters from sample to sample, the data points will be widely scattered, and it will be impossible to determine the slope of the transmission versus length curve with acceptable accuracy. However, data points which fall nicely onto a straight line constitute a priori proof that sufficient consistency has been achieved, even though the absolute magnitude of the coupling loss may not be known.

Any departure of the data from the straight line can be taken as an indication of experimental error, assuming that the waveguide is homogeneous. For example, in Fig. 5.7, the increasing differential loss for lengths greater than 4mm was caused by aperturing of the output beam, which occurred when the light spread laterally in the planar waveguide to an extent that all of it could not be collected by the output lens.

The method of loss measurement described above is most advantageous because of its simplicity and accuracy. However, it does have some inherent disadvantages. Perhaps the most important of these is that it is generally destructive, with the waveguide being chopped to bits in the measurement process. Also, no discrimination is made between the losses in different modes. For these reasons, the technique is most often used for semiconductor waveguides, which are usually single-mode due to the relatively small index difference involved, and which can be easily cleaved to produce end faces of consistent quality.

5.4.2 Prism-Coupled Loss Measurements

In order to determine the loss associated with each mode of a multimode waveguide, the basic measurement technique described in the preceding section can be modified by using prism couplers [5.25] as shown in Fig. 5.8. Because light can be selectively coupled into each mode by properly choosing the angle of incidence of the laser beam, the loss for each mode can be measured separately. Generally, the position of the input prism is kept fixed while the output prism is moved after each measurement to change the effective sample length. Data are plotted and analyzed just as in the end-fire coupling method. For best accuracy, the output detector should be positioned and masked so as to collect only light from the "*m-*" line corresponding to the desired mode. If there is no mode conversion, all of the transmitted light should be contained in this one *m*-line, and the output prism could be replaced by a simpler lens-coupled detector in the end-fire configuration to collect all of the light emerging from the waveguide. In that case, the input prism would have to be moved successively to change the effective sample length. Unless one is certain that mode conversion is insignificant, two prisms should be used. In fact, the use of an output prism is the best way to detect the presence of mode conversion, since multiple *m*-lines will be observed in that case.

While prism coupling is a more versatile technique for loss measurement than end-fire coupling, it is generally less accurate, because it is very difficult to reproduce the same coupling loss each time the prism is moved to a new

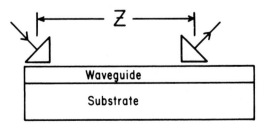

Fig. 5.8. Experimental set-up for measurement of waveguide attenuation employing prism coupling

position. As will be discussed in more detail in Chap. 6, the coupling efficiency is a strong function of the pressure used to hold the prism in contact with the waveguide; this parameter is difficult to control reproducibly. Nevertheless, the prism coupling method has been used successfully to determine losses as low as 0.02 dB/cm, by employing a small, precisely held prism, with index matching oil between the prism and waveguide [5.26].

5.4.3 Scattering Loss Measurements

All of the methods of loss measurement that have been discussed thus far determine the *total* loss resulting from scattering, absorption and radiation, with no distinction being made between the three mechanisms. In cases in which absorption and radiation are known to be negligible compared to scattering, such as in straight glass waveguides well above cutoff, it is possible to measure scattering loss conveniently. Since scattered light leaves the waveguide in directions other than the original direction of propagation in the guide, it is possible to determine scattering loss by collecting and measuring the scattered light with an apertured directional detector. One of the most convenient detectors for this purpose is a p-n junction photodiode, coupled to an optical fiber which is used as a probe to collect light scattered from the waveguide [5.27]. Usually the fiber is held at right angle to the waveguide and scanned along its length so that a plot of relative scattered optical power versus length can be made. The loss per unit length can be determined from the slope of this curve as was done in the case of the previously described methods. This technique of loss measurement implicitly assumes that the scattering centers are uniformly distributed, that the intensity of the scattered light in the transverse direction is proportional to the number of scattering centers, and that radiation and absorption losses are negligible. In that case, it is not necessary to collect all of the scattered light. It is required only that the detector aperture be constant. When an optical fiber is used to probe the waveguide, the spacing between the fiber end-face and the waveguide surface must be kept constant in order to satisfy the requirement of constant detector aperture.

As in the case of loss measurements made with either prism or end-fire coupling, the extent of scatter of the data points about a straight line variation is a reliable indicator of the accuracy of the measurements. If randomly located large scattering centers are present in the waveguide, or if scattering is nonuniform, the loss data points will not fall neatly into line. The transverse optical fiber probe method is most accurate for waveguides with relatively large scattering loss. When scattering loss is less than about 1 dB/cm, measurements are difficult because of the low intensity that must be detected.

Since scattering loss is usually the dominant mechanism in dielectric thin-film waveguides, this method is used most often to measure losses in such waveguides. In fact, it is usually assumed that the loss measured by the transverse fiber probe method is the total loss in the waveguide. Absorption and radiation loss are assumed to be negligible. In semiconductor wave-guides, where absorption loss is more significant, the transverse fiber probe method can be used to determine total loss. However, this will not provide any indication of the relative magnitudes of the three loss mechanisms in the waveguide.

In order to explain the various techniques of waveguide loss measure-ment, it has been necessary to briefly describe some of the methods of coupling light into and out of waveguides. In the next chapter, these methods are described in greater detail, and some additional coupling techniques, such as grating coupling, are reviewed.

Problems

5.1   If $P = P_0 \exp(-\alpha z)$, where P_0 is the power at the input end of a waveguide and P is power as a function of distance traveled in the propagation direction (z), show that

$$\alpha = \frac{\text{power lost per unit length}}{\text{power transmitted}}.$$

5.2   Show that the relationship between attenuation coefficient α (in cm^{-1}) and loss $\mathscr{L}$ (in dB/cm) is given by

$$\mathscr{L} = 4.3\alpha.$$

5.3   In the optical integrated circuit shown below, all of the waveguides have the same cross sectional dimensions and loss per unit length due to scattering and absorption. However, the curved waveguides have an additional loss per

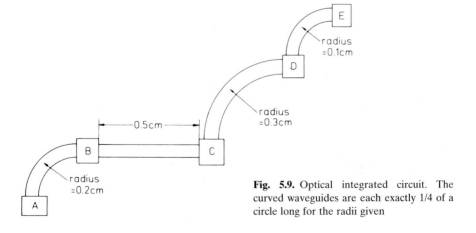

Fig. 5.9. Optical integrated circuit. The curved waveguides are each exactly 1/4 of a circle long for the radii given

unit length due to radiation.

If the total loss between the following elements is:

Between D and E $L_T = 1.01\,\mathrm{dB}$
Between C and D $L_T = 1.22\,\mathrm{dB}$
Between B and C $L_T = 1,00\,\mathrm{dB}$.

What is the total loss L_T between elements A and B? (Neglect coupling losses – consider only waveguide loss as above.)

5.4   Describe the physical reason why radiation loss from a guided optical mode in a curved waveguide increases as the radius of curvature is reduced.

5.5   A certain ribbed channel waveguide, 1-μm deep, is used in an OIC for guiding light of vacuum wavelength $\lambda_0 = 6328\,\text{Å}$. Loss measurements made on test sections of the guide have shown that the loss coefficient in a straight sample is $\alpha = 0.3\,\mathrm{cm}^{-1}$, while in a curved section with radius of curvature $R = 0.5\,\mathrm{mm}$ it is $\alpha = 1.4\,\mathrm{cm}^{-1}$, and in a curved section with $R = 0.3\,\mathrm{mm}$ it is a $\alpha = 26.3\,\mathrm{cm}^{-1}$.

What is the minimum raidus of curvature that can be used if α must be less than $3\,\mathrm{cm}^{-1}$ at all points in the circuit?

5.6   a) Sketch the index of refraction profile in a strip-loaded waveguide, indicating the relative magnitudes of the index in different regions.

b) III–V semiconductor waveguide research has progressed to different materials in the order shown in the list below. What reason (i.e., advantage) motivated each step in this progression?

LPE GaAs waveguides on GaAs substrates
LPE $Ga_{(1-x)}Al_xAs$ waveguides on GaAs substrates
LPE $Ga_{(1-x)}In_xAs$ waveguides on GaAs substrates
MBE $Ga_{(1-x)}In_xAs_yP_{(1-y)}$ waveguides on GaAs substrates
(LPE – Liquid phase epitaxial; MBE – Molecular beam epitaxial)

5.7   A certain uniform waveguide 2-cm long is transmitting an optical signal, the power of which is measured at the output of the waveguide to be 1 Watt. If the waveguide is cut so that its length is reduced by 10%, the optical power at the output is found to be 1.2 Watts. What is the attenuation coefficient (in cm^{-1}) in the waveguide?

6. Waveguide Input and Output Couplers

Some of the methods of coupling optical energy into or out of a waveguide were mentioned briefly in Chap. 5. In this chapter, we shall consider in more detail the various coupling techniques that can be used. The methods that are employed for coupling an optical beam between two waveguides are different from those used for coupling an optical beam in free space to a waveguide. Also, some couplers selectively couple energy to a given waveguide mode, while others are multimode. Each type of coupler has its attendant set of advantages and disadvantages; none is clearly best for all applications. Hence, a knowledge of coupler characteristics is necessary for the OIC user, as well as for the designer.

Coupler fabrication is generally accomplished by using techniques such as photoresist masking, thin film deposition and epitaxial growth, which have been described in Chap. 4; thus, a separate chapter devoted to coupler fabrication is not required. However, certain specialized methods, such as holographic exposure of photoresist to make grating couplers, are discussed in this chapter.

6.1 Fundamentals of Optical Coupling

The principal characteristics of any coupler are its efficiency and its mode selectivity. Coupling efficiency is usually given as the fraction of total power in the optical beam, which is coupled into (or out of) the waveguide. Alternatively, it may be specified in terms of a coupling loss in dB. For a mode-selective coupler, efficiency can be determined independently for each mode, while multimode couplers are usually described by an overall efficiency. However, in some cases it is possible to determine the relative efficiencies for the various modes of a multimode coupler. Thus, the basic definition of coupling efficiency is given by

$$\eta_{cm} \equiv \frac{\text{power coupled into (out of) the } m\text{th order mode}}{\text{total power in optical beam prior to coupling}} \qquad (6.1.1)$$

and coupling loss (in dB) is defined as

$$\mathscr{L}_{cm} \equiv 10\log \frac{\text{total power in optical beam prior to coupling}}{\text{power coupled into (out of) the } m\text{th order mode}}. \qquad (6.1.2)$$

If the power in each mode cannot be separately determined, overall values of η_{Cm} and $\mathscr{L}_{Cm}$ are used.

Coupling efficiency depends most strongly on the degree of matching between the field of the optical beam and that of the waveguided mode. This principle can be best illustrated by considering the case of the transverse coupler.

6.2 Transverse Couplers

Transverse couplers are those in which the beam is focused directly onto an exposed cross-section of the waveguide. In the case of a free space (air) beam, this may be accomplished by means of a lens. Transverse coupling of two solid waveguides may be done by butting polished or cleaved cross-sectional faces together.

6.2.1 Direct Focusing

The simplest method of transverse coupling of a laser beam to a waveguide is the direct focusing or *end-fire* approach shown in Fig. 6.1. The waveguide may be of either the planar or channel type, but we assume a planar waveguide for the moment. The transfer of beam energy to a given waveguide mode is accomplished by matching the beam-field to the waveguide mode field. The coupling efficiency can be calculated from the overlap integral [6.1] of the field pattern of the incident beam and the waveguide mode, given by

$$\eta_{cm} = \frac{\left[\int A(x)B_m^*(x)dx\right]^2}{\int A(x)A^*(x)dx \int B_m(x)B_m^*(x)dx}, \tag{6.2.1}$$

where $A(x)$ is the amplitude distribution of the input laser beam, and $B_m(x)$ is the amplitude distribution of the mth mode.

The end-fire method is particularly useful for coupling gas-laser beams to the fundamental waveguide mode because of the relatively good match

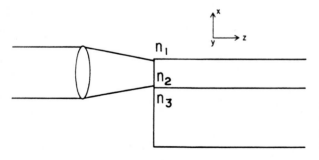

Fig. 6.1. The transverse coupling method, which is sometimes referred to as *end-fire* coupling

between the Gaussian beam profile and the TE_0 waveguide mode shape. Of course, the beam diameter must be closely matched to the waveguide thickness for optimum coupling. In principle, coupling efficiency could be nearly 100% if field contours were carefully matched. However, in practice, efficiencies of about 60% are usually achieved, because film thicknesses are on the order of 1 µm, and thus alignment is very critical. End-fire coupling is often used in the laboratory because of its convenience. However, the difficulty of maintaining alignment without an optical bench limits its usefulness in practical applications.

6.2.2 End-Butt Coupling

Transverse coupling does have a practical application in the case of coupling a waveguide to a semiconductor laser, or to another waveguide. A parallel end-butt approach [6.2] can be used, as shown in Fig. 6.2. Very efficient coupling can be achieved, since the thickness of the waveguide can be made approximately equal to that of the light emitting layer in the laser, and since the field distribution of the fundamental lasing mode is well matched to the TE_0 waveguide mode. The method is especially useful for coupling a laser diode to a planar waveguide, because efficient coupling of an injection laser to a thin film waveguide is difficult to achieve by using either a prism, grating, or tapered film coupler. The reason for this is that the injection laser has a relatively uncollimated emitted beam which diverges at a half-angle of typically 10 to 20°. Prism, grating and tapered film couplers are all very sensitive to the angle of incidence of the light beam, requiring collimation to better than 1° for efficient coupling, as will be explained later in this chapter.

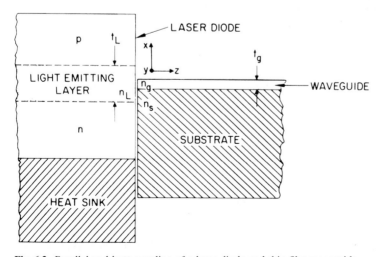

Fig. 6.2. Parallel end-butt coupling of a laser diode and thin-film waveguide

For the case of a laser diode, operating in the fundamental TE_0 mode and coupled to a planar waveguide, as shown in Fig. 6.2, the coupling efficiency for the TE modes is given by [6.2]

$$
\eta_{cm} = \underbrace{\frac{64}{(m+1)^2 \pi^2}}_{\text{normalization}} \cdot \underbrace{\frac{n_L n_g}{(n_L + n_g)^2}}_{\text{reflection}} \cdot
$$

$$
\cos^2\left(\frac{\pi t_g}{2 t_L}\right) \cdot \underbrace{\frac{1}{\left[1 - \left(\frac{t_g}{(m+1)t_L}\right)^2\right]^2}}_{\text{overlap}} \cdot \underbrace{\frac{t_g}{t_L}}_{\substack{\text{area} \\ \text{mismatch}}} \cdot \cos^2\left(\frac{m\pi}{2}\right),
$$

$$
m = 0,1,2,3,\ldots \qquad (6.2.2)
$$

The above expression is based on the assumptions that all waveguide modes are well confined, and that $t_g \leq t_L$. It is interesting to note from the last factor of (6.2.2) that there is no coupling to odd-order waveguiding modes. This is because the field distributions have cancelling lobes when their overlap integrals are taken with the even ($m = 0$) laser mode. The first factor of (6.2.2) is just a normalization term, while the second factor arises from reflections at the laser-waveguide interface. The other terms account for mismatch in the field distributions in the laser and waveguide.

Calculated curves of η_{Cm} as a function of relative waveguide thickness are plotted in Fig. 6.3, along with experimental data [6.2] for the case of a GaAs laser diode coupled to a Ta_2O_5 waveguide on a glass substrate. If $t_g \cong t_L$, coupling efficiency can theoretically approach 100% for the lowest-order waveguide mode. In that case, coupling into higher-order waveguide modes is nearly zero. *Hammer* et al. have reported using this method to couple as much as 27 mW of optical power from a single-mode diode laser ($\lambda_0 = 0.84$ μm) into a Ti diffused $LiNbO_3$ waveguide with a coupling efficiency of 68% [6.3].

The coupling efficiencies shown in Fig. 6.3 are optimum values, corresponding to perfect alignment of the laser and waveguide. Coupling efficiency is most sensitive to transverse lateral misalignment in the x direction. A displacement X of the waveguide relative to the laser, as shown in Fig. 6.4, reduces the coupling efficiency according to the relation [6.2]

$$
P/P_0 = \cos^2\left(\frac{\pi X}{t_L}\right), \qquad (6.2.3)
$$

where P_0 is the coupled power fox $X = 0$. The above expression assumes that $t_g < t_L$ and $X \leq (t_L - t_g)/2$. The dashed curve in Fig. 6.4 is the theoretically calculated P/P_0 for the case of $t_L = 5.8$ μm and $t_g = 2.0$ μm, while the solid curves

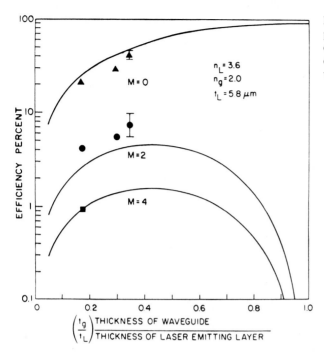

Fig. 6.3. Comparison of experimental coupling efficiency data with theoretical curves as a function of waveguide thickness [6.2]

represent experimentally measured data. A prism output coupler was used on the waveguide to determine the relative power that was butt coupled into each of the three modes that were observable.

The spacing between the laser and the waveguide in the z direction is also very critical, and must be controlled to a precision on the order of a wavelength for optimum coupling. Figure 6.5 shows the experimentally measured variation of coupled power as a function of z displacement. The oscillatory shape of the curve results from modulation of the effective reflectivity of the laser output face by resonance in the Fabry-Perot etalon formed by the plane parallel faces of the laser and waveguide. In principle, this effect could be eliminated by using an index matching fluid between the laser and waveguide, so that the coupled power would vary smoothly with z displacement, as shown by the dashed curve in Fig. 6.5.

The results presented so far demonstrate that end-butt coupling can be a very efficient means of coupling a diode laser to a waveguide. *Enochs* [6.4] has found similar results for the case of butt coupling a laser to an optical fiber. Yet, it is also obvious that submicrometer alignment tolerances are required if optimum efficiency is to be obtained. Alignment to such tolerances can be achieved by using piezoelectrically driven micrometer heads, which feature a small piezoelectric crystal stage bonded onto the end of a conventional screw micrometer. Coarse alignment is established with the micrometer screw, and then final alignment is produced by applying a voltage to the piezoelectric stage to move the laser (or waveguide), while coupled optical power is moni-

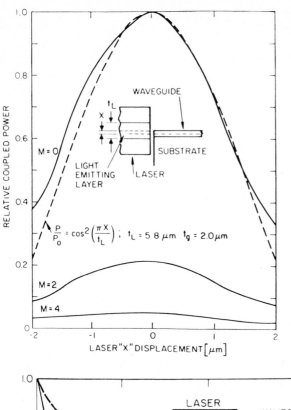

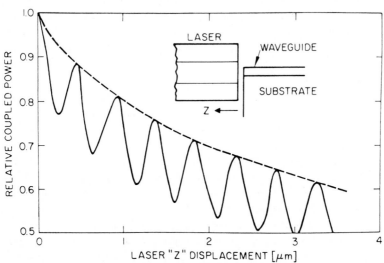

Fig. 6.4. Comparison of experimental coupling efficiency data (*solid line*) with theoretical curve (*dashed*) as a function of lateral misalignment of laser and waveguide [6.2]

Fig. 6.5. Experimentally measured dependence of coupling efficiency on spacing between laser and waveguide [6.2]

tored by means of an appropriate output coupler and photodetector. Piezo-electric micrometers with a sensitivity of better than 40 Å per volt, over a 2000 V range, are commercially available. Thus, alignment to better than 0.1 μm can be obtained rather easily. Once alignment has been established, the laser heatsink can be permanently bonded to the waveguide support structure with epoxy, or with a metallic bond. Because the size and mass of both the laser and the OIC are relatively small, vibration sensitivity is not a significant problem, and reliable alignment can be maintained.

While the above-described alignment technique is an effective method of coupling a laser diode to a waveguide it is relatively time consuming, and hence expensive in a production line setting. To solve this problem, techniques of both hybrid and monolithic integrated fabrication have been developed which take advantage of automated, batch-fabrication procedures. For ex-ample, *Yanagisawa* et al. [6.5] have used a hybrid-integration process to couple an AlGaAs laser diode to a glass waveguide on a silicon substrate. Vertical alignment is achieved by using the surface of the silicon substrate as a refer-ence plane, while lateral alignment is done with conventional photolitho-graphic techniques. Alignment accuracy is within 1 μm, leading to a coupling loss of approximately 3 dB.

An example of monolithic integration is the waveguide coupling of an InGaAsP distributed-feedback laser and electro-absorption modulator on an InP substrate described by *Aoki* et al. [6.6]. Their method is based on carefully controlled selective-area *M*etal-*O*rganic-*V*apor-*P*hase *E*pitaxy (MOVPE) of *M*ultiple-*Q*uantum-*W*ell (MQW) structures. (See Chap. 16 for a discussion of MQW structures.) The integrated laser/modulator was employed to transmit data over an 80 km single-mode fiber at a data rate of 2.5 Gbit/s.

6.3 Prism Couplers

Transverse coupling can be used only when a cross-sectional end face of the waveguide is exposed. In many cases, it is necessary to couple light into a waveguide that is buried within an OIC, with only the surface exposed. One could envision focusing the light onto the surface of the waveguide at an oblique angle, as shown in Fig. 6.6, but a fundamental problem is encountered in that case. For coupling to occur, it is necessary that the components of the phase velocities of the waves in the z direction be the same in both the waveguide and the beam. Thus, a phase-match condition must be satisfied, which requires

$$\beta_m = kn_1\sin\theta_m = \frac{2\pi}{\lambda_0}n_1\sin\theta_m. \tag{6.3.1}$$

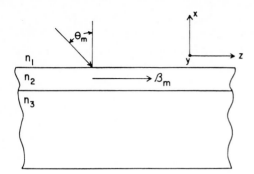

Fig. 6.6. Diagram of an attempt to obliquely couple light into a waveguide through its surface

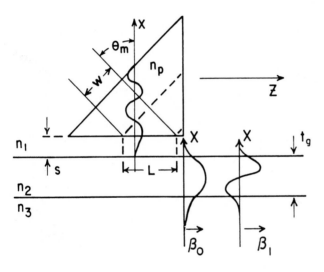

Fig. 6.7. Diagram of a prism coupler. The electric field distributions of the prism mode and the $m = 0$ and $m = 1$ waveguide modes in the x direction are shown

However, in Chap. 2 it was shown that, for a waveguided mode,

$$\beta_m > kn_1. \tag{6.3.2}$$

Combining (6.3.1) and (6.3.2) leads to the result that sin $\theta_m > 1$, which is, of course, impossible.

One solution to the problem of phase matching is to use a prism, as shown in Fig. 6.7. A beam of light of width W is directed into the face of the prism, which has $n_p > n_1$. The beam is totally internally reflected at the $n_p - n_1$ interface, setting up a standing wave mode in the prism, as shown in Fig. 6.7. This mode is stationary in the x direction, but moves in the z direction with a phase constant β_p. In the waveguide, various guided modes can exist, moving in the z direction with phase constants β_m. All of these guided modes have an evanescent tail extending slightly beyond the $n_1 - n_2$ interface. If the prism spacing s is small enough so that the tails of the waveguide modes overlap the tail of the prism mode, there is coherent coupling of energy from the prism mode to the mth waveguide mode when θ_m is chosen so that $\beta_p = \beta_m$. The condition for matching of the β terms is given by

$$\frac{2\pi n_p}{\lambda_0} \sin\theta_m = \beta_m. \tag{6.3.3}$$

Although θ_m must be carefully chosen in order to couple to a given mode, a single prism can be used to couple to many different modes by merely changing the angle of incidence of the optical beam. The beam need not be perpendicular to the prism surface, as shown in Fig. 6.7. However, if the beam is not perpendicular to the prism surface, refraction at that interface will require a modification of the expression for θ_m given by (6.3.3). See Problem 6.2 for an example of this effect.

The process of coupling energy via the overlapping mode tails, while the incident beam tends to be totally internally reflected in the prism, is sometimes called *optical tunneling*, because it is analogous to the quantum mechanical tunneling of a particle through an energy barrier. The modes in the waveguide are only weakly coupled to the mode in the prism. Hence, negligible perturbation of the basic mode shapes occurs. Of course, the condition

$$\theta_m > \theta_c = \sin^{-1}\left(\frac{n_1}{n_p}\right) \tag{6.3.4}$$

must also be satisfied if total internal reflection is to occur in the prism, where θ_c is the critical angle.

Because of the size of the prism, the interaction between prism and waveguide modes can occur only over the length L. The theory of weakly coupled modes [6.1] indicates that a complete interchange of energy between phasematched modes occurs if the interaction length in the z direction satisfies the relation

$$\kappa L = \pi/2, \tag{6.3.5}$$

where κ is the coupling coefficient. The coefficient κ depends on n_p, n_1 and n_2, which determine the shape of the mode tails, and on the prism spacing s. From (6.3.5), the length required for complete coupling is given by

$$L = \frac{W}{\cos\theta_m} = \frac{\pi}{2\kappa}. \tag{6.3.6}$$

For a given L, the coupling coefficient required for complete coupling is thus given by

$$\kappa = \frac{\pi\cos\theta_m}{2W}. \tag{6.3.7}$$

This condition for complete coupling assumes that the amplitude of the electric field is uniform over the entire width W of the beam. In a practical case this is never true. For a Gaussian beam shape, it results that the maximum coupling efficiency is about 80%. For a more detailed discussion of the effect of beam width and shape on coupling efficiency see *Tamir*, [6.7] *Pasmooli* et al. [6.8] and *Klimov* et al. [6.9]. Also, it can be seen that, in order to get 100% coupling even with a uniform beam, the trailing edge of the beam must exactly intersect the right-angle corner of the prism. If it intersects too far to the right, some of the incident power will be either reflected or transmitted directly into the waveguide and will not enter the prism mode. If the beam is incident too far to the left, some of the power coupled into the waveguide will be coupled back out into the prism.

The prism coupler is frequently used in integrated optics applications because of its versatility. It can be used as either an input or output coupler. When used as an output coupler, the prism is arranged exactly as in Fig. 6.7, except that the direction of travel of the waveguided light would be in the negative z direction. If more than one mode is propagating in the guide, light is coupled out at specfic angles corresponding to each mode. Because of this characteristic, the prism coupler can be used as an analytical tool to determine the relative power in each waveguide mode, as described in Chap. 5. The prism can also be moved along the length of the waveguide to determine losses. However, care must be taken to apply the same mechanical pressure to the prism during each measurement of coupled power so that the spacing, and hence the coupling coefficient, will be constant.

One disadvantage of the prism coupler is that n_p must be not only greater than n_1 but also greater than n_2. This is true because the waveguide index n_2 is generally close to the substrate index n_3, which leads to the result that

$$\beta_m \cong kn_2 = \frac{2\pi}{\lambda_0} n_2. \tag{6.3.8}$$

Since $\sin \theta_m \leq 1$, (6.3.8) coupled with (6.3.3) implies that $n_p > n_2$. In the case of glass waveguides, with indices $\cong 1.5$, it is easy to find a suitable prism material with $n_p > n_2$. However, semiconductor waveguides, which typically have indices $\cong 3$ or 4, are more difficult to couple with prisms. Both the index and the transparency of the prism material must be considered at the wavelength of interest. Table 6.1 gives a number of different prism materials that are avail-

Table 6.1. Practical prism materials for beam couplers

Material	Approximate refractive index	Wavelength range
Strontium titanate	2.3	visible – near IR
Rutile	2.5	visible – near IR
Germanium	4.0	IR

able in good optical quality, along with their indices of refraction at various wavelengths.

Another disadvantage of the prism coupler is that the incident beam must be highly collimated because of the critical angular dependence of coupling efficiency into a given mode. Due to this problem, prism couplers cannot be used effectively with semiconductor lasers, which have a beam divergence half angle of 10–20° unless a lens is used to collimate the beam.

Prism couplers are very useful in laboratory applications where flexibility is desired regarding the position of the incident beam. However, the requirement of a stable mechanical pressure to hold the prism in place makes it less useful in practical applications, in which vibration and temperature variations are usually encountered. The grating coupler can be used to avoid this problem, without giving up the advantage of mode selectivity.

6.4 Grating Couplers

The grating coupler, like the prism coupler, functions to produce a phase matching between a particular waveguide mode and an unguided optical beam which is incident at an oblique angle to the surface of the waveguide, as shown in Fig. 6.8. It will be recalled that, without the grating, the phase matching conditon is given by (6.3.1), and cannot be satisfied for any θ_m.

6.4.1 Basic Theory of the Grating Coupler

Because of its periodic nature, the grating perturbs the waveguide modes in the region underneath the grating, thus causing each one of them to have a set of spatial harmonics [6.10] with z-direction propagation constants given by

$$\beta_v = \beta_0 + \frac{v\,2\pi}{\Lambda},\tag{6.4.1}$$

where $v = 0, \pm 1, \pm 2, \ldots$, and where Λ is the periodicity of the grating. The fundamental factor β_0 is approximately equal to the β_m of the particular mode in the waveguide region not covered by the grating. Because of the negative values of v, the phase matching condition (6.3.1) can now be satisfied so that

$$\beta_m = kn_1 \sin\theta_m,\tag{6.4.2}$$

even though $\beta_m > kn_1$.

Since all of the spatial harmonics of each mode are coupled to form the complete surface wave field in the grating region, energy introduced from the beam into any one of the spatial harmonics is eventually coupled into the fundamental ($v = 0$) harmonic as it travels to the right and past the grating region. This fundamental harmonic is very close to, and eventually becomes, the β_m mode outside of the grating region. Thus, the grating coupler can be

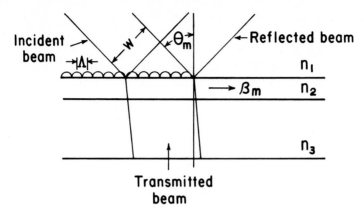

Fig. 6.8. Diagram of a grating coupler

used to selectively transfer energy from an optical beam to a particular waveguide mode by properly choosing the angle of incidence. The grating can also be used as an output coupler, because, by reciprocity, energy from waveguide modes will be coupled out at specific angles, θ_m, corresponding to a particular mode.

The preceding paragraphs have described the operation of the grating coupler in relatively simple terms. However, the details of the coupling phenomena are quite complex, and depend very strongly on the cross-sectional shape of the grating bars as well as on their spacing. For a detailed discussion of the effect of grating shape, see [Ref. 6.7 pp. 110–118]. As in the case of the prism coupler, an optimum coupling efficiency of approximately 80% is theoretically possible when coupling a Gaussian beam with a grating. However, typical unblazed gratings (with symmetric profiles) generally have efficiencies of 10 to 30%. The principal reason for this is that much of the incident energy is usually transmitted through the guide and lost in the substrate, because, unlike a prism, the grating does not operate in a total internal reflection mode. Power can also be coupled into higher-order diffracted beams produced by the grating, unless the ratio of grating periodicity to guide wavelength is approximately 1.

The efficiency of a grating coupler can be greatly improved by shaping its profile asymmetrically to 'blaze' it for optimum performance at the coupling angle and wavelength of interest. For example, *Tamir* and *Peng* [6.11] have shown that the theoretical maximum efficiency for coupling either the TE_0 or TM_0 mode to an air beam is roughly 50% for symmetric grating profiles, while an asymmetric saw-tooth profile can produce an efficiency greater than 95%. These theoretical predictions are supported by experimental results indicating very high efficiencies with blazed gratings [6.12, 13] and by further theoretical work [6.14, 16].

The principal advantage of the grating coupler is that, once fabricated, it is an integral part of the waveguide structure. Hence, its coupling efficiency remains constant and is not altered appreciably by vibration or ambient conditons. Also, the grating coupler can be used on high-index semiconductor waveguides for which it is difficult to obtain a suitable prism material. However, since it is highly angle dependent, the grating coupler cannot be used effectively with the relatively divergent beam of a semiconductor laser. Perhaps the greatest disadvantage of the grating coupler is that it is very difficult to fabricate, requiring the use of sophisticated masking and etching techniques.

6.4.2 Grating Fabrication

A grating structure may be formed either by masking and etching the waveguide surface [6.12, 17, 18] or by masking the surface and depositing a thin film grating pattern [6.19, 20]. In either case, the most difficult part of the process is defining the pattern of the closely spaced grating bars. The spacing should be on the order of a wavelength in the waveguide material. Hence, for visible and near ir wavelengths, spacing of 1000–3000 Å are typical, for indices in the range from 1.4 to 4. Conventional photoresists, used in the microelectronics industry, have adequate resolution, but the practical limit of conventional photomasks is approximately 1 micrometer. Therefore, gratings are generally produced by using an optical interference exposure process [6.19, 20], sometimes called a *holographic* process [6.21].

In this process the substrate containing the waveguide is first coated with photoresist using any of the methods described in Chap. 4. Then the resist is exposed using an interference pattern generated by combining coherent laser beams directed at the surface as shown in Fig. 6.9. (Both beams are usually obtained from one laser by using a beam splitter.) Simple geometric considerations show that the relationship between grating periodicity Λ and the beam angle α is given by

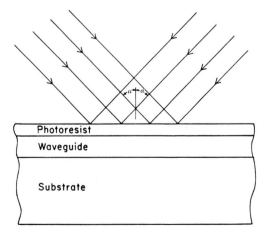

Fig. 6.9. Holographic photoresist exposure. The interference pattern produced by two obliquely intersecting coherent laser beams is used to expose the photoresist to produce a grating pattern

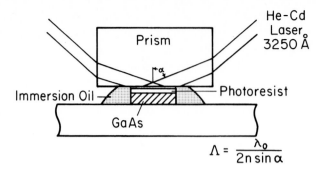

$$\Lambda = \frac{\lambda_0}{2n \sin \alpha}$$

Fig. 6.10. Diagram of prism method of holographic grating exposure as applied to the fabrication of a first order grating in a GaAs waveguide [6.22]

$$\Lambda = \frac{\lambda_0}{2 \sin\alpha}. \tag{6.4.3}$$

From (6.4.3) it is obvious that Λ is limited to values greater than $\lambda_0/2$. However, by using a rectangular prism, as shown in Fig. 6.10, it is possible to expose grating patterns that are more closely spaced than would ordinarily be possible using a given laser. In this case, (6.4.3) becomes

$$\Lambda = \frac{\lambda_0}{2n \sin\alpha}, \tag{6.4.4}$$

where n is the index of refraction of the prism material. Using a quartz prism and a He-Cd laser, *Yen* et al. [6.22] have produced grating spacings as small as 1150 Å.

Once the photoresist has been exposed, it is developed using standard methods to produce the required mask on the waveguide surface. Either chemical or ion beam etching can be used to produce the grating, following etching procedures described in Chap. 4. In general, ion beam etching produces more uniform grating structures, but chemical etching produces less damage in the waveguide material. Chemical etching is also capable of penetrating more deeply without undercutting, if the etchant is chosen carefully. For example, a buffered solution of NH_4F can be used to etch GaAs to depths greater than 100 μm [6.23]. Another advantage of chemical etching is that, by proper choice of substrate orientation and etchant, one can obtain non-isotropic etch rates, which yield an asymmetric grating blazed for a desired wavelength [6.12].

As an alternative to etching the grating structure into the waveguide surface, it can also be produced by depositing thin-film bars on the waveguide surface, using a photoresist mask to define the shape. In fact, the photoresist itself can even be used as a deposited grating material [6.19, 20]. These methods generally yield a grating coupler with greater scattering and absorption losses than those obtained with an etched grating, because a portion of the beam is blocked from the waveguide.

6.5 Tapered Couplers

The tapered coupler [6.20] is based on the principle that a waveguide which is below cuttoff transfers energy into radiation modes. The waveguide thickness is tapered in the coupler region to produce a reduced-height waveguide with a decreasing cuttoff wavelength, as shown in Fig. 6.11. The coupling mechanism can be conveniently visualized by using the ray-optic approach. A guided wave incident on the tapered coupler undergoes zig-zag bounces with the angle of incidence to the guide-substrate interface (measured from the normal) steadily decreasing. When the angle of incidence becomes less than the critical angle for total internal reflection, energy is refracted into the substrate. The energy from subsequent rays is also refracted out of the waveguide in a like manner, so that up to 70% coupling efficiency is obtainable [6.24]. The 30% loss is associated mostly with scattering into air radiation modes when the thickness of the guide reaches the appropriate cutoff point. The waveguide mode is totally co-upled out within about 8 vacuum wavelengths after the cutoff point is reached [6.10].

The greatest advantage of the tapered coupler is that it is simple to fabricate and functions reasonably well as an output coupler. However, it forms a divergent beam, as shown in Fig. 6.11, which spreads over an angle of 1–20°, depending on the taper. This divergent beam is somewhat inconvenient to use, but can be tolerated in many applications in which the form of the output beam is not a critical factor.

In principle, the tapered coupler can also be used as an input coupler. However, in order to obtain high efficiency, one would have to construct a converging input beam which was the reciprocal of the diverging beam shown in Fig. 6.11. Since this is practically an impossibility, only very low efficiencies are usually observed when tapered waveguides are used as input couplers. One practical application of the tapered coupler may be in coupling a thin film waveguide to an optical fiber, since the end of the fiber can be located very close to the waveguide and the end face can be shaped to improve the coupling efficiency.

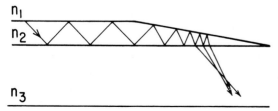

Fig. 6.11. Diagram of a tapered coupler

6.6 Fiber to Waveguide Couplers

A key element of an integrated optic system is the coupler that transfers the optical wave between a fiber, used for long distance transmission, and a waveguide of the OIC used for signal processing. Research in recent years has resulted in the development of a number of different types of fiber-to-waveguide couplers.

6.6.1 Butt Coupling

The fiber may be directly butted in contact with the waveguide, without any interfacing device, in an end-on alignment. If the cross-sectional area of the fiber core and the waveguide are closely matched, high efficiency coupling can be achieved, as in the case of butt-coupled channel waveguides or a laser diode and a channel waveguide [6.25]. An index matching fluid can be used to reduce reflection loss at the interface. The greatest problem with the butt coupling approach is that it is extremely difficult to establish and maintain correct alignment, since both the fiber core and waveguide typically have micron sized dimensions.

A mechanical arrangement for effectively coupling a single-mode fiber to a laser diode (or waveguide) which has been demonstrated by *Enoch* [6.4] is shown in Fig. 6.12. The coupling problem is difficult because the fiber core

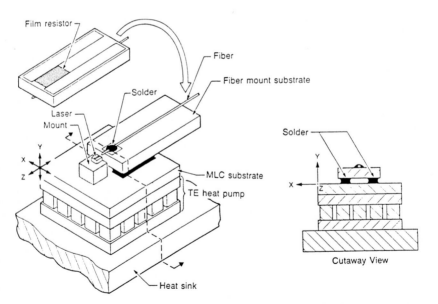

Fig. 6.12. Optical fiber interconnect to a single-mode laser diode [6.4] (Diagram courtesy of Tektronix, Inc.)

diameter is only a few micrometers and the laser light emitting region is less than 1 micrometer thick. Also, it is necessary to maintain the alignment while the fiber is permanently bonded in place. Piezoelectrically driven micromanipulators can be used to align the fiber, as mentioned previously in this chapter. To provide permanent bonding of the fiber and laser, the fiber is metallized and soldered to a fiber mount substrate. That substrate is, in turn, attached to a thermoelectric heat sink with solder. On the underside of the fiber mount substrate (shown in the upper left corner of Fig. 6.12) are metallization leads and a thin film resistor which are used to melt the solder by passage of an electrical current. A computerized feedback control system is used to move the laser and fiber into optimum alignment while current passing through the thin film resistor keeps the solder fluid. When the desired alignment has been achieved, the computer turns off the current, allowing the solder to cool and set. Thus the laser and fiber remain permanently bonded to the thermoelectric heat sink and alignment is maintained.

More recently, the silicon V-groove and flip-chip techniques, which have been used for years in the electrical integrated circuits industry [6.26], have been applied to the problem of aligning a channel waveguide with a single-mode fiber [6.27]. *Sheem* et al. [6.28] have used a Si V-groove/flip-chip coupler to couple a 3 μm wide diffused LiNbO$_3$ waveguide to a 4.5 μm core single-mode fiber. They obtained a coupling efficiency of 75% into both TE and TM modes. *Murphy* and *Rice* [6.29] have used the overlap between a precision-etched V-groove silicon chip and the top surface of a Ti-diffused LiNbO$_3$ waveguide substrate to simultaneously align an array of single mode fibers with corresponding channel waveguuides. The coupling structure, shown in Fig. 6.13, aligns all but one of the six degrees of freedom automatically. Only

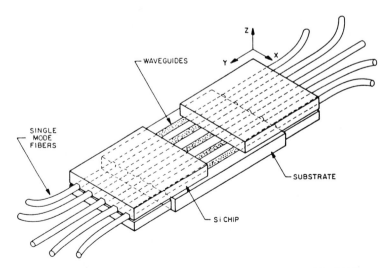

Fig. 6.13. Structure for self-aligned coupling of fibers and waveguides [6.29]

the transverse alignment in the x direction must be adjusted for maximum coupling. Although only five fibers are shown in the diagram for simplicity, the method has been used to attach arrays of twelve fibers with average excess loss of 0.9 dB at a wavelength of $\lambda_0 = 1.3$ μm. More recent work by *Sugita* et al. [6.30] has reduced loss to 0.6 dB in an 8 groove interface.

Etched grooves can also be utilized to align optical fibers with either planar or channel waveguides in polymers [6.31]. Precisely oriented and sized slots are generated by eximer-laser ablation to accept the core of an optical fiber from which the cladding has been stripped. Angular precision of 0.3° and translational accuracy of 0.5 μm have been achieved, resulting in coupling losses less than 0.5 dB.

The optimization of coupling efficiency has remained a topic of continuing interest. Detailed theoretical models have been developed to describe the coupling between optical fibers and both planar [6.32] and channel [6.33] waveguides, and modeshape adapters have been implemented [6.34] to improve the match between circular cross-section fibers and rectangular waveguides.

To conclude this section on butt coupling of fibers, we note that light-emitting diodes (LED's), which emit over a relatively large surface area, can be coupled to optical fibers with comparative ease. For a thorough discussion of this topic, the reader is referred to a review by *Barnoski* [6.35].

6.6.2 Tapered Film Fiber Couplers

Tien et al. [6.36] have proposed using a tapered section of waveguide to couple to a fiber as shown in Fig. 6.14a. The fiber is inserted into a cylindrical hole, drilled into the substrate just below the waveguide. The hemispherical end of this hole is filled with a high index material to produce a lens. Light waves coupled out of the thin-film waveguide into the substrate are collected by the fiber. Reasonably good efficiency is obtainable in the case of multimode fibers

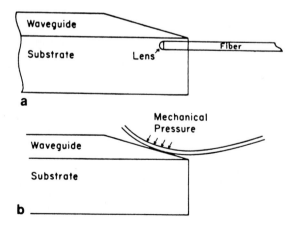

Fig. 6.14a, b. Tapered film fiber couplers; a coupling through substrate as proposed by *Tien* [6.36]; b coupling through surface as proposed by *Teh* and *Stegeman* [6.37]

with a large acceptance angle. For example, in one experiment in which the refractive indices of the film, substrate, fiber and lens were, respectively, 1.53, 1.49, 1.51 and 1.55, light from a 6328 Å He-Ne laser was coupled with 60% efficiency.

Another approach to coupling a fiber and a waveguide by using a tapered film has been demonstrated by *Teh* and *Stegeman* [6.37]. Instead of inserting the fiber into the substrate beneath the tapered waveguide section, they stripped the cladding from the fiber, and then pressed a short length of the exposed core against the surface of the taper, as shown in Fig. 6.14b. Since a highly multimode 60 μm fiber was used, it collected about 90% of the light radiated from the waveguide.

Problems

6.1   A strontium titanate prism ($n_p = 2.32$) is used as an output coupler to couple light out of a Ta$_2$O$_5$ waveguide ($n_2 = 2.09$). Three "*m*- lines" are visible at angles of 36.5°, 30.2°, and 24.6° from the waveguide surface. The output face of the prism makes an angle of 60° with the waveguide surface and the wavelength λ_0 is 9050 Å. What are the β's of the three modes?

6.2   If a rutile prism ($n_p = 2.50$) is used as an input coupler to the same waveguide as in Problem 6.1, what angle should the incident light beam make with the waveguide surface to efficiently couple into the lowest order mode? Assume the input face of the prism makes an angle of 60° with the waveguide surface.

6.3   In the case of a prism coupler, explain why the condition

$$\beta_m = \frac{2\pi n_p}{\lambda_0} \sin\theta_m$$

must be satisfied in order for coupling to occur. Derive the equation from geometric considerations (n_p: index of refraction of prism, β_m: phase constant of *m*th mode in the waveguide, θ_m: angle between normal to waveguide surface and incident ray of light within the prism, and λ_0: wavelength of light in air).

6.4   A strontium titanate right-angle prism ($n_p = 2.32$) is used as an input coupler to couple light of 9050 Å (vacuum) wavelength into a Ta$_2$O$_5$ waveguide (n = 2.09) from an air beam. The face of the prism makes an angle of 45° with the surface of the waveguide. The β of the lowest-order mode in the waveguide is 1.40×10^5 cm^{-1}.

a) Make a sketch of the coupler labeling the significant angles Φ_1, Φ_2, Φ_3, etc. Use these labels in your calculations in part b.

b) What angle must the input laser beam (in air) make with the surface of the waveguide in order to be coupled into the lowest-order mode?

6.5   A grating situated on a planar waveguide can act as a 180° reflector for the waves within the guide. If the propagation constant of the guided mode is $\beta = 1.582\ k$ and $\lambda_0 = 0.6328$ μm, find the smallest grating spacing Λ that will cause the mode to be reflected.

6.6   A rutile prism ($n_p = 2.50$) is used to couple light of vacuum wavelength $\lambda_0 = 0.9050$ μm to the fundamental mode in a waveguide which has a refractive index $n_g = 2.09$. Given that the phase constant of the fundamental mode is $\beta_0 = 1.44 \times 10^5$ cm^{-1} what angle γ must the input face of the prism make with the waveguide surface and what angle ϕ must the optical beam within the prism make with the waveguide surface in order to obtain the most efficient coupling.

6.7   A grating with spacing $\Lambda = 0.4$ μm, situated on a GaAs planar waveguide, is to be used for coupling a beam of light from a He-Ne laser ($\lambda_0 = 1.15$ μm) into the waveguide. If the propagation constant for the lowest-order mode in the guide is $\beta_0 = 3.6\ k$, what angle must the laser beam make with the surface of the waveguide in order to couple to this mode? Assume first-order coupling, i.e. $|v| = 1$.

6.8   A thin film waveguide has $n_1 = 1$, $n_2 = 1.5$ and $n_3 = 1.462$. The waveguide thickness is 0.9 μm. Light from a He-Ne laser ($\lambda_0 = 6328$ Å) is being guided in the (fundamental) TE$_0$ mode, for which the effective refractive index of the guide is $n_{\text{eff}} = 1.481$, If a 45°–45°–90° prism with index $n_p = 2.25$ is used as an output coupler, what angle will the exiting beam make with the surface of the waveguide?

6.9   A prism coupler with index $n_p = 2.2$ is used to observe the modes of a waveguide as shown below. The light source is a He-Ne laser with $\lambda_0 = 6328$ Å. If the light from a particular mode is seen at an angle of 26.43° with the normal to the prism surface, what is the propagation constant β for that mode?

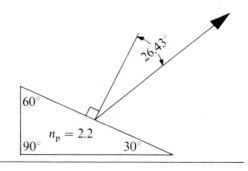

7. Coupling Between Waveguides

The phenomenon of optical tunneling can be used not only to couple energy from a fiber or a beam to a waveguide, as described in Chap. 6, but also to couple one waveguide to another. Couplers of this type are usually called directional couplers because the energy is transferred in a coherent fashion so that the direction of propagation is maintained. Directional couplers have been fabricated in two basic geometries: multilayer planar structures, and dual side-by-side channel waveguides. In this chapter the different types of waveguide to waveguide couplers are described and a concise theory of operation is developed. For a thorough mathematical treatment of these devices the reader is referred to the work of *Burns and Milton* [7.1].

7.1 Multilayer Planar Waveguide Couplers

While butt coupling can be used to couple two planar waveguides, as mentioned in Chap. 6, the more common method is to bring the guides into close proximity and allow coupling to occur through phase coherent energy transfer (optical tunneling), as shown in Fig. 7.1. The indices n_0 and n_2 in the guiding layers must be larger than n_1 and n_3, and the thickness of the confining layer 1 must be small enough that the evanescent tails of the guided modes overlap. In order for energy transfer to occur between the two guides, they must have identical propagation constants. Thus, the indices and the thicknesses of the waveguiding layers must be very carefully controlled to provide matching propagation constants. This is difficult to do but has been accomplished with excellent results. As in the case of other devices embodying the synchronous coupling principle, such as the prism coupler, the interaction length must be carefully chosen for optimum coupling. The condition for total transfer of energy is again given by (6.3.5), but the value of κ would obviously be different for a pair of overlapped planar waveguides than for a prism coupler.

This method of coupling is diffcult to use with deposited thin film waveguides because thicknesses and indices of refraction cannot be conveniently controlled. However, in the case of epitaxially grown waveguides, the superior control of thicknesses and index (by controlling composition) make the problem of matching propagation constants much easier to solve. The method seems to be particularly well suited to coupling an integrated laser

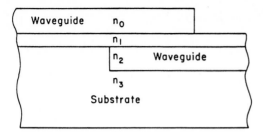

Fig. 7.1. Coupling between two planar waveguides by optical tunneling. Transfer of energy occurs by phase coherent synchronous coupling through the isolation layer with index n_1

diode to a waveguide. For example *Vawter* et al. [7.2] have used a multilayer planar structure to couple a GaAlAs transverse junction stripe (TJS) laser to a waveguide of the same material. In that case the intervening isolation layer was composed of lower refractive index GaAlAs. The method has also been used by *Utaka* et al. [7.3] to couple lasers and waveguides composed of GaInAsP, separated by a thin layer of InP. This twin-guide structure [7.4] can be used to fabricate an efficient single-mode laser with distributed feedback gratings. A very efficient laser can be obtained by using two separate, but coupled, waveguides to isolate the active region, where photons are generated, from the distributed feedback region. Such devices will be discussed in more detail in Chap. 13.

7.2 Dual-Channel Directional Couplers

The dual-channel directional coupler, which is analogous to the microwave dual-guide multihole coupler [7.5], consists basically of parallel channel optical waveguides sufficiently closely spaced so that energy is transferred from one to the other by optical tunneling, as shown in Fig. 7.2. This energy is transferred by a process of synchronous coherent coupling between the overlapping evanescent tails of the modes guided in each waveguide. Photons of the driving mode, say in guide 0, transfer into the driven mode in guide 1, maintaining phase coherence as they do. This process occurs cumulatively over a significant length; hence, the light must propagate with the same phase velocity in each channel in order for this synchronous coupling to occur. The fraction of the power coupled per unit length is determined by the overlap of the modes in the separate channels. Thus, it depends on the separation distance s, the interaction length L, and the mode penetration into the space between channels, which can be characterized by the extinction coefficients p and q. (See Chaps. 2 and 3 for a discussion of the mode shapes.)

7.2.1 Operating Characteristics of the Dual-Channel Coupler

In a dual-channel coupler, the energy transfers alternately from one waveguide to the other, and then back again if the interaction legnth is sufficient. If one were to measure the optical energy density while moving in the z

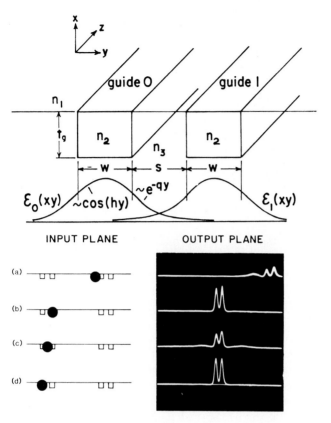

Fig. 7.2. Diagram of dual-channel directional coupler. The amplitudes of the electric field distributions in the guides are shown below them

Fig. 7.3. Optical power distribution at the output of a 3 dB dual-channel directional coupler, for various input conditions (as explained in the text). The oscillographs of output power were made using a scanning system like that shown in Fig. 2.3. The waveguides, which were formed by proton bombardment of the GaAs substrate, had 3 μm × 3 μm cross-section and were separated by 3 μm. The interaction length was 1 mm

direction along one channel of a directional coupler, a sinusoidal variation with distance would be observed. For a coupler to transfer any given fraction of the energy, it is necessary only to bend away the secondary channel at the proper point. In this way, for example, either a 10 dB coupler for measurement padding, a 3 dB coupler for beam splitting, or a 100% coupler for beam switching can be made.

The experimentally measured transmission characteristics of a dual-channel 3 dB coupler are shown in Fig. 7.3. The diagram at the left shows the position of the input laser beam, and the photograph on the right shows an oscillograph scan (in the y direction) of the optical power densities in the waveguides at the output plane. The waveguides forming this particular coupler were fabricated by 300 keV proton bombardment of a GaAs substrate [7.6]. The waveguides had a 3 μm square cross-section and were separated by approximately 3 μm. A 1.15 μm wavelength He-Ne laser beam was focused onto the cleaved input plane of the sample, which happened to contain two

INPUT PLANE   OUTPUT PLANE

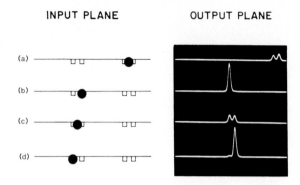

Fig. 7.4. Optical power distribution at the output of a 100% dual-channel directional coupler for various input conditions. The waveguides were like those of Fig. 7.3, except the interaction length was 2.1 mm

pairs of coupled channel waveguides. When the laser beam was focused directly on one guide, as in cases b and d of Fig. 7.3, the optical energy at the output plane was equally divided between both of the coupled guides, i.e. a 3 dB coupling was evident. When the laser beam covered only a portion of one of the guides, as in case a, or a portion of both guides as in case c, a reduced amount of light was coupled, and appeared at the output still roughly equally divided between the two guides of the dual-channel pair.

The coupling effect is even more striking in the transmission characteristics of a 100% coupler, as shown in Fig. 7.4. This device was identical to the 3 dB coupler except that it was 2.1 mm long rather than 1 mm. This length was chosen to produce total transfer of energy at the output plane. In this case, when the input laser beam was focused on either channel of a dual pair, as in (b) and (d), the light was contained in the opposite channel of the pair at the output plane. When the input beam was spread across both channels, as in (a) or (c), a reduced amount of light was coupled into both channels.

The results shown in Figs. 7.3 and 7.4 demonstrate that dual-channel couplers can produce a very efficient transfer of optical energy from one guide to another over a relatively short interaction length. The embedded waveguides produced by ion implantation, which lie beneath the surface of the substrate and have an index difference ~0.005 from the substrate material, are particularly useful for dual-channel coupler fabrication, because the mode tails extend significantly into the space between guides. This is not the case, for example, with a ridged guide formed on the surface of the substrate and surrounded on three sides by air. Nevertheless, effective dual-channel couplers can be made in glass if the critical dimensions and indices are carefully controlled [7.7]. Various techniques of fabricating dual-channel couplers will be described and compared in more detail in a later section of this chapter, following a discussion of the theory of operation.

7.2.2 Coupled-Mode Theory of Synchronous Coupling

A concise theory of operation of the dual-channel directional coupler can be developed by following the coupled mode theory approach of *Yariv* [7.8, 9].

This model has been compared with experimental measurements of three-arm couplers fabricated in Ti:LiNbO$_3$ by *Peall* and *Syms* [7.10], and excellent qualitative agreement was obtained. The electric field of the propagating mode in the waveguide is described by

$$\overline{E}(x, y, z) = A(z)\overline{\mathscr{E}}(x, y), \tag{7.2.1}$$

where $A(z)$ is a complex amplitude which includes the phase term $\exp(i\beta z)$. The term $\overline{\mathscr{E}}(x, y)$ is the solution for the field distribution of the mode in one waveguide, assuming that the other waveguide is absent. By convention, the mode profile $\overline{\mathscr{E}}(x, y)$ is assumed to be normalized to carry one unit of power. Thus, for example, the power in guide number 1 is given by

$$P_1(z) = \left| A_1(z) \right|^2 = A_1(z)A_1^*(z). \tag{7.2.2}$$

The coupling between modes is given by the general coupled mode equations for the amplitudes of the two modes. Thus,

$$\frac{dA_0(z)}{dz} = -i\beta_0 A_0(z) + \kappa_{01} A_1(z), \tag{7.2.3}$$

and

$$\frac{dA_1(z)}{dz} = -i\beta_1 A_1(z) + \kappa_{10} A_0(z), \tag{7.2.4}$$

where β_0 and β_1 are the propagation constants of the modes in the two guides, and κ_{01} and κ_{10} are the coupling coefficients between modes.

Consider the guides shown in Fig. 7.2. Assume that the guides are identical and that they both have an exponential optical loss coefficient α. Thus,

$$\beta = \beta_r - i\frac{\alpha}{2}, \tag{7.2.5}$$

where $\beta = \beta_0 = \beta_1$, and β_r is the real part of β. For the case of identical guides, it is obvious from reciprocity that

$$\kappa_{01} = \kappa_{10} = -i\kappa, \tag{7.2.6}$$

where κ is real. Then, using (7.2.5) and (7.2.6), Eqs. (7.2.3) and (7.2.4) can be rewritten in the form

$$\frac{dA_0(z)}{dz} = -i\beta A_0(z) - i\kappa A_1(z), \tag{7.2.7}$$

and

$$\frac{dA_1(z)}{dz} = -i\beta A_1(z) + i\kappa A_0(z). \tag{7.2.8}$$

If it is assumed that light is coupled into guide 0 at the point $z = 0$, so that boundary conditions for the problem are given by

$$A_0(0) = 1 \text{ and } A_1(0) = 0, \tag{7.2.9}$$

then the solutions are described by

$$A_0(z) = \cos(\kappa z)e^{i\beta z} \tag{7.2.10}$$

and

$$A_1(z) = -i \sin(\kappa z)e^{i\beta z}. \tag{7.2.11}$$

Thus, the power flow in the guides is given by

$$P_0(z) = A_0(z)A_0^*(z) = \cos^2(\kappa z)e^{-\alpha z} \tag{7.2.12}$$

and

$$P_1(z) = A_1(z)A_1^*(z) = \sin^2(\kappa z)e^{-\alpha z}. \tag{7.2.13}$$

From (7.2.12) and (7.2.13), it can be seen that the power does indeed transfer back and forth between the two guides as a function of length, as shown experimentally in Figs 7.3 and 7.4. Note also, in (7.2.10) and (7.2.11), the distinct phase difference that exists between the amplitude of the fields in the two guides. The phase in the driven guide always lags 90° behind the phase of the driving guide. Thus, initially at $z = 0$, the phase in guide 1 lags 90° behind that in guide 0. That lagging phase relationship continues for increasing z, so that at a distance z that satisfies $\kappa z = \frac{\pi}{2}$, all of the power has been transferred to guide 1. Then, for $\frac{\pi}{2} \leqq \kappa z \leqq \pi$, the phase in guide 0 lags behind that in guide 1, and so on. This phase relationship results from the basic mechanism which produces the coherent transfer of energy. The field in the driving guide causes a polarization in the dielectric material which is in phase with it, and which extends into the region between guides because of the mode tail. This polarization then acts to generate energy in the mode of the driven guide. It is a basic principle of field theory that generation occurs when polarization leads the field, while dissipation occurs when polarization lags the field [7.11]. Thus, the lagging field in the driven guide is to be expected. Because of this definite phase relationship, the dual-channel coupler is a directional coupler. No energy can be coupled into a backward wave traveling in the $-z$ direction in the driven waveguide. This is a very useful feature in many applications.

From (7.2.12) and (7.2.13), it can be seen that the length L necessary for complete transfer of power from one guide to the other is given by

$$L = \frac{\pi}{2\kappa} + \frac{m\pi}{\kappa}, \tag{7.2.14}$$

where $m = 0, 1, 2, \ldots$ In a real guide, with absorption and scattering losses, β is complex. Hence, the total power contained in both guides decreases by a factor $\exp(-\alpha z)$. The theoretical power distribution as a function of z from (7.2.12) and (7.2.13) is plotted in Fig. 7.5.

The coupling coefficient κ is a strong function of the shape of the mode tails in the guides. For well confined modes, in which the overlapping of the tails causes only a negligible perturbation of the basic mode shape, it can be shown that the coupling coefficient is given by [7.12]

$$\kappa = \frac{2h^2 q e^{-qs}}{\beta W(q^2 + h^2)}, \tag{7.2.15}$$

where W is the channel width, s is the separation, h and β are the propagation constants in the y and z directions, respectively, and q is the extinction coefficient in the y direction. It will be recalled that these parameters have been assumed to be identical for both waveguides.

In a practical situation, it may be difficult to fabricate two identical waveguides to form a coupler. If, for example, the guides do not have exactly the same thickness and width, the phase velocities will not be the same in both. This will not necessarily destroy the coupling effect entirely. If the difference in phase constants $\Delta\beta$ is small, it can be shown [7.6] that the power distributions in the two guides are given by

$$P_0(z) = \cos^2(gz)e^{-\alpha z} + \left(\frac{\Delta\beta}{2}\right)^2 \frac{\sin^2(gz)}{g^2}e^{-\alpha z}, \tag{7.2.16}$$

and

$$P_1(z) = \frac{\kappa^2}{g^2}\sin^2(gz)e^{-\alpha z}, \tag{7.2.17}$$

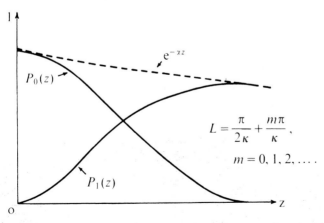

Fig. 7.5. Theoretically calculated power distribution curves for a dual-channel directional coupler. The initial condition of $P_0(0) = 1$ and $P_1(0) = 0$ has been assumed

where

$$g^2 \equiv \kappa^2 + \left(\frac{\Delta\beta}{2}\right)^2. \tag{7.2.18}$$

It can be seen from (7.2.16–18) that, in the presence of a phase constant difference $\Delta\beta$, transfer of power will still occur. However, the transfer will be incomplete, since (7.2.16) has no zeros for any z.

The preceding equations can be used to calculate the expected performance of dual-channel directional couplers for the case of slightly non-identical guides, with a non-zero $\Delta\beta$. We will return to these equations in Chap. 8 to consider the case in which $\Delta\beta$ is made electrically controllable, to produce an optical modulator and switch.

7.2.3 Methods of Fabricating Dual-Channel Directional Couplers

One of the most convenient methods of fabricating the closely-spaced waveguides of the dual-channel directional coupler, in the case of GaAs (or GaP) substrates, is masked proton bombardment. The proton bombardment step is diagrammed in Fig. 7.6. A vapor-deposited gold mask, one or two micrometer thick, is used to block the protons in the regions surrounding the waveguides. In between the bars of the mask, the protons penetrate the semiconductor, producing a carrier-compensated waveguide as described in Chap. 4. A 1.5 μm thick layer of gold is sufficient to block 300 keV protons.

The mask is formed by first depositing a gold layer over the entire substrate surface; then applying a layer of photoresist on top of the gold by using the standard thin film spinning process described in Sect. 4.1.2. The photoresist is then exposed and developed, down to the gold using standard photolithographic techniques, to produce a striped pattern of resist with a ridged shape. This photoresist pattern is in turn used to mask the gold layer during ion-beam micromachining [7.13]. The ion-beam sputtering process can be easily controlled to remove the gold just down to the surface of the GaAs. If an ion such as Ar⁺ or Kr⁺ is used, the sputtering yield for gold will be relatively high, compared to that for either GaAs or photoresist. Thus, a 2 μm thick photore-

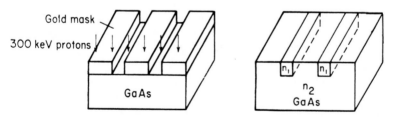

Fig. 7.6. Diagram of the masked proton implantation method of dual-channel coupler fabrication. For a proton energy of 300 keV, a 3 μm thick waveguide is formed in GaAs

Fig. 7.7. Scanning-electron-microscope photograph of the evaporated gold mask used in coupler fabrication as in Fig. 7.6. Spacing between mask bars equals 3 μm. (Remains of the photoresist mask can be seen above the gold stripes). (Photo courtesy of Hughes Research Laboratories, Malibu, CA)

sist mask will adequately mask a 2 μm thick gold layer, and also very little GaAs will be removed, even if the process is not halted instantly when the GaAs surface is first reached. This is important because it provides a built-in compensation for any possible variations in thickness of the photoresist. Figure 7.7 shows a scanning-electron-microscope photograph of a mask formed on a GaAs substrate by the process which has been described. It can be seen that edge variations where the gold mask meets the GaAs are less than about 0.05 μm, so that scattering loss should be negligible. Absorption loss in proton implanted waveguides is on the order of $1 \, cm^{-1}$ [7.14].

After the gold mask is in place on the substrate, channel waveguides are formed by proton implantation as described in Chap. 4. Couplers formed by 300 keV H^+ bombardment of GaAs to produce 2.5 μm wide guides spaced 3.9 μm apart, have exhibited a coupling coefficient κ equal to $0.6 \, mm^{-1}$ [7.15]. In this case, the increase in index of refraction within the guides due to proton bombardment Δn was calculated to be 0.0058. The coupling coefficient κ is a strong function of Δn. For example, a 10% reduction of Δn from the above value would increase κ to $0.74 \, mm^{-1}$.

As an alternative to proton bombardment, diffusion of a compensating dopant atom [7.16, 17] could be used to form the waveguides. Also, masked diffusion of a third element into a binary compound can be used to form a channel waveguide. For example, *Martin* and *Hall* [7.18] used an SiO_2 mask on ZnSe to form Cd diffused channel waveguides that were 10 μm wide. In $LiNbO_3$ and $LiTiO_3$, diffusion of metal atoms can be used to produce the required waveguide pair. For example, *Kaminow* et al. [7.19] have used Ti diffusion into $LiNbO_3$ to produce 4.6 μm wide waveguides. In most cases, diffusion doping is much less predictable than ion implantation doping, exhibiting greater variations in dopant concentration profile. This makes diffusion doping more difficult to use for coupler fabrication, since the coupling coefficient is such a strong function of the index profiles in the guides. However, evanescent coupling between two indiffused waveguides has been mathematically modeled [7.17, 20].

Dual-channel couplers can also be fabricated by masking the substrate and depositing metallic strips on the surface to produce a closely spaced pair of strip-loaded guides. *Campbell* et al. [7.21] have used that approach to form a dual-channel coupler in GaAs. They began with an epitaxially grown, carrier-concentration-reduction type, planar waveguide on a heavily doped ($N_d \sim 10^{18}$ cm^{-3}) n-type substrate. Thin (5–25 μm wide) metallic strips were produced by using standard photoresist masking and etching techniques, and vapor depositing either Au, Ni or Pt. Separation between the waveguides was on the order of a few micrometers. This type of coupler is particularly useful in electro-optic modulator or switch applications, since the metal loading strips can also be used as electrodes to control the phase matching between the two channels. Devices of that type are discussed in more detail in Chap. 8.

All of the methods of dual-channel coupler fabrication that have been described thus far produce imbedded waveguides, which lie below the surface of the substrate. However, couplers can also be made by beginning with a substrate containing a planar waveguide over its entire surface and then removing superfluous regions by masking and etching to produce raised ridged channel guides. The planar waveguide is usually made by epitaxial growth. The masking is accomplished with standard photoresist techniques, and either chemical or ion-beam etching can be used. However, in general, chemical etchants are difficult to control to produce the required degree of sidewall smoothness. If ion-beam machining is used, the photoresist pattern is replicated in the waveguide layer just as it was in the gold layer used as a mask for proton bombardment, as previously described. Very smooth sidewalls can be produced using this method, as shown in Fig. 7.8. Edge variations are seen in the photograph to be less than a few hundred Angstroms. However, because of the larger index difference at the sidewall interface, scattering losses are particularly sensitive to wall smoothness. Hence, care must be exercised to minimize variations, limiting them to at most about 500 Å.

Another problem inherent in fabricating couplers with ridged channel guides is that the larger index difference confines the modes so well to their respective channels that insufficient coupling is obtained. To increase the coupling, only a partial removal of the material between channels can be performed, as shown in Fig. 7.9. The strength of coupling then depends on the thickness of the waveguide bridge between channels [7.6]. Alternatively, the coupling between two ridged channels can be increased by filling the region between them with another material having a slightly lower refractive index.

Ridged channel waveguides are most useful in materials such as LiNbO$_3$ and glasses, in which imbedded waveguides are difficult to produce. For example, *Kaminow* et al. [7.22] have used ion-beam machining to produce 19 μm

Fig. 7.8. Scanning electron microscope photograph of an ion-beam machined ridged channel waveguide in GaAs. The channel is 1.4 μm high and 2.0 μm wide [7.6]

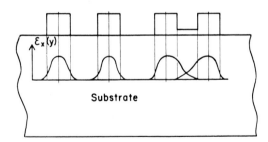

Fig. 7.9. Diagram of dual-channel couplers fabricated by masked etching of a planar waveguide layer. The electric field distributions are shown for the case of total etched isolation on the left and for bridged isolation on the right

wide ridged guides in LiNbO₃. *Kawabe* et al. [7.23] have made ridged-waveguide directional couplers in LiNbO₃ by employing an ion-bombard-ment-enhanced etching technique. This method utilizes the fact that a material damaged by ion bombardment generally has a higher chemical etch rate than unbombarded material. The waveguide pattern was defined by masked 60 keV Ar⁺ bombardment, and unwanted material was then removed from around the waveguides by etching in dilute HF [7.24]. Resulting edge roughness from this process was estimated to be less than 300 Å.

7.2.4 Applications Involving Directional Couplers

The dual-channel directional couplers that have been described in this chapter are among the more useful of integrated optic devices. They can be used as power dividers, input or output couplers, and directionally selective taps on an optical data bus. The foregoing are all examples of passive applications, in that the fraction of power coupled in each case is a constant. However, the most important application of the dual-channel coupler is an active modulator or switch, in which the coupled power is electrically controlled. It has been shown in (7.2.16) and (7.2.17) that the coupled power is a strong function of any phase mismatch in the two guides. In the next chapter, techniques will be described for producing a change in index of refraction by application of an electric field. The combination of this electro-optic effect with the dual-channel coupler yields a very efficient high-speed modulator, which is described in Chap. 8.

7.3 Butt-Coupled Ridge Waveguides

Techniques of butt-coupling similar to those described for fiber-waveguide couplers in Chap. 6 can also be used to couple two ridge waveguides. Once again, the coupling efficiency depends on the overlap integral of the modes in the waveguides. *Koshiba* et al. [7.25] have used the scalar finite element method to evaluate the effective index, modal field profile, and far-field pattern of the guided mode of a ridge waveguide, and to determine the coupling efficiency of a butt-joint structure of two single-mode rib waveguides. Two ridge waveguides of identical cross-section theoretically can have 100% coupling efficiency if they are perfectly aligned and interface reflections are eliminated.

The alignment of coupled ridge waveguides is a critical problem. In the laboratory it can accurately be accomplished by using stages positioned by piezoelectric micrometers, but this is not a satisfactory solution for application in the field. In that case, accurately formed slots or grooves are frequently used to produce the desired alignment. For example, *Booth* [7.26] has reported polymer waveguide to waveguide interconnects created by employing laser ablation to micromachine interlocking slots in the waveguides and substrates. Both single-mode and multi-mode coupling have been achieved, with single-mode loss less than 0.2 dB and multi-mode loss of 0.5 dB. The loss has been attributed mostly to overlap mismatch of the abutting guides. Multilayer packaging structures, typically on the order of 200 μm or more in thickness, can be used to increase the strength of the slotted connectors to permit connect/ disconnect cycling. Greater than 100 cycles have been tested without any measurable increase in loss.

7.4 Branching Waveguide Couplers

The branching waveguide coupler, or confluent coupler as it is sometimes called, is a straight forward passive combining of a multiplicity of waveguides with a single waveguide at a branching junction. It may be used as a combiner, as shown in Fig. 7.10, or as a splitter, if the direction of propagation is reversed from that shown in the figure. The dimensions and refractive index profiles in the waveguides must be carefully controlled in order to obtain eiter equal splitting or uniform combining. The theory of coupled local normal modes can be used to design branching waveguide couplers. *Burns* and *Milton* [7.27] have applied this approach to both two-branch and three-branch couplers.

Branching waveguide couplers can be produced by any of the usual waveguide fabrication methods. For example, *Haruna* and *Koyama* [7.28] have made them in Ti-diffused Z-cut $LiNbO_3$ with the input and output waveguide branches 4 µm wide to support only the fundamental guided mode for 6328 Å wavelength. *Beguin* et al. [7.29] have fabricated branching waveguide couplers by ion exchange in glass. The ion exchange process has two steps. First thallium ions are diffused into the glass at an elevated temperature to cause the exchange of Tl^+ ions with Na^+ and K^+ ions of the substrate, producing an

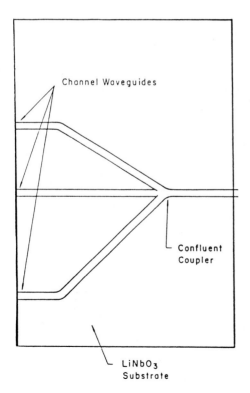

Channel Waveguides

Confluent
Coupler

$LiNbO_3$
Substrate

Fig. 7.10. Three-arm branching waveguide coupler

increase in the index of refraction. The guides are then buried and symmetrized by a second ion exchange in the presence of an electric field to prevent out diffusion of the Tl$^+$ ions. In 1×2 couplers fabricated by this method, splitting uniformity was measured to be ± 0.1 dB and excess loss was less than 0.2 dB. The corresponding values for 1×16 couplers were splitting uniformity $= \pm 0.3$ dB and excess loss < 1.0 dB.

The key feature of branching waveguide couplers which makes them attractive is that they can be relatively inexpensive when produced in large quantities in a low-cost substrate such as glass. For example, they have been proposed for use in the subscriber loops of telecommunication systems [7.30], where many such devices would be required.

The dual-channel directional couplers and branching waveguide couplers which have been described in this chapter are passive devices that have no controlled switching characteristics. However, by the addition of properly placed control electrodes it is possible to turn these couplers into electro-optic switches and modulators. Devices of this type are described in Chap. 8.

Problems

7.1 A dual-channel-coupler type of modulator has been designed so that

$$\kappa L = \frac{\pi}{2} + m\pi, \qquad m = 0, 1, 2, \ldots,$$

where κ is the coupling coefficient and L is the length. Thus, complete transfer of light will occur from channel 0 to channel 1. If we now apply a voltage to produce a $\Delta\beta = \beta_0 - \beta_1$, the condition for complete cancellation of the transfer is

$$gL = \pi + m\pi, \qquad m = 0, 1, 2, \ldots,$$

where

$$g^2 = \kappa^2 + \left(\frac{\Delta\beta}{2}\right)^2.$$

Derive an expression for $\Delta\beta$ required to produce this complete cancellation in terms of the length L.

7.2 In order to produce a value of $\Delta\beta$ (as determined in Problem 7.1) that is necessary for complete cancellation, what change in index of refraction in one waveguide (Δn_g) would be required? Give your answer in terms of an expression in which the parameters are wavevectors (k) and coupling length (L).

7.3   Using the result of Problem 7.2, what would be the change of index (Δn_g) required for the case of light with a vacuum wavelength $\lambda_0 = 9000\,\text{Å}$ and a coupling length $L = 1\,\text{cm}$?

7.4   If two dual-channel waveguide directional couplers of identical channel geometry and spacing are formed in the same substrate material, except that coupler A has an index of refraction n_A in the channels and coupler B has an index of refraction n_B in the channels, which coupler has the larger coupling coefficient κ if $n_A > n_B$?

Fig. 7.5   A dual-channel directional coupler has $\kappa = 4\,\text{cm}^{-1}$, $\alpha = 0.6\,\text{cm}^{-1}$, and $\Delta\beta = 0$. What length should it be to produce a 3 dB power divider? If that length is doubled, what fraction of the input power is in each channel at the output?

8. Electro-Optic Modulators

This chapter begins the discussion of optical-signal modulation and switching In many cases, the same device can function as either a modulator or a switch depending on the strength of the interaction between the optical waves and the controlling electrical signal, as well as on the arrangement of input and output ports. The device is considered to be a modulator if its primary function is to impress information on a light wave by temporaly varying one of its properties. A switch, on the other hand, changes the spatial position of the light, or else turns it off and on. Many of the same factors must be considered in designing or evaluating both modulators and switches. Hence, it is logical to discuss them together.

8.1 Basic Operating Characteristics of Switches and Modulators

8.1.1 Modulation Depth

One important characteristic of modulators and switches is the modulation depth,or modulation index, η. In the case of an intensity modulator in which the applied electrical signal acts to decrease the intensity of the transmitted light, η is given by

$$\eta = (I_0 - I)/I_0, \tag{8.1.1}$$

where I is the transmitted intensity and I_0 is the value of I with no electrical signal applied. If the applied electrical signal acts no increase the transmitted light intensity. η is given by

$$\eta = (I - I_0)/I_m, \tag{8.1.2}$$

where I_m is the transmitted intensity when maximum signal is applied. The maximum modulation depth, or extinction ratio, is given by

$$\eta_{max} = (I_0 - I_m)/I_0 \quad \text{for} \quad I_m \leqq I_0, \tag{8.1.3}$$

or by

$$\eta_{max} = (I_m - I_0)/I_m \quad \text{for} \quad I_m \geqq I_0. \tag{8.1.4}$$

It is also possible to define the modulation depth for phase modulators, as long as the phase change can be functionally related to an equivalent intensity change. For the case of interference modulators, it can be shown that the modulation depth is given by [8.1, 2]

$$\eta = \sin^2(\Delta\varphi/2), \tag{8.1.5}$$

where $\Delta\varphi$ is the phase change.

Modulation depth has been defined for intensity modulators (and indirectly for phase modulators); however, an analogous figure. of merit, the maximum deviation of a frequency modulator, is given by

$$D_{\max} = |f_{\mathrm{m}} - f_0|/f_0, \tag{8.1.6}$$

where f_0 is the optical carrier frequency, and f_{m} is the shifted optical frequency when the maximum electrical signal is applied.

8.1.2 Bandwidth

Another important characteristic of modulators and switches is the bandwidth, or range of modulation frequencies over which the device can be operated. By convention, that bandwidth of a modulator is usually taken as the difference between the upper and lower frequencies at which the modulation depth falls to 50% of its maximum value. In the case of a switch, frequency response is usually given in terms of the switching speed, or switching time. The switching time T is related to the bandwidth Δf by the expression

$$T = 2\pi/\Delta f. \tag{8.1.7}$$

Minimizing switching time is most important when large-scale arrays of switches are used to route optical waves over desired paths. Similarly, modulation bandwidth is a critical factor when many information channels are to be multiplexed onto the same optical beam. Thus, the unusually fast switching speed and wide bandwidth of waveguide switches and modulators, which will be discussed later in this chapter, make them particularly useful in large telecommunications systems.

8.1.3 Insertion Loss

Insertion loss is another important characteristic of optical switches and modulators that must be known for system design. Insertion loss is generally stated in decibels, and for the case in which the modulating signal acts to decrease the intensity, it is given by

$$\mathscr{L}_{\mathrm{i}} = 10\log(I_{\mathrm{t}}/I_0), \tag{8.1.8}$$

where I_{t} is the optical intensity that would be transmitted by the waveguide if the modulator were absent, and I_0 is the intensity transmitted with the modulator in place, but with no applied signal. For a modulator in which the applied signal acts to increase the transmitted intensity, the insertion loss is given by

$$\mathscr{L}_i = 10\log(I_t/I_m), \tag{8.1.9}$$

where I_m is the transmitted intensity when maximum signal is applied. Insertion loss is, of course, an optical power loss. However, it ultimately increases the amount of electrical power that must be supplied to the system, since higher power optical sources must be used.

8.1.4 Power Consumption

Electrical power must also be supplied to drive the modulator or switch. In the case of modulators, the required drive power increases with modulation frequency. Hence, a useful figure of merit is the drive power per unit bandwidth, $P/\Delta f$, usually expressed in milliwatts per megahertz. As is discussed in more detail in Sect. 8.7, a key advantage of channel-waveguide modulators is that they have a significantly lower $P/\Delta f$ than that required for bulk modulators.

The power requirements of optical switches operating at high clock rates, for example to time-division multiplex a number of different signals, can be evaluated in much the same way as is used for modulators. Hence, $P/\Delta f$ would still be a useful figure of merit in that case. However, if switching is done at relatively slow rates, a more important quantity is the amount of power required to hold the switch in a given state. An ideal switch would consume significant power only during the change of state; holding power would be negligible. Since electro-optic switches require the presence of an electric field to maintain at least one state, they could not be called ideal in that respect. However, except for leakage current, little power is needed to maintain a field in the small volume of a waveguide switch.

8.1.5 Isolation

The degree of isolation between various inputs and outputs of a switch or modulator is a major design consideration. In a modulator, the isolation between input and output is merely the maximum modulation index, as defined previously. However, it is usually expressed in decibels when used to specify isolation. In the case of a switch, the isolation between two ports (either input or output) is given by

$$\text{isolation [dB]} = 10\log\frac{I_2}{I_1}, \tag{8.1.10}$$

where I_1 is the optical intensity in the driving port, and I_2 is the intensity at the driven port when the switch is in the *off* state with respect to port 1 and 2. Thus, a switch with a signal leakage, or crosstalk, of 1% with respect to two ports would have -20 dB isolation.

8.2 The Electro-Optic Effect

The fundamental phenomenon that accounts for the operation of most electrooptic modulators and switches is the change in index of refraction produced by the application of an electric field. In the most general case, this effect is nonisotropic, and contains both linear (Pockels effect) and nonlinear (Kerr effect) components. In crystalline solids, the change in index produced by the linear electro-optic effect, can be most conveniently characterized by the change in the components of the optical indicatrix matrix [8.3]. The equation of the index ellipsoid in the presence of an electric field is

$$\left(\frac{1}{n^2}\right)_1 x^2 + \left(\frac{1}{n^2}\right)_2 y^2 + \left(\frac{1}{n^2}\right)_3 z^2 + 2\left(\frac{1}{n^2}\right)_4 yz + 2\left(\frac{1}{n^2}\right)_5 xz + 2\left(\frac{1}{n^2}\right)_6 xy = 1.$$

(8.2.1)

If x, y, and z are chosen to be parallel to the principal dielectric axes of the crystal, the linear change in the coefficients due to an electric field E is given by

$$\Delta\left(\frac{1}{n^2}\right)_i = \sum_{j=1}^{3} r_{ij} E_j,$$

(8.2.2)

where $i = 1, 2, 3, 4, 5, 6$, and where $j = 1, 2, 3$ are associated with x, y, z, respectively. If (8.2.2) is written in matrix form, the 6×3 $[r_{ij}]$ matrix is called the electro-optic tensor. It can be shown that the linear electro-optic effect exists only in crystals that do not possess inversion symmetry [8.4]. Even in the case of these noncentrosymmetric crystals, for most symmetry classes, only a few elements of the electro-optic tensor are nonzero [8.3]. Hence, in the design of an electro-optic modulator or switch, both the waveguide material and its orientation with respect to the applied electric field must be chosen carefully. Nevertheless, many materials that can be used to make low loss waveguides, such as GaAs, GaP, $LiNbO_3$, $LiTaO_3$ and quartz, also have significantly large Pockels coefficients, for certain orientation. Thus, the linear electro-optic effect is widely used in integrated optic applications.

The nonlinear (quadratic) Kerr electro-optic coefficient is relatively weak in commonly used waveguide materials. Also, a nonlinear dependence on electric field introduces unwanted modulation crossproducts (distortion) into the modulated signal. Consequently, it is not particularly useful in most integrated optic applications.

8.3 Single-Waveguide Electro-Optic Modulators

There are several different types of electro-optic modulators and switches that can be fabricated in a single-waveguide structure. The waveguide may be

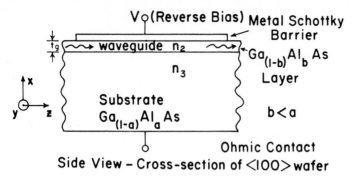

Fig. 8.1.   Basic electro-optic modulator structure

either a planar, or a channel guide. For example, the relatively simple planar waveguide shown in Fig. 8.1 can function as either a phase modulator, an amplitude (intensity) modulator, or as an optical switch. The waveguide shown in Fig. 8.1 is formed in $Ga_{1-x}Al_xAs$ using the heteroepitaxial growth technique described in Chap. 4. However, any electro-optic semiconductor such as $GaAs_xP_{(1-x)}$, GaAs, or GaP could be used, and also the waveguide could alternatively be formed by the carrier-concentration-reduction method.

8.3.1 Phase Modulation

The waveguide of Fig. 8.1 is asymmetric, since the indices of the air and the metal at the top surface are both much less than n_2, while n_3 is relatively close to n_2. The total change in index between the guide and the substrate is given by

$$\Delta n_{23} = n_2 - n_3 = \Delta n_{\text{chemical}} + \Delta n_{\text{CCR}} + \Delta n_{\text{EO}}. \tag{8.3.1}$$

where $\Delta n_{\text{chemical}}$ is the index change due to the differing aluminium concentrations a and b, Δn_{CCR} is the change resulting from decreased carrier concentration (if any) in the guide, and Δn_{EO} is the index change caused by the electro-optic effect. To make a phase modulator, the dimensions and doping of the waveguide are chosen so that it is above cutoff for the $m = 0$ mode at the wavelength desired, but is below cutoff for the $m = 1$ mode. Thus, from (3.1.31)

$$\frac{1}{32n_2}\left(\frac{\lambda_0}{t_g}\right)^2 < \Delta n_{\text{chemical}} + \Delta n_{\text{CCR}} < \frac{9}{32n_2}\left(\frac{\lambda_0}{t_g}\right)^2. \tag{8.3.2}$$

When a voltage V is applied with reverse bias polarity to the Schottky barrier diode, as shown in Fig. 8.1, the waveguide becomes part of the depletion layer of the diode, and the electric field causes a change in the phase of light waves traveling along the guide that is proportional to V. For the crystal orientation shown, the change in index of refraction caused by the field for a TE wave (polarized along the y axis) is given by [8.5]

$$\Delta n_{EO} = n_2^3 r_{41} \frac{V}{2t_g},$$ (8.3.3)

while there is no field-induced index change for TM waves (polarized in the x direction). By definition of terms,

$$\Delta n = \Delta\beta/k = (\Delta\beta\lambda_0)/2\pi.$$ (8.3.4)

Hence, if (8.3.4) is substituted into (8.3.3), the phase change produced by the electric field is given by

$$\Delta\varphi_{EO} = \Delta\beta L = \frac{\pi}{\lambda_0} n_2^3 r_{41} \frac{VL}{t_g},$$ (8.3.5)

where L is the length of the modulator in the z direction. The above equations assume that both the optical and electrical fields are uniform and occupy the same volume. If that is not the case, a correction factor Γ can be added to allow for an overlap integral which is less than 1 [8.6].

Many variations of the basic single-waveguide phase modulator structure have been demonstrated. For example, *Hall* et al. [8.7] have made a planar modulator structure like that of Fig. 8.1 in GaAs, in which $\Delta n_{chemical}$ was zero but doping concentration changes in the layers produced a Δn_{CCR} sufficient for waveguiding. *Kaminow* et al. [8.8] have made a modulator of this type in an outdiffused LiNbO₃ waveguide structure that was shaped by ion beam etching to produce a 19 µm wide ridged channel waveguide modulator. The required modulating power was only 20 µW/MHz/rad. In that case, the electric field was introduced into the waveguide by means of two evaporated metal strip electrodes on either side of the waveguide. A voltage of 1.2 V produced a phase change of 1 rad. A reduction of required power to 1.7 µW/MHz/rad was obtained in a 5 µm wide Ti-diffused LiNbO₃ single-channel waveguide modulator [8.9].

Electro-optic modolators can also be made in polymer materials. *Shuto* et al., [8.10] have reported electro-optic light modulation and second-harmonic. generation in diazo-dye-substituted poled polymers. A poled dimenthyl-substituted diazo dye (3RDCVXY) film exhibited a linear EO coefficient of 40 pm/V at 0.633 µm wavelength. An EO channel-waveguide modulator was made in this film with a half-wave voltage of 5 V.

8.3.2 Polarization Modulation

Phase modulators are limited in their usefulness by the fact that a phase coherent detection system must be used. To avoid this complication, the simple modulator structure of Fig. 8.1 can be used in a slightly different fashion by introducing the linearly polarized optical beam at 45° to the x and y axes. Because the phase change occurs only for waves polarized in the y direction and not for those polarized in the x direction, a rotation of the polarization vector will result as the waves propagate in the z direction. This change of polarization can be detected with a polarization sensitive detector, or by placing a polarization-seletive filter (usually called an analyzer) ahead of the detector. In the case of a discrete waveguide modulator, used for an air beam, a conventional wiregrid polarizer or an absorptive polarizing filter can be used as the analyzer. An analogous system can be implemented for the optical integrated circuit. For example, both grating and prism couplers are polarization sensitive and, hence, can be used as analyzers. However, the difficulty of fabricating an effective analyzer monolithically has limited the use of polarization modulators in OIC's, and has led to a preference for intensity modulation.

8.3.3 Intensity Modulation

Since polarization and phase modulation are difficult to detect compared to intensity modulation, the device of Fig. 8.1 is most often used in an intensity modulation mode. To make an intensity modulator, the difference in index of refraction at the waveguide-substrate interface must be carefully tailored to make the waveguide be just at the threshold for guiding the lowest order mode with no electric field present. Then, when an electric field is created by applying a voltage to the electrodes, it produces a slight additional change in index which causes the waveguide to become transmissive. The total change of index is given by (8.3.1). Hence, the zero field threshold condition for an intensity modulator of this type is given by

$$\Delta n_{23} = \Delta n_{\text{chemical}} + \Delta n_{\text{CCR}} = \frac{1}{32n_2}\left(\frac{\lambda_0}{t_{\text{g}}}\right)^2, \tag{8.3.6}$$

where (3.1.31) has been used for the cutoff condition of an asymmetric guide. An intensity modulator of this type was first implemented by *Hall* et al. [8.7], in a GaAs carrier-concentration-reduction planar waveguide, for light of 1.15 µm wavelength. A voltage of 130 V was required to bring the guide above cutoff for the TE$_0$ mode. *Campbell* et al. [8.11] have made a channel-waveguide modulator of this type in GaAs, with a 95% maximum modulation depth, a 150 MHz bandwidth, and a power requirement of less than 300 µW/MHz. *Kawabe* et al. [8.12] have made an intensity modulator for 6328 Å light in a 2.4 µm wide Ti-diffused LiNbO$_3$ channel waveguide. With an applied voltage of – 10 V (15 kV/cm) the $E^z{}_{11}$ mode was clearly guided, and with an applied voltage of + 10 V it was cutoff. The extinction ratio between the two extremes was – 19 dB. Obviously, intensity modulators such as have been described in this section, can also function as effective optical switches, as long as the magnitude of ($\Delta n_{\text{chemical}} + \Delta n_{\text{CCR}}$) is chosen to place the waveguide just below threshold and the applied field V/t_{g} is large enough to bring the guide well above cutoff at the desired wavelength.

8.3.4 Electro-Absorption Modulation

The modulators discussed thus far have all depended on the linear electro-optic effect for their operation. However, there is another type of modulator, the electro-absorption modulator, which must be classed as an electro-optic modulator in that it uses an electric field to produce intensity modulation, yet it does not employ the Pockels effect. The electro-absorption modulator is, instead, based on the Franz-Keldysh effect [8.13]. In the presence of a strong electric field, the absorption edge of a semiconductor is shifted to longer wavelength, as shown in Fig. 8.2 for the case of GaAs in a field of 1.3×10^5 V/cm. Because of the steepness of the absorption edge in a direct bandgap material such as GaAs, very large changes in absorption of wavelength near the band edge can be produced by application of an electric field. In the specific example shown in Fig. 8.2, for light of 9000 Å wavelength, the absorption coefficient α increases from 25 cm^{-1} to 10^4cm^{-1} when the field is applied.

The mechanism responsible for the Franz-Keldysh effect can be described straightforwardly with reference to the semiconductor energy band diagram shown in Fig. 8.3. In the presence of a strong electric field, the band edges bend. The left-hand limit of the diagram represents the surface of the semiconductor, at which a Schottky barrier contact or shallow p-n junction has been formed. Application of a reverse-bias voltage to this rectifying junction causes a charge depletion layer to form, extending to depths x within the semiconduc-

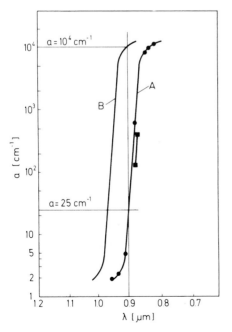

Fig. 8.2. Franz-Keldysh shift of the absorption edge of GaAs. *Curve A* is the zero field absorption curve for GaAs. (Circular dots represent experimental data points for n-type material with carrier concentration $n = 3 \times 10^{16}$ cm^{-3}. For the square dots $n = 5.3 \times 10^{16}$ cm^{-3}.) *Curve B* shows the shifted absorption edge for a field of 1.3×10^5 V/cm

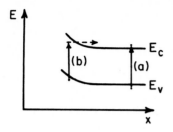

Fig. 8.3. Energy band diagram of a semi-conductor exhibiting the Franz-Keldysh effect in the presence of a strong electric field. The parameter x represents the distance from the surface of the semiconductor, and E is the electron energy. E_c and E_v are the conduction and valence bandedges, respectively

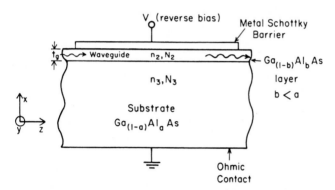

Fig. 8.4. Basic electro-absorption modulatore. Aluminum concentration b in the waveguide should be chosen so that the absorption edge wavelength is just slightly shorter than the guided wavelength, i.e., the guide is transparent with $V = 0$. concentrations should be chosen so the $N_3 \gg N_2$; thus, a relatively large electric field is produced in the guide when V is applied

tor. A nonuniform electric field is present within the depletion layer, with the largest amplitude existing at the surface. The resulting band bending is also greatest at the surface. (For a detailed discussion of depletion layers and band bending see, for example, [8.14].) Outside of the depletion layer where the field exists, the bands are flat, as at the right side of Fig. 8.3. In this region, a photon can be absorbed only if it has enough energy to lift an electron across the bandgap, as in transition (a). Closer to the surface, where the bands have been bent by the field, a transition (b) can occur in which photon energy is sufficient only to lift the electron partway across the gap. Ordinarily such a transition could not occur because there would be no allowed electron state within the bandgap. However, the electric field effectively broadens the states of the conduction band so that there is a finite probability of finding the electron in the gap. This, of course, reduces the effective bandgap and thereby shifts the absorption edge to longer wavelength. It can be shown that the effective change in bandgap energy ΔE is given by [8.13, 15]

$$\Delta E = \frac{3}{2}(m^*)^{-1/3}(q\hbar\mathscr{E})^{2/3},$$

(8.3.7)

where m^* is the effective mass of the carrier, q is the magnitude of the electrical charge of a carrier, h is Planck's constant divided by 2π, and $\mathscr{E}$ is the electric field strength.

Since ΔE depends on the electric field strength, and since α is a very strong function of ΔE, as shown in Fig. 8.2, a very effective electro-absorption modulator can be made for light of slightly less than bandgap wavelength. A basic electro-absorption modulator structure is shown in Fig. 8.4. The surface electrode can be either a Schottky barrier contact or a shallow p-n junction. In either case the electric field is produced in the depletion layer. Ideally the dopant concentration in the waveguide N_2 should be small enough that the depletion layer extends all the way through the guide in the x direction. The length of the modulator, and the applied voltage are chosen by using absorption curves such as those of Fig. 8.2 in order to establish a desired minimum insertion loss and maximum modulation depth for a given wavelength.

Although any type of waveguide can be used, the waveguide modulator structure can be improved by using a GaAlAs heterostructure guide, as shown in Fig. 8.4. In that case, the aluminum concentration in the guide can be adjusted to produce optimum performance at a given wavelength. For example. *Reinhart* [8.16] has produced such structures for use at 9000 Å which exhibit a change in transmission by a factor of 100 for – 8 V applied bias. The power necessary for 90% modulation was on the order of 0.1mW/MHz. By using multilayer waveguide structures of the quaternary compound GaInAsP, electroabsorption modulators for use at 1.3 or 1.55 µm wavelength can be made [8.17, 18].

In Sect. 8.3, electro-optic modulators employing a single waveguide have been discussed. In the next section, we consider modulators involving the transfer of optical energy between two waveguides. Such devices are inherently switches as well as modulators.

8.4 Dual-Channel Waveguide Electro-Optic Modulators

In Chap. 7, we saw that two closely spaced channel waveguides could function as a directional coupler, in which optical energy was synchronously transferred from one guide to the other. Such a coupler can be made into an electro-optic modulator by merely adding two electrodes, as shown in Fig. 8.5.

8.4.1 Theory of Operation

If a modulating signal voltage is applied to the electrodes, as shown in Fig. 8.5, it produces a slight difference in the indices of refraction in the guides, which results in a propagation constant difference $\Delta\beta$. By once again following the coupled mode theory approach used in Chap. 7 [8.19], the coupling equations can be shown to be given by

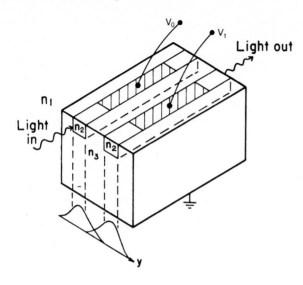

Fig. 8.5. Basic dual-channel modulator structure

$$\frac{dA_0(z)}{dz} = -i\beta_0 A_0(z) - ikA_1(z), \tag{8.4.1}$$

and

$$\frac{dA_1(z)}{dz} = i\beta_1 A_1(z) - ikA_0(z), \tag{8.4.2}$$

where β_0 and β_1 are the propagation constants in the two guides, and the other terms have already been defined previously. The solution of (8.4.1) and (8.4.2), subject to the boundary conditions that

$$A_0(0) = 1 \quad \text{and} \quad A_1(0) = 0, \tag{8.4.3}$$

yields the following expressions for $A_0(z)$ and $A_1(z)$.

$$A_0(z) = \left(\cos gz - i\frac{\Delta\beta}{2g} \sin gz \right) \exp\left[-i\left(\beta_0 - \frac{\Delta\beta}{2} \right)z \right], \tag{8.4.4}$$

and

$$A_1(z) = -\left(\frac{-ik}{g} \sin gz \right) \exp\left[-i\left(\beta_1 + \frac{\Delta\beta}{2} \right)z \right], \tag{8.4.5}$$

where

$$\Delta\beta = \beta_0 - \beta_1,$$

and where

$$g^2 \equiv k^2 + \left(\frac{\Delta\beta}{2} \right)^2. \tag{8.4.6}$$

Thus, in this case of imperfect phase match, the power flow in the two guides is given by

$$P_0(z) = A_0(z)A_0^*(z) = \cos^2(gz)e^{-\alpha z} + \left(\frac{\Delta\beta}{2}\right)^2 \frac{\sin^2(gz)}{g^2}e^{-\alpha z},$$ (8.4.7)

and

(8.4.8)

$$P_1(z) = A_1(z)A_1^*(z) = \frac{\kappa^2}{g^2}\sin^2(gz)e^{-\alpha z},$$

where α is the exponential loss coefficient in the guides. Note that (8.4.7) and (8.4.8) become identical to (7.2.12) and (7.2.13) when $\Delta\beta$ equals zero. Thus, the condition for total transfer of power for zero applied voltage is given once again by (7.2.14), which states that

$$\kappa L = \frac{\pi}{2} + m\pi,$$ (8.4.9)

where $m = 0, 1, 2, \ldots$. Similarly, it can be seen from (8.4.7) and (8.4.8) that, when a modulating voltage is applied to produce a $\Delta\beta$, the coupling is completely cancelled [i.e., $P_1(L) = 0$ and $P_0(L) = 1$] if

$$gL = \pi + m\pi, \quad \text{where} \quad m = 0,1,2,\ldots.$$ (8.4.10)

From (8.4.9) and (8.4.10) it can be shown that the value of $\Delta\beta$ required for 100% modulation is given by

$$(\Delta\beta)L = \sqrt{3}\pi.$$ (8.4.11)

The effective index of refraction in a guide is given by

$$n_g \equiv \frac{\beta}{k}.$$ (8.4.12)

Thus, the change in effective index needed for 100% modulation is given by

$$\Delta n_g = \frac{\sqrt{3}\pi}{kL}.$$ (8.4.13)

In typical cases, the magnitude of Δn_g required for 100% modulation is surprisingly small. For example, in a GaAlAs 3 μm × 3 μm dual-channel waveguide modulator such as that of Fig. 8.5, of length equal to 1 cm, (8.4.13) predicts that light of 9000 Å vacuum wavelength can be totally switched by producing a Δn_g of approximately 1×10^{-4}. From (8.3.3) it can be determined that the magnitude of the required electric field is about 3×10^4 V/cm, corresponding to a voltage of 10 V across the 3 μm thick channels.

The particular modulator geometry shown in Fig. 8.5 is probably the simplest arrangement that can be envisioned, so that it serves as a good example to illustrate the principles of dual-channel modulator operation.

However, many other electrode configurations are possible, some of them having distinct advantages for given applications.

8.4.2 Operating Characteristics of Dual-Channel Modulators

The concept of using a dual-channel directional coupler as a modulator was proposed as early as 1969 by *Marcatili* [8.20], but quite a few years passed before a working device was realized, largely because of the difficulty of fabricating the dual-channel structure to the required close tolerances. *Somekh* et al. [8.21, 22] fabricated dual-channel directional couplers in GaAs with 100% coupling, and theoretically analyzed the case of non-zero $\Delta\beta$. *Taylor* [8.23] theoretically analyzed the performance of a dual-channel coupler with three electrodes, as shown in Fig. 8.6; and *Campbell* et al. [8.24] produced the first operational device of this type, in 1975. At about the same time *Papuchon* et al. [8.25, 26] reported the successful operation of the first two-electrode type modulator, such as that shown in Fig. 8.5.

The three-electrode device of *Campbell* et al. was fabricated in GaAs, by using the metal electrodes themselves to create a pair of strip-loaded waveguides, as shown in Fig. 8.7. Ninety-five percent amplitude switching, with a maximum extinction ratio of 13 dB, was observed for the light from a 1.06 μm Nd: YAG laser. For the case of 6 μm wide waveguides separated by 7 μm, the maximum switching condition occurred for an applied voltage of 35 V. A 7 ns rise-time was measured, implying a 3 dB bandwidth of 100 MHz. The power-bandwidth ratio was determined to be approximately 180 μW/MHz.

The frequency response of modulators and switches of this type is generally limited by the capacitance of the electrodes. For example, *Campbell* et al. estimate that the above mentioned 7 ns rise-time could be reduced by an order of magnitude by reducing the width of the Schottky barrier contacts from 100 to 10 μm, thus giving a projected bandwidth of 1 GHz. The two-electrode modulator of *Papuchon* et al. [8.25, 26] offers the potential advantage of lower capacitance by eliminating the center electrode. They called their modulator *commutateur optique binaire rapide*, leading to the acronym COBRA which is often used to label the two-electrode type of dual-channel modulators. Typical coupling lengths observed for couplers formed from 2 μm wide Ti diffused strips, in LiNbO$_3$, and used to guide 5145 Å light, were 500 μm for the case of 2 μm separation, and 1 mm for 3 μm spacing. A voltage of 6 V was found to be

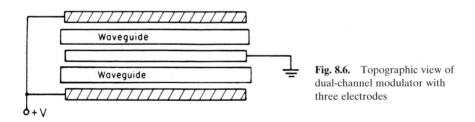

Waveguide

Waveguide

$+V$

Fig. 8.6. Topographic view of dual-channel modulator with three electrodes

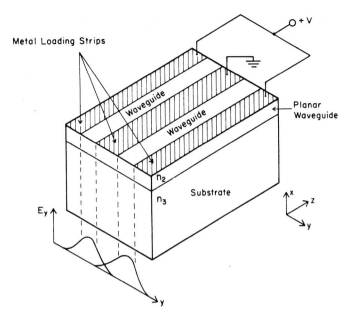

Fig. 8.7. Dual-channel modulator featuring strip loaded waveguides [8.22]. The three loading strips also function as the Schottky Barrier electrodes of the modulator. Typical electric field distributions for the TE_0 modes are plotted beneath the modulator structure

sufficient to *switch off* the coupling that was present for 0 V applied. The effect was about three times stronger for TM guided waves than for TE because of the particular orientation of the field and the $LiNbO_3$ substrate, and the strong anisotropy of the electro-optic tensor.

A problem that is shared by both the two-electrode and three-electrode modulators discussed thus far is that the length of the device must be carefully chosen to establish maximum coupling with no applied bias, as per (8.4.9); the *on* state cannot be electrically adjusted. *Kogelnik* and *Schmidt* [8.27] have demonstrated a dual-channel modulator in which three basic electrodes are split in half, as shown in Fig. 8.8. The polarity of the applied voltage is reversed in the two halves to produce two sections with $\Delta\beta$ of equal magnitude, but opposite sign. The effect of this stepped $\Delta\beta$ reversal is to yield a device in which both the *off* and *on* states can be electrically adjusted for a relative wide range of lengths. Obviously this allows one to maximize the extinction ratio and minimize crosstalk. For example, *Kogelnik* and *Schmidt* [8.27] have shown that providing sections with alternating $\Delta\beta$ makes it possible to cause a complete transfer of the light from one waveguide to another by electrical adjustment of the on and off states, as long as the ratio L/l is greater than unity, where L is the total length of the modulator, and l is the effective interaction length, defined by $l = \pi/2\kappa$. The frequency response of dual-channel modulators is also improved by stepped $\Delta\beta$ reversal [8.28]. For a detailed discussion of the theory of reversed $\Delta\beta$ modulators see *Alferness* [8.29]. During the decade

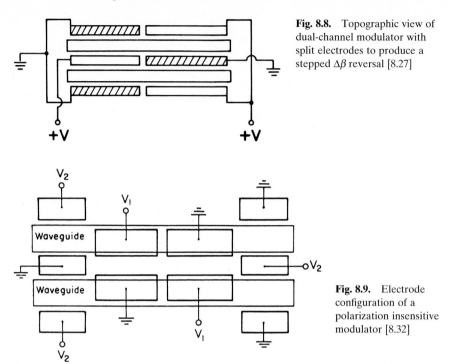

Fig. 8.8. Topographic view of dual-channel modulator with split electrodes to produce a stepped $\Delta\beta$ reversal [8.27]

Fig. 8.9. Electrode configuration of a polarization insensitive modulator [8.32]

of the 1980's the effectiveness of stepped $\Delta\beta$ reversal modulators has been greatly improved by the use of more sophisticated device structures. For example, *Veselka* et al. [8.30] have produced a 1×4 switch array of these devices (intended for time-division-multiplexing applications) which switches at a frequency of 4 GHz through the use of traveling wave electrodes.

Because of the anisotropic nature of the electro-optic tensor, electro-optic modulators are generally sensitive to the polarization of the light waves. In many applications this is not a problem because the proper polarization can be chosen to maximize the desired interaction. However, in the case of modulators to be used in conjunction with fiber optic waveguides, polarization sensitivity becomes an important problem to be dealt with. Linearly polarized light coupled into circular, single-mode fibers undergoes conversion to elliptical polarization, which excites both TE and TM waves in the rectangular guide of an OIC. Thus, in order to obtain maximum extinction ratio and minimum crosstalk, switches and modulators must be carefully designed to act on both polarization components equally. *Steinberg* and *Giallorenzi* [8.31] have shown that this can be done for most types of electro-optic modulators in LiNbO$_3$ by properly selecting the crystal orientation. *Steinberg* et al. [8.32] have also shown that a polarization-insensitive modulator can be made by using the unique electrode configuration of Fig. 8.9, which combines both between-guide electrodes and over-guide electrodes in a stepped $\Delta\beta$ reversal pattern. Since the between-guide electrodes produce electric field lines parallel to the substrate plane, while the over-guide electrodes produce lines perpendicular to it, the designer is given an additional degree of freedom that can be used to cancel polarization sensitivity.

8.5 Mach-Zehnder Type Electro-Optic Modulators

Not all two-channel modulators employ the synchronous coupling of energy between overlapping mode tails. Another class of modulators is based on a waveguide version of the Mach-Zehnder interferometer, in which interference is produced between phase coherent light waves that have traveled over different path lengths. The basic modulator structure is shown in Fig. 8.10. Light input to the modulator is via a single-mode waveguide. A beam splitter divides the light into two equal beams that travel through guides a and b, respectively. By applying a voltage to the electrodes, the effective path lengths can be varied. In an ideally designed modulator of this type, the path lengths and guide characteristics are identical, so that with no applied voltage the split beams recombine in the output waveguide to produce the lowest-order mode once more. If an electric field is applied so as to produce a phase change of π radians between the two arms, then the recombination results in an optical field that is zero at the center of the output waveguide, corresponding to the first order ($m = 1$) mode. If the output waveguide is a single mode guide, identical to the input guide, the first order mode is cut off, and rapidly dissipates over a short length by substrate radiation. Thus, the modulator can be switched from a transmitting to a nontransmitting state by application of a voltage.

Various embodiments of the basic Mach-Zehnder interferometer structure of Fig. 8.10 have been demonstrated to be effective. *Zernike* [8.33] first proposed this type of modulator in integrated optic form, using 3-dB directional couplers as the beam splitter and recombiner, rather than the prisms used in a conventional discrete component interferometer. *Martin* [8.34] fabricated a Mach-Zehnder modulator in ZnSe, employing diffused waveguides

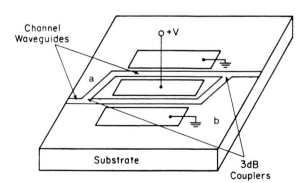

Fig. 8.10. Mach-Zehnder type modulator

that were formed into single mode branching "Y"s to obtain the desired beam splitting and recombining. Operating at a wavelength of 0.63 μm, this device could be switched from a state of 60% transmission to about 1% transmission by application of approximately 25 V. Mach-Zehnder modulators can be fabricated in LiNbO$_3$ substrates by forming either Ti indiffused or proton exchanged waveguides [8.35]. One problem with all of the Mach-Zehnder modulators described thus far is that even minute variations in fabrication parameters result in a device which is not in the *on* state for zero applied voltage. Thus, careful control of the applied voltage must be maintained for both the *off* and *on* states. *Ramaswamy* et al. [8.36] have gone one step further by using electrically switched dual-channel directional couplers as the beam splitter and recombiner. This modification permits electrical adjustment of both the splitting fractions and the phase change in the interferometer arms. Thus, the *off* and *on* states can be electrically selected for optimum extinction ratio. Once the bias voltages on the directional couplers have been adjusted, only the voltage on the interferometer arm must be changed to control switching. An off-on ratio of 22 dB was observed in a 38 mm long device of this type formed in Ti diffused LiNbO$_3$.

Electro-optic modulators of the single-waveguide (Sect. 8.3), dual-channel (Sect. 8.4) and Mach-Zehnder types have become the first integrated optic devices to be offered for sale as commercial products on an "off-the-shelf" basis. The operating characteristics of these commercially available devices are discussed in Chap. 17.

8.6 Electro-Optic Modulators Employing Reflection or Diffraction

A number of different modulators and switches have been demonstrated that utilize electro-optic control of either reflection or diffraction of the waveguided light. Diffraction modulators are generally based on the Bragg effect [8.37], which involves distributed interactions with multiple *reflecting* elements, usually in the form of an optical grating.

8.6.1 Bragg-Effect Electro-Optic Modulators

A typical Bragg-effect modulator is shown in Fig. 8.11, which consists of an interlaced, comb like, pair of electrodes. A voltage applied to the interdigitated surface electrodes perturbs the index of refraction beneath them thus forming an effective optical grating pattern in the waveguide. This grating causes a change in the direction of propagation of the light beam. If the direction of the light beam in the waveguide is adjusted so it is incident onto the grating bars at an angle equal to the Bragg angle θ_B, the light is diffracted

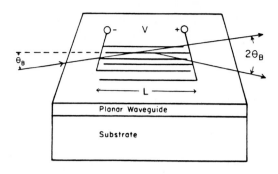

Fig. 8.11. Bragg-effect electro-optic modulator

with maximum efficiency at an angle $2\theta_B$ with respect to the input beam. It can be shown [8.38] that θ_B is given by

$$\sin \theta_B = \lambda_0 / 2 \Lambda n_g, \tag{8.6.1}$$

where Λ is the grating spacing, and n_g is the effective guide index (β/k).

The derivation of (8.6.1) is based on the *thick grating* assumption that

$$2\pi \lambda_0 L \gg \Lambda^2. \tag{8.6.2}$$

If the input beam strikes the grating at an angle different from the Bragg angle, diffraction still occurs over a limited range of $\Delta\theta_B$, but with reduced efficiency. The angular range for a 50% reduction is given by [8.35]

$$\Delta\theta_B = \frac{2\Lambda}{L}, \tag{8.6.3}$$

for small θ_B such that $\sin \theta_B \cong \theta_B$.

The intensity of light diffracted is dependent on the applied voltage, and has the general form [8.35]

$$\frac{I}{I_0} = \sin^2 VB, \tag{8.6.4}$$

where I is the intensity of the diffracted beam with V applied, I_0 is the transmitted intensity with $V = 0$, and B is a constant dependent on the effective guide index and on the applicable element of the electro-optic tensor.

Electro-optic, Bragg diffraction, modulators were first proposed by *Hammer* [8.39], and independently by *Giarusso* and *Harris* [8.40]. *Hammer* et al. demonstrated efficient grating modulation in epitaxial ZnO waveguides on sapphire substrates [8.41] and in LiNb$_x$Ta$_{1-x}$O$_3$ waveguides on LiTaO$_3$ substrates [8.42]. A detailed theoretical model for this type of modulator has been developed by *Lee* and *Wang* [8.43]. Experimental work [8.44, 45] has shown that very efficient Bragg diffraction electro-optic modulators can be made if grating geometry and tolerances are carefully controlled. For example, *Tangonan* et al. [8.45] formed modulators in Ti diffused LiNbO$_3$ having a multiple, series grating structure designed for high efficiency. They measured

diffraction efficiencies of 98%, with an extinction ratio of 300:1 (24.7 dB) at
1.06 μm wavelength, and a diffraction efficiency of 98%, with an extinction
ratio of 250:1 (24 dB) at 6328 Å.

8.6.2 Electro-Optic Reflection Modulators

It is possible to use the linear electro-optic effect to reduce the index of
refraction in a layer, thereby bringing about the total internal reflection of an
optical beam. A device of this type, which has been proposed by *Tsai* et al.
[8.46], is shown in Fig. 8.12. Four horn-shaped tapered channel waveguides
form the input and output ports to a planar waveguide modulator which
contains a region in which the refractive index can be reduced by application
of an electric field. If there is no applied voltage, an incident light beam from,
for example, port 1 will encounter no index interface and will pass freely to
port 4. If the horn tapers are carefully designed and fabricated to minimize
scattering and mode conversion, very little crosstalk will occur at port 3.
However, when a voltage is applied with the proper polarity to reduce the
index between the electrodes, two interfaces are created between regions of
different index. Total internal reflection may occur at the first interface if the
angle of incidence is greater than the critical angle, thereby causing partial (or
possible total) switching of the light beam to port 3. It can be shown for the
arrangement diagrammed in Fig. 8.12 that the critical angle is given by [8.46]

$$\theta_c = \sin^{-1}\left[1 - \frac{1}{2}n_1^2 r_{33}\left(\frac{V}{d}\right)\right], \tag{8.6.5}$$

where n_1 is the effective index outside of the electric field region, and d is the
electrode separation. In terms of the voltage required to switch a beam incident at a given angle θ_i, (8.6.5) can be written as

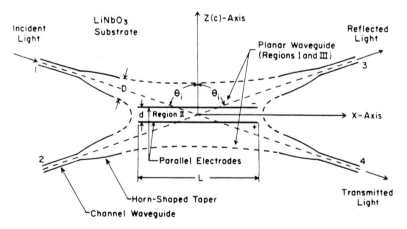

Fig. 8.12. A total internal reflection (TIR) electro-optic modulator and switch [8.46]

$$\frac{V}{d}\bigg|_{\text{TIR}} = \frac{2(1 - \sin\theta_i)}{n_1^2 r_{33}} \cong \frac{1}{n_1^2 r_{33}}\left(\frac{\pi}{2} - \theta_i\right)^2.$$

(8.6.6)

A TIR switch of the type which has been described has been fabricated by *Tsai* et al. [8.43] by Ti diffusion of Y cut LiNbO$_3$. The input/output horns were 4.7 mm long, tapering from 4 to 40 μm in width. The length of the electrode pair was $L = 3.4$ mm and d equalled 4 μm. Complete switching of a 6328 Å beam was observed for V approximately equal to 50 V. Cross-talk to port 3 in the absence of an applied voltage was measured to be –15 dB. A detailed theoretical model for the type of TIR switch has been developed by *Sheem* [8.47]. Estimated switching speeds exceed 6 GHz because of the relatively low device capacitance. Oh et al. [8.48] have fabricated a TIR optical switch like that of Fig. 8.12 in which the controllable reflecting interface is provided by p/n/p/n current blocking layers. This device, fabricated in InGaAsP waveguides on an InP substrate, had a very low operating current of 20 mA.

8.7 Comparison of Waveguide Modulators to Bulk Electro-Optic Modulators

At several places in this chapter, the relatively low drive power required by a waveguide modulator has been noted. To quantitatively compare the power requirements of a waveguide modulator to those of a bulk electro-optic modulator, it is convenient to develop a simple, yet general, expression for the average external power P_e needed to operate the modulator at a maximum frequency equal to its bandwidth (Δf). For the case of 100% modulation, this power is given by

$$P_e = (\Delta f)\mathcal{W}.$$

(8.7.1)

where $\mathcal{W}$ is the energy supplied from an external source to switch the device on or off. For an ideal electro-optic modulator with no Ohmic losses, all of this energy goes into the stored electric field between the electrodes. Hence, we can take

$$\mathcal{W} = \frac{1}{2}\int \varepsilon E_a^2 dv,$$

(8.7.2)

where E_a is the peak amplitude of the applied field, and ε is the permittivity. The integral is to be taken over the entire volume occupied by the field. If we assume, for convenience, that all of the electric field is confined to the modulator volume, and additionally that E_a is uniform over that volume, (8.7.2) becomes

$$\mathcal{W} = \frac{\varepsilon WtLE_a^2}{2},$$

(8.7.3)

where W is the width, t is the thickness and L is the length of the modulator active volume. Hence, the external drive power, from (8.7.1), is given by

$$P_e = \frac{(\Delta f)\varepsilon WtLE_a^2}{2}. \tag{8.7.4}$$

The key feature of (8.7.4) is that the modulating power required is proportional to the active volume. Thus, if we compare a bulk electro-optic modulator, such as that shown in Fig. 8.13a, to the planar waveguide modulator of Fig. 8.13b, it is obvious that significantly less power is required by the planar waveguide device. Still greater power reduction is obtained by going to a channel waveguide modulator, such as that shown in Fig. 8.13c. Consider the following numerical example.

For the specific case of an electro-optic modulator formed in GaAs with the orientation shown in Fig. 8.1, we find by using (8.3.3) that the applied field and the resulting change in index of refraction are related by

$$E_a = \frac{2\Delta n}{n_2^3 r_{41}}. \tag{8.7.5}$$

Therefore, combining (8.7.5) and (8.7.4) yields

$$P_e = \frac{2(\Delta f)\varepsilon WtL}{n_2^6 r_{41}^2} \Delta n^2. \tag{8.7.6}$$

For the special case of the dual-channel 100% modulator, it has been shown in (8.4.13) that

$$\Delta n = \frac{\sqrt{3}\pi}{kL} = \frac{\sqrt{3}\lambda_0}{2L}. \tag{8.7.7}$$

Substituting (8.7.7) into (8.7.6) gives the expression

$$\frac{P_e}{(\Delta f)} = \frac{3\varepsilon Wt\lambda_0^2}{2n_2^6 r_{41}^2 L}. \tag{8.7.8}$$

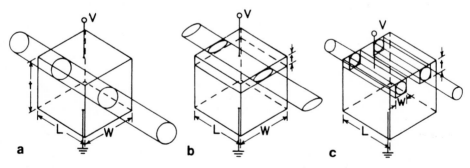

Fig. 8.13a–c. Basic electro-optic modulator structures; **a** bulk; **b** planar waveguide; **c** channel waveguide

If we take the following typical values for GaAs: $W = 6 \times 10^{-6}$m, $t = 3 \times 10^{-6}$m, $\lambda_0 = 0.9 \times 10^{-6}$m, $n_2 = 3.6$, $r_{41} = 1.2 \times 10^{-12}$m/V, $\varepsilon/\varepsilon_0 = 12$, and $L = 0.5$cm, the result is that $P_e/(\Delta f) = 0.148$ mW/MHz. The comparable value of $P_e/(\Delta f)$ for a planar waveguide modulator would typically be on the order of ten times larger because of the increase in W, while that for a bulk modulator would be 100 to 1000 times larger because of corresponding increases in both W and t. It should be noted that the values of $P_e/(\Delta f)$ calculated using (8.7.6) are based on the assumption that the optical fields and the electric field are both uniformly confined to a volume WtL. If this is not the case, a slightly modified relation can be used, which is given by

$$P_e = \frac{2(\Delta f)\varepsilon \left(\dfrac{W}{c_1}\right)\left(\dfrac{t}{c_2}\right)L}{n_2^6 r_{41}^2} \Delta n^2, \tag{8.7.9}$$

where c_1 and c_2 are constants, having a value less than 1, to account for electric field and optical field not being perfectly confined to the same volume.

8.8 Traveling Wave Electrode Configurations

The various electro-optic modulators shown in the preceding figures of this chapter all have "lumped-element" electrodes. In this case resistance, capacitance and inductance are all assumed to be non-distributed, and the electrode is considered as an equipotential element. While this model is accurate at lower modulation frequencies, it cannot be used above about 1 GHz. At these higher frequencies the magnitude of the wavelength of the modulating signal approaches the dimensions of the electrode, and such parameters as resistance and capacitance must be considered to be distributed. The modulating signal produces a voltage wave, with peaks and troughs along the length of the electrode. Thus for efficient coupling of the modulating signal the electrodes should be designed so as to form an appropriate microwave transmission line. Traveling wave electrode configurations for a phase modulator and a Mach-Zehnder modulator are shown in Fig. 8.14. In these devices the electrodes have been patterned to form a microwave transmission line with a characteristic impedance of 50 Ω The modulation signal input is at the left end of the transmission line and the modulation signal output at the right end of the line would be terminated in a matching impedance to avoid reflections. As the microwave modulation signal travels along the transmission line the traveling voltage wave produces a similarly moving pattern of changed index of refraction in the underlying waveguide. This change in index causes a phase shift in the optical waves in the waveguide. Because of efficient coupling between the traveling modulating voltage wave and the optical waves the modulators shown in Fig. 8.14 have a 3 dB bandwidth of 4 to 6 GHz. Similar devices made with conventional electrodes have a bandwidth of only 2 to 3 GHz.

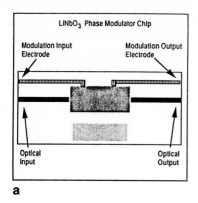
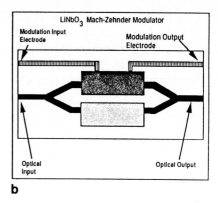

a   **b**

Fig. 8.14 a, b.   Waveguide modulators with traveling wave electrodes in LiNbO₃. **a** phase modulator, **b** Mach-Zehnder modulator

The characteristic impedance (z) of a traveling wave electrode is given by

$$\frac{1}{z} = \frac{c}{\sqrt{\varepsilon_{\text{eff}}}}\left(\frac{C}{L}\right),$$

where c is the speed of light in vacuum, (C/L) is the capacitance per unit length and ε_{eff} is the effective microwave dielectric constant (= 35.8 for lithium niobate) [8.6]. For a thorough discussion of traveling wave electrode design and performance see *Alferness* [8.6].

By carefully designing the traveling-wave electrode configuration it is possible to produce electro-optic modulators with very high operating frequencies. The key to success is to match the velocity of the light waves in the optical waveguide to the velocity of the microwaves on the electrode transmission line. For example, *Gopalkrishnan* et al. [8.49] have reported a Ti diffused LiNbO₃ EO modulator with electrodes configured as a microwave coplanar transmission line which has a 3 dB cutoff frequency of 20 GHz. Millimeter wave coplanar slow-wave structures operating in the frequency range of 35 to 50 GHz have been produced in GaAs [8.50] and InP [8.51].

In this chapter, a number of different types of electro-optic modulators have been discussed. One of these, the Bragg diffraction type, uses an electro-optically induced grating to change the path of the optical beam. In the next chapter, we will see how surface acoustic waves can be used for the same purpose.

Problems

8.1   We wish to design a GaP electro-optic phase modulator as shown below for operation at 6300 Å wavelength.

a) What is the minimum thickness (t) required in the waveguiding layer if the carrier concentrations are $N_2 = 1 \times 10^{15} \text{cm}^{-3}$ and $N_3 = 3 \times 10^{18} \text{cm}^{-3}$?

b) How large a voltage (V) can be applied without producing electrical breakdown?

c) If this voltage is applied, how long (L) must the device be to produce a phase shift of π radians in the transmitted light wave? Assume the incident light is polarized in the Y direction.

For GaP:

E_c (critical electric field for breakdown): $5 \times 10^5 \text{V/cm}$,

r_{41} (electro-optic coefficient for the above orientation): $5 \times 10^{-11} \text{cm/V}$,

m^* (effective mass): $0.013 m_0$,

$n = 3.2$.

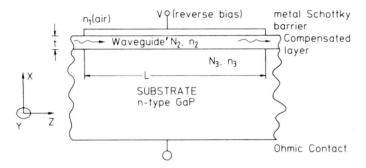

8.2 In the electro-optic waveguide switch shown in the figure, how large a voltage (V) must be applied to the electrode to turn the waveguide on – i.e., to increase the index of refraction in the waveguide sufficiently so as to bring it above cutoff for the lowest order waveguide mode?

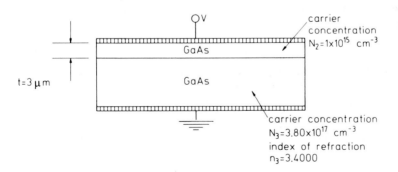

Assume: Figure

1) Wavelength of light in air $\lambda_0 = 1.0 \mu\text{m}$.
2) Effective mass of electron $m^* = 0.08 m_e$.
3) All of the voltage is dropped over the waveguide thickness rather than in the substrate.

4) Crystal orientation is such that r_{41} is the appropriate electro-optic tensor element to be used, and positive voltage (V) produces an increase in index of refraction.

8.3   a) A Schottky barrier type electro-optic modulator can be used to produce which of the following types of light modulation? – phase, polarization, frequency, intensity, pulse code?
    b) Describe the Franz-Keldysh effect.

8.4   In an electro-optic modulator, does the polarization of the light wave generally have an important effect on the modulation depth? Why?

8.5   What are the two basic conditions that must be met if optical modes in two different waveguides are to be coupled by synchronous coupling or "optical tunneling"?

8.6   A dual-channel GaAs waveguide modulator of the type shown in Fig. 8.5, with waveguides of $5\mu m \times 5\mu m$ cross-section, is used to modulate light of wavelength (in air) $\lambda_0 = 1.06\mu m$. It has a coupling coefficient $\kappa = 1\,cm^{-1}$ for the coupling of optical energy from one channel to the other when the voltage applied to the field plates is zero. (Assume absorption loss is negligible).
    a) What length L must the device have to completely couple the light from channel 0 to channel 1? (Assume lossless propagation of the lowest order mode only).
    b) What is the minimum voltage V_1 which must be applied to cause the light output from channel 1 to go to zero if $V_0 = 0$ and if the effective electro-optic coefficient for the geometry shown is $r = 1 \times 10^{-12} m/V$. Assume that the substrate conductivity is so large that all of the electric field is confined to the waveguide channel.

8.7   In the waveguide modulator structure of Problem 8.1, with all parameters being as given in that problem:
    a) What is the index of refraction difference at the interface between layers 2 and 3 due to the difference in carrier concentrations?
    b) What applied voltage would be required to produce a Δn of the same magnitude due to the electro-optic effect? (Assume the minimum waveguide thickness as calculated in Problem 8.1a).
    c) For the special case of $N_2 = N_3 = 1 \times 10^{15} cm^{-3}$ (i.e. a Schottky barrier formed directly on a uniformly doped substrate), would the answer to part (b) be the same?

8.8   a) Sketch the shape of the electric-field distribution versus depth for the waveguide modulator structures of Probs. 8.1 and 8.7c, assuming that the applied voltages have been adjusted to produce equal waveguide thickness in each case.

b) Would the difference in these two field distributions have a more pronounced effect on the profile of Δn versus depth in a lightly doped or more heavily doped semiconductor substrate.

8.9   If the maximum modulation bandwidth of an electro-optic modulator is given by

$$\Delta f = (2\pi R_L C)^{-1}$$

where R_L is the load resistance of the circuit and C is the device capacitance, show that the power to produce a given phase change $\Delta\Phi_{EO}$ is given by

$$P = \frac{(\Delta\Phi_{EO})^2 \lambda_0^2 \varepsilon A(\Delta f) t_g}{\pi L^2 n^6 r^2}$$

where λ_0 denotes the wavelength in air, ε is the permittivity of the material, A the cross-sectional area, L the optical path length in modulator, n the index of refraction of material, r the appropriate element of the E-O tensor, and t_g the waveguide thickness.

9. Acousto-Optic Modulators

In the preceding chapter, we have shown that modulators and switches can be made by using the electro-optic effect to produce a grating-shaped variation of the index of refraction within the waveguide. This grating structure causes diffraction of the guided optical waves, resulting in modulation or switching.

Acoustic waves can also be used to produce the desired grating pattern in the index profile. The acousto-optic effect is the change in the index of refraction caused by mechanical strain which is introduced by the passage of an acoustic-strain wave. The resulting index variation is periodic, with a wavelength equal to that of the acoustic wave. Two basic types of acousto-optic modulators are discussed in this chapter: the Bragg and the Raman-Nath configurations, which differ mainly in the interaction length between the acoustic and optical waves.

9.1 Fundamental Principles of the Acousto-Optic Effect

Mechanical strain in a solid causes a change in the index of refraction which can affect the phase of a light wave traveling in the strained medium. This photoelastic effect, as it is called, can be characterized by a fourth rank tensor (the strain-optic, or photoelastic tensor) that relates the strain tensor to the optical indicatrix, just as the electro-optic tensor characterizes changes in the indicatrix produced by an electric field. For a thorough discussion of the strain-optic tensor, see *Iizuka* [9.1].

In the case of the acousto-optic effect, mechanical strain is produced in a material by the passage of an acoustic wave. That strain therefore causes a change in the index of refraction via the photoelastic effect. *Pinnow* [9.2] has shown that the change in index of refraction Δn is related to the acoustic power P_a by the expression

$$\Delta n = \sqrt{n^6 P^2 10^7 P_a / 2\rho v_a^3 A},$$ (9.1.1)

where n is the index of refraction in the unstrained medium, p is the appropriate element of the photoelastic tensor, P_a is the total acoustic power in Watts, ρ is the mass density, v_a is the acoustic velocity, and A is the cross-sectional area through which the wave travels. All quantities in (9.1.1) are given in cgs units except for P_a. In terms of the commonly used acousto-optic figure of merit M_2, (9.1.1) can be written as

$$\Delta n = \sqrt{M_2 10^7 P_a / 2A} \, ,$$
(9.1.2)

where M_2 is defined by

$$M_2 \equiv n^6 p^2 / \rho v_a^3 \, .$$
(9.1.3)

In crystalline solids, such as are used most often for substrates in OIC's, the acousto-optic effect depends strongly on the orientation, i.e. on p. However, this effect is relatively small even for optimum choices of material and orientation. For example at a wavelength of 6328 Å, for fused quartz M_2 equals $1.51 \times 10^{-18} s^3/gm$, and for $LiNbO_3$ it is $6.9 \times 10^{-18} s^3/gm$ [9.2]. Thus, it can be determined from (9.1.2) that Δn is on the order of 10^{-4} in these materials, even for an acoustic power density of 100 W/cm². Despite the small value of Δn that can be produced by an acoustic wave, the overall effect on an optical beam can be quite significant, because each small Δn produced by a strain peak of the acoustic wave can result in optical interactions that can be accumulated constructively (or destructively) if proper phase matching is provided. Thus, substantial diffraction effects can be produced.

The acousto-optic modulators and switches used in optical integrated circuits generally employ traveling acoustic waves. Thus, the grating structure that is produced in the optical index profile is actually in motion with respect to the optical beam. However, this motion has an insignificant effect on the operation of most devices. The average effect of the moving grating structure is identical to that of a stationary one, except that the mth order of diffracted light is shifted in frequency by an amount equal to $\pm m f_0$, where f_0 is the frequency of the acoustic wave. Since the acoustic frequency is typically many orders of magnitude less than the optical frequency, the effect is generally negligible in acousto-optic phase or intensity modulators and beam deflectors. However, the effect has been used to produce optical frequency division multiplexing of signals [9.3].

Optical wave diffraction can be produced by interaction with either bulk acoustic waves traveling in the volume of the medium, or with surface acoustic waves (SAW), traveling within roughly an acoustic wavelength of the surface [9.4]. Since optical waveguides are typically only a few micrometers thick, SAW modulators and switches are compatible with most OIC applications.

Regardless of whether bulk or surface acoustic waves are used, two fundamentally different types of modulation are possible. In the Raman-Nath type of modulator the optical beam is incident transversely to the acoustic beam, and the interaction length of the optical path (i.e., the width of the acoustic beam) is relatively short, so that the optical waves undergo only a simple phase grating diffraction, thus producing a set of many interference peaks in the far field pattern. If the acoustic beam is much wider, so that the optical waves undergo multiple rediffraction before leaving the acoustic field, a much different diffraction pattern is produced. The diffraction in this case is similar to the volume diffraction of x-rays by multiple atomic planes in a crystal, which was first observed by Bragg. In Bragg-type acousto-optic modulators, the optical beam is incident at a specific angle (the Bragg angle) to the *bars* of the acoustically-produced index grating structure, and only one diffraction lobe is observed in the farfield pattern. A more detailed description of both Raman-Nath and Bragg-type modulators and switches is given in Sect. 9.2 and 9.3.

9.2 Raman-Nath-Type Modulators

The basic structure of a Raman-Nath modulator is shown in Fig. 9.1. Light passing through this device in the z-direction undergoes a phase shift given by [9.5]

$$\Delta\varphi = \frac{\Delta n 2\pi l}{\lambda_0}\sin\left(\frac{2\pi y}{\Lambda}\right), \tag{9.2.1}$$

where Δn is the acoustically produced index change, l is the interaction length, and Λ is the acoustic wavelength. The zero of the y axis is taken at the center of the incident beam. Combining (9.2.1) with (9.1.2) yields the expression

$$\Delta\varphi = \frac{2\pi}{\lambda_0}\sqrt{M_2 10^7 P_a l/2a}\sin\left(\frac{2\pi y}{\Lambda}\right), \tag{9.2.2}$$

where the fact that the area A equals l multiplied by the thickness of the acoustic beam a has been used.

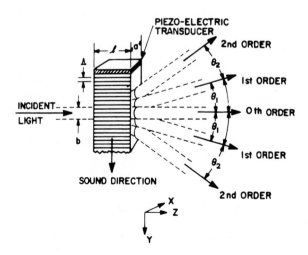

Fig. 9.1. Basic structure of a Raman-Nath type acoustooptic modulator [9.5]

For Raman-Nath type diffraction to take place, the interaction length must be so short that no multiple diffraction can occur. Such is the case when the condition

$$l \ll \frac{\Lambda^2}{\lambda} \qquad (9.2.3)$$

is satisfied, where λ is the optical wavelength within the modulator material. Incident light is then diffracted into a set of different orders at angles given by

$$\sin\theta = \frac{m\lambda_0}{\Lambda}, \qquad m = 0, \pm 1, \pm 2, \dots . \qquad (9.2.4)$$

The intensity of these orders is given by the relation [9.5]

$$I / I_0 = \begin{cases} [\mathrm{J_m}(\Delta\varphi')]^2 / 2, & |m| > 0 \\ [\mathrm{J_0}(\Delta\varphi')]^2, & m = 0 \end{cases} \qquad (9.2.5)$$

where the J's are the ordinary Bessel functions, I_0 is the intensity of the optical beam that is transmitted in the absence of an acoustic wave, and $\Delta\varphi'$ is the maximum value of the $\Delta\varphi$ given by (9.2.1), i.e.,

$$\Delta\varphi' = \frac{2\pi l \Delta n}{\lambda_0} = \frac{2\pi}{\lambda_0} \sqrt{M_2 10^7 P_a l / 2a} . \qquad (9.2.6)$$

The output channel of a Raman-Nath modulator is usually taken to be the zeroth-order mode. In that case, the modulation index equals the fraction of the light diffracted out of the zeroth order, and is given by

$$\eta_{\mathrm{RN}} = \frac{[I_0 - I(m = 0)]}{I_0} = 1 - [\mathrm{J_0}(\Delta\varphi')]^2 . \qquad (9.2.7)$$

Raman-Nath modulators generally have a smaller modulation index than that of comparable Bragg modulators. Also, the Raman-Nath modulator cannot be conveniently used as an optical switch, because the diffracted light is distributed over many orders, at different angles. Because of these disadvantages, the Raman-Nath modulator, while it is interesting from a theoretical standpoint, has not been often used in OIC applications. By contrast, the Bragg modulator has been widely used, as an intensity modulator, a beam deflector, and an optical switch.

9.3 Bragg-Type Modulators

For Bragg-type diffraction to take place, the interaction length between the optical and acoustic beams must be relatively long, so that multiple diffraction can occur. A quantitative relation that expresses this condition is given by

$$l \gg \frac{\Lambda^2}{\lambda}. \tag{9.3.1}$$

By comparing (9.3.1) and (9.2.3) one can see that there is a transitional range of l/λ for which a composite of both Bragg and Raman-Nath diffraction occurs. However, it is usually desirable to design a modulator so that it will clearly operate in either the Bragg or Raman-Nath regime, thus enabling the input and output angles of the optical beam to be chosen for maximum efficiency.

In the case of a Bragg-type modulator, the input angle of the optical beam should optimally be the Bragg angle θ_B, given by

$$\sin\theta_B = \frac{\lambda}{2\Lambda}. \tag{9.3.2}$$

The diffracted (1st order) output beam emerges at an angle of $2\theta_B$ with respect to the undiffracted (0th order) beam, as shown in Fig. 9.2. Generally the output of the modulator is taken to be the zeroth order beam. In that case, the modulation depth is given by [9.5]

$$\frac{I_0 - I}{I_0} = \sin^2\left(\frac{\Delta\varphi'}{2}\right), \tag{9.3.3}$$

where I_0 is the transmitted intensity in the absence of the acoustic beam, and I is the 0th order intensity in the presence of the acoustic beam. The maximum modulation depth, or modulation index, can be obtained by combining (9.3.3) and (9.2.6), which yields

$$\eta_B = \frac{(I_0 - I)}{I_0} = \sin^2\left[(\pi/\lambda_0)\sqrt{10^7 M_2 P_a l / 2a}\right]. \tag{9.3.4}$$

The basic modulator structures shown in Figs. 9.1 and 9.2 could be either bulk or waveguide modulators, depending on the ratio of the thickness a to the wavelength of light in the material. For a/λ very much greater than one, a bulk

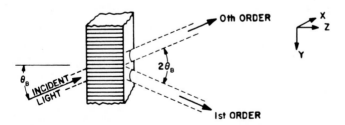

Fig. 9.2. Basic structure of a Bragg-type acousto-optic modulator [9.5]

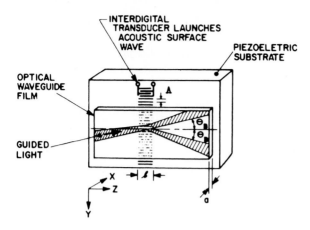

INTERDIGITAL
TRANSDUCER LAUNCHES
ACOUSTIC SURFACE
WAVE

PIEZOELETRIC
SUBSTRATE

OPTICAL
WAVEGUIDE
FILM

GUIDED
LIGHT

Fig. 9.3. Waveguide acousto-optic Bragg modulator structure [9.9]

modulator would result. Bulk acousto-optic modulators were widely used even before the advent of integrated optics, and their use continues today. For a review of typical applications see, for example, the articles by *Adler* [9.6], *Dixon* [9.7], and *Wade* [9.8]. Waveguide acousto-optic modulators, such as are used in OIC's, function in essentially the same manner as their bulk counterparts. However, they have the advantage of reduced drive power requirements because the optical and acoustic waves can both be confined to the same small volume.

The predicted performance of either bulk or waveguide modulators can be calculated by using (9.2.5) and (9.3.3). However, the calculation of the phase shift $\Delta\theta$ is more complicated for the case of waveguide modulators, because the optical and acoustic fields are generally not uniform over the active volume. For the case of non-uniform fields, (9.2.6) cannot be used to accurately determine the phase shift. Instead, an overlap integral must be calculated [9.9, 10]. However, (9.2.6) is adequate if only approximate results are desired.

The first waveguide acousto-optic modulator, as reported by *Kuhn* et al. [9.9], was a hybrid device, with the basic geometric configuration shown in Fig. 9.3. A y-cut α quartz substrate ($n = 1.54$) was used because of its relatively large piezoelectric coefficient, while a 0.8 μm thick glass film of larger index of refraction ($n = 1.73$) was sputter-deposited to form the waveguide. An interdigitated pattern of metal film electrodes was used as a transducer to launch an acoustic surface wave in the y direction. The strain pattern produced at the surface of the substrate by this wave was coupled into the optical waveguiding glass film by mechanical contact. Guided light ($\lambda_0 = 6328$ Å), introduced at the Bragg angle by means of a grating coupler, was observed to be diffracted through an angle $2\theta_B$. A modulation index of $\eta_B = 70\%$ was measured for an acoustic surface wave of frequency $f_a = 191$ MHz and wavelength $\Lambda = 16$ μm.

Kuhn et al. [9.9] chose a hybrid approach in order to use the strong piezoelectric effect in α quartz to launch the acoustic waves. *Wille* and *Hamilton* [9.11] also used this approach, obtaining a Bragg modulation index of 93% at a wavelength of 6328 Å for a sputter-deposited Ta_2O_5 waveguide on an α quartz substrate, driven by a 175 mW acoustic wave with $f_a = 290$ MHz. *Omachi* [9.12] obtained similar results with an As_2S_3 waveguide film on a $LiNbO_3$ substrate. At a wavelength of 1.15 μm, he observed a Bragg modulation index of 93% for an acoustic power of 27 mW, at $f_a = 200$ MHz. More recently, an acoustoroptic modulator has been designed using a piezoelectric film covering a multiple quantum-well structure to take advantage of the quantum-confined Stark effect [9.13]. (See Chap. 16 for a discussion of this phenomenon.

While the hybrid circuit approach is obviously an effective means of obtaining high-efficiency acousto-optic modulation, it is not essential in all applications. Monolithic OIC's, in which the acoustic wave is launched directly in the waveguide material, are also possible. For example, *Chubachi* et al. [9.14] have demonstrated a Bragg acousto-optic modulator featuring a ZnO sputtered-film waveguide on a fused quartz substrate, in which the acoustic waves were launched directly into the ZnO. They obtained a modulation index of greater than 90%. *Schmidt* et al. [9.15] fabricated a Raman-Nath type modulator in an outdiffused $LiNbO_3$ waveguide structure. They have reported that 25 mW/MHz of acoustic power was required to completely extinguish the zeroth order transmitted beam. Also, *Loh* et al. [9.16] have made guided wave Bragg-type modulators in GaAs. These examples demonstrate that monolithic acousto-optic modulators, in which the acoustic and optical waves are launched and travel in the same waveguide material, are feasible. However, much optimization must be achieved before they will be competitive with hybrid types.

To conclude the discussion of basic types of acousto-optic modulators, it is appropriate to note that there are alternative methods available, in addition to the Bragg and Raman-Nath techniques. Some of these methods involve the use of acoustic waves to induce coupling between various guided modes of the waveguide, or between guided and radiation modes. *Chu* and *Tamir* [9.17] have developed a detailed theory of acoustic-optic interactions using the coupledmode approach, and a number of different modulators have been demonstrated that employ intermode coupling [9.18–20]. Another alternative is to somehow use acoustic waves to induce a change in the optical transmission of a material, thus producing an intensity modulator. *Karapinar* and *Gunduz* [9.21] have done this in a homeotropically oriented nematic liquid crystal. *Gryba* and *Lefebvre* [9.22] have utilized the electro-absorptive effects in quantum wells and the electric field induced by surface acoustic waves to make a modulator. (See chap. 16 for further discussion of absorption in quantum wells.)

9.4 Bragg-Type Beam Deflectors and Switches

Bragg diffraction modulators are inherently capable of functioning as beam deflectors and optical switches, since the output beam exits at a different angle from the input beam. If the frequency of the acoustic wave is held constant, the optical beam can be switched through the angle $2\theta_B$ by supplying sufficient acoustic power to produce 100% diffraction from the zeroth order beam to the first-order beam. Alternatively, the frequency (wavelength) of the acoustic wave can be varied to deflect the optical beam through different angles, as given by (9.3.2).

In beam deflection applications, an important parameter is the number N of resolvable spots. A theoretical expression that can be used to determine N can be derived as follows. Assume that an optical beam of width b is incident on an acoustic wave over an interaction length l, as shown in Fig. 9.4. If b is much greater than Λ, so that the optical beam covers a number of of acoustic wave periods, we know from the fundamental diffraction theory of a beam incident on a grating [9.23], that the far-field pattern contains a set of diffraction maxima having a half-power width given by

$$\Delta\theta_1 = \lambda/b , \tag{9.4.1}$$

with angular separation of the peaks given by

$$\Delta\theta_2 = \lambda/\Lambda . \tag{9.4.2}$$

When the optical beam is incident at the Bragg angle, only the first-order lobe of the diffraction pattern is significant, the other higher order have negligible intensity. The diffracted beam is thus concentrated in this peak and has a

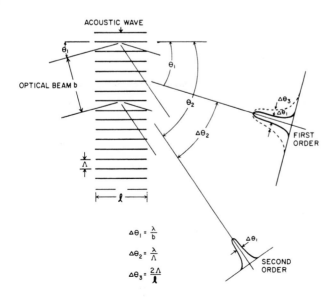

Fig. 9.4. Diffraction patterr of an optical beam incident on an acoustic wave grating The second order (and higher) lobes would be absent in the case of $\theta_1 = \theta_B$

maximum intensity. If the frequency (wavelength) of the acoustic wave is varied from that for which the Bragg condition is exactly satisfied, the diffracted beam is angularly scanned, but its intensity is reduced. It can be shown that the intensity follows a bell-shaped pattern with a half-power width give by [9.24]

$$\Delta\theta_3 = \frac{2\Lambda}{l}. \tag{9.4.3}$$

The number of resolvable spots is given by the ratio of the envelope width $\Delta\theta_3$ to the spot width $\Delta\theta_1$; thus, from (9.4.3) and (9.4.1)

$$N = \frac{\Delta\theta_3}{\Delta\theta_1} = \frac{2\Lambda b}{\lambda l}. \tag{9.4.4}$$

It must be emphasized with regard to (9.4.1–4) that angular differences $\Delta\theta_i$ ($i = 1, 2, 3$) and wavelengths Λ and λ are those measured within the medium in which the modulator is fabricated. Also, it has been assumed that the medium is optically isotropic.

It is interesting to note the equivalence between (9.4.3), which gives the angular deviation of the diffracted beam (from the Bragg angle) when the acoustic frequency is changed by a Δf sufficient to produce a 50% reduction in optical intensity, and (8.6.3), which gives the angular deviation of the input beam (from the Bragg angle) required to produce a 50% reduction in the intensity of the diffracted beam.

Another important operating characteristic of a Bragg-type beam scanner is the bandwidth Δf of the acoustic signal that it can accept. The decrease in diffracted optical beam intensity associated with angular deviation from the Bragg condition, as described above, implies an upper limit to this bandwidth. It can be shown that the maximum value of Δf corresponding to the half-power point of intensity is given by [9.24]

$$\Delta f_0 \cong \frac{2\upsilon_a\Lambda}{\lambda l}, \tag{9.4.5}$$

where υ_a is the acoustic velocity. In terms of this bandwidth, the number of resolvable spots can be shown from (9.4.4) and (9.4.5) to be given by

$$N = (\Delta f_0)t, \tag{9.4.6}$$

where t is the transit time of the acoustic wave across the optical beam width, given by

$$t = b/\upsilon_a. \tag{9.4.7}$$

The limitation of bandwidth by optical factors as expressed in (9.4.5–7) is not the only effect that must be considered. Very often the bandwidth is limited to significantly lower frequencies by the response time of the piezoelectric transducer that is used to launch the acoustic waves. The overall response time τ of an acoustic beam deflector is thus given by

$$\tau = \frac{1}{\Delta f_0} + \frac{1}{\Delta f_a} + t,\qquad(9.4.8)$$

where Δf_a is the bandwidth of the acoustic transducer, and Δf_0 and t have been defined previously.

If the number of resolvable spots is desired to be much greater than one, it can be seen from (9.4.6) the $1/\Delta f_0$ will be much less than t. Hence, in that case, (9.4.8) reduces to

$$\tau \cong \frac{1}{\Delta f_a} + t = \frac{1}{\Delta f_a} + \frac{b}{v_a}.\qquad(9.4.9)$$

In order to obtain maximum speed of operation, τ must be minimized by reducing t and increasing Δf_a. Note from (9.4.9) and (9.4.6), however, that there is an unavoidable trade-off between the number of resolvable spots and speed, since the transit time can be reduced by making b smaller, but that results in fewer resolvable spots. Of course, the bandwidth of the acoustic transducer Δf_a may be the limiting factor, in any case. A number of different transducer geometries designed for small τ are described in the next section.

9.5 Performance Characteristics of Acoustic-Optic Modulators and Beam Deflectors

The simplest type of acoustic transducer that can be used in a waveguide modulator consists of an interdigitated, single-periodic pattern of metal fingers deposited directly onto the waveguide surface, as shown in Fig. 9.5. Such a transducer launches a surface acoustic wave in the direction normal to its fingers. Patterns suitable for frequencies up to 1 GHz can be fabricated using standard photolithographic techniques described in Chap. 4, while interdigitated structure for higher frequencies may require the use of electron beam lithography [9.25, 26]. this relatively simple transducer, however, has a relatively small bandwidth compared to that obtainable with more sophisticated structures. For example, *Tsai* [9.27] has studied the case of a modulator for 6328 Å light, composed of a 2 µm thick in-diffused waveguide formed in a Y-cut LiNbO$_3$ substrate. A transducer such as that shown in Fig. 9.5 was used, which has a center frequency of 700 MHz and an interaction length (acoustic aperture) $l = 3$ mm. The 3 dB modulator bandwidth in this case is only 34 MHz, limited mostly by the 3 mm acoustic aperture. A larger bandwidth can be produced by reducing l. For example, a modulator such as the above, but with a waveguide thickness of 1 µm, a SAW center frequency of 1 GHz, and an acoustic aperture of 0.2 mm has a bandwidth of 380 MHz [9.27]. However, this increased bandwidth resulting from a smaller value of l is gained at the ex-

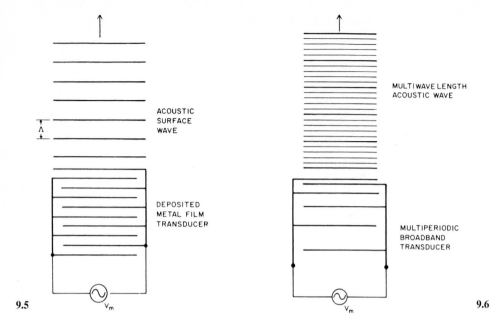

Fig. 9.5. Single-periodic interdigitated transducer. Application of a modulation voltage V_m induces a surface acoustic wave through the piezoelectric effect

Fig. 9.6. Multi-periodic "chirp" interdigitated transducer

pense of significantly reduced diffraction efficiency, see (9.3.4). Reduced diffraction efficiency, of course, implies that a larger acoustic drive power is required for operation.

If both wide bandwidth and high diffraction efficiency are required, more sophisticated transducer structures must be employed. One such structure is the aperiodic, or *chirp* transducer, shown in Fig. 9.6 [9.28, 29], in which the interdigitated spacing is gradually changed along the length of the transducer in the direction of propagation of the acoustic wave. Since the transducer is most efficient in generating acoustic waves when the spacing is one-half wavelength [9.27], different acoustic wavelengths are optimally generated at different positions along the transducer comb, thus increasing the overall bandwidth.

Rather than using a single chirp-transducer to obtain wide bandwidth, one can use a multiple array of tilted transducers, as shown in Fig. 9.7 [9.30]. The individual transducers are staggered in periodicity (center frequency), and are tilted with respect to the optical beam. The tilt angle between each adjacent pair of transducers just equals the difference in the Bragg angles that correspond to their center frequencies. Thus, the composite of surface acoustic waves launched by such a transducer satisfies the Bragg condition in multiple frequency ranges, and thereby yields a much wider bandwidth than that which could be obtained with a single transducer. The electrical drive power must be coupled to each transducer through a suitable matching network, but the transducers are then driven in parallel through a power divider.

Lee et al. [9.31] have fabricated a modulator with multiple tilted transducers in a Ti-diffused waveguide in Y-cut LiNbO$_3$. The transducers had center frequencies of 380, 520, 703 and 950 MHz. The TE$_0$ mode from a He-Ne laser at 6328 Å was modulated over a 680 MHz bandwidth with a diffraction efficiency of 8% at a total rf drive power of 0.8 W. The overall drive power of 1.17 mW/MHz was reasonably low, even though the conversion efficiency of the 950 MHz transducer was very poor (– 15 dB). *Tsai* [9.27] predicted that, when more efficient high frequency transducers have been fabricated, modulators of this type should be capable of a 1 GHz bandwidth with 50% diffraction efficiency at an electrical drive power of 1 mW/MHz.

Multiple tilted transducers are very effective because they permit the Bragg condition to be satisfied over a large acoustic aperture and over a wide frequency range. Another approach for achieving the same objective is to use a phased-array transducer set, as shown in Fig. 9.8 [9.32]. In this case, the individual transducer elements have identical center frequencies and parallel propagation axes. However, they are arranged in a stepped configuration that results in a variable phase shift between adjacent surface acoustic waves as the SAW frequency is varied. Because of this phase shift and resultant wave interference, a scanning acoustic wavefront is generated, in the same manner that a scanning radar beam is generated by a phased-array antenna. Scanning of the wavefront produces a wide aperture acoustic beam that tracks the Bragg condition over a large frequency range. *Nguyen* and *Tsai* [9.32] have fabricated a modulator of this type in a 7 μm thick outdiffused LiNbO$_3$ waveguide. A six element array with a center frequency of 325 MHz and a total acoustic aper-

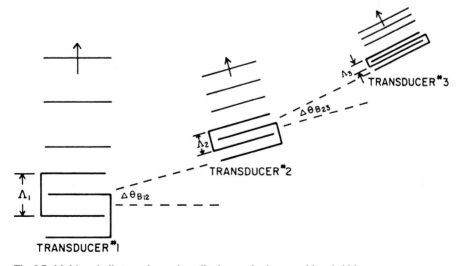

Fig. 9.7. Multi-periodic transducers in a tilted array for increased bandwidth

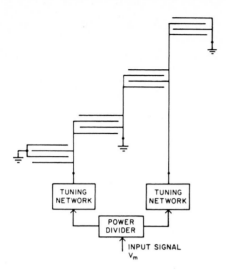

Fig. 9.8. Phased array transducers [9.32]

ture of 10.44 mm required only 3.5 mW of acoustic drive power (68 mW of electrical power) to diffract 50% of the light, over a bandwidth of 112 MHz. As in the case of other multi-element wide bandwith acousto-optic modulators that have been described, Δf was limited by the transducer bandwidth rather than by any optical phenomena.

Acousto-optic modulators and beam deflectors are among the more advanced integrated optic devices. They have not only been developed to a high degree of sophistication in the laboratory, but have also been employed in a number of practical applications, such as the rf spectrum analyzer [9.33], which is described in Chap. 17, and the DOC II 32-bit digital optical computer [9.34], which employs multichannel acousto-optic modulators.

9.6 Accusto-Optic Frequency Shifters

In addition to their use in modulators and beam deflectors, acousto-optic devices can also be employed to shift the frequency of an optical wave. As mentioned in Sect. 9.1, the frequency shifting effect of an acousto-optic modulator is so small that it is usually neglected when considering modulators and beam shifters. However, even this slight shift in frequency can be significant enough to be usful in certain applications.

For example, the standing wave surface acoustic wave optical modulator (SWSAWOM) which is shown in Fig. 9.9 can be used to produce frequency shifts of an optical carrier wave that are sufficiently large to permit frequency division multiplexing of a number of information signal channels onto an optical beam [9.3]. In this device the light from a semiconductor laser is butt coupled into a Ti diffused waveguide formed in a LiNbO$_3$ substrate and is then

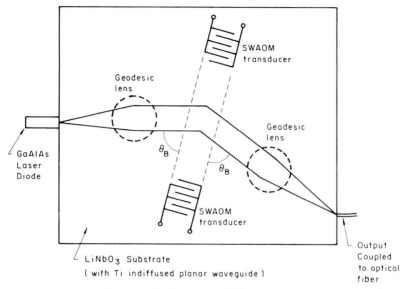

Fig. 9.9. A SWSAWOM for optical frequency shifting

collimated by a geodesic lens before passing through a SWSAWOM. The SWSAWOM features two surface wave acousto-optic modulator (SWAOM) transducers which launch surface waves in opposite directions. The acoustic waves traveling in one direction cause an upward shift in the frequency of the optical waves while the acoustic waves traveling in the opposite direction produce a downward shift in the frequency of the optical waves. Thus the emerging optical beam has frequency components at $f_0 + f_m$ and also $f_0 - f_m$, where f_m is the frequency at which the transducers are driven. Thus, when the optical beam is refocused by another geodesic lens and directed to a square law detector (possibly over an interconnecting optical fiber), the two components of the optical beam beat together to produce a beam with a modulated subcarrier frequency of $f_c = 2f_m$. A number of such modulators can be combined in an optical integrated circuit as shown in Fig. 9.10 to produce a frequency division multiplexed optical beam with a number of subcarrier channels at frequencies fc_1, fc_2, fc_3, etc. Each one of these channels separately carries the particular information signal that is impressed on its diode laser by modulating the laser drive current. These signals can be sorted out at the receiving end by conventional microwave filtering methods. A SWSAWOM of the type shown in Fig. 9.9 has been fabricated to operate at a subcarrier frequency of $f_c = 600\,\text{MHz}$ [9.35]. In this case the transducers were driven at a frequency of 300 MHz.

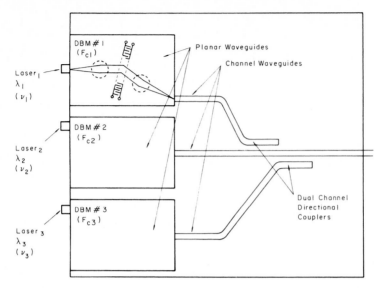

Fig. 9.10. An optical frequency division multiplexer

Problems

9.1 Show that, in general, a properly designed waveguide light modulator requires less electrical driving power than a bulk modulator of the same type (Bragg, Raman-Nath, etc.) in the same material.

9.2 What change in index of refraction can be induced in a quartz waveguide 3 μm thick and 100 μm wide by an acoustic power of 1 W?

9.3 A Bragg modulator is formed in a LiNbO$_3$ planar waveguide which is capable of propagating only the lowest order mode for light of 6328 Å (vacuum) wavelength. This mode has $\beta = 2.085 \times 10^5$ cm^{-1}. The bulk index of refraction for LiNbO$_3$ at this wavelength is 2.295. The acoustic wavelength in the waveguide is 2.5 μm, the optical beam width 4.0 mm, and the interaction length 2.0 mm. Obtain the following:

a) The angle (in degrees, with respect to the acoustic wave propagation direction) at which the optical beam must be introduced so as to obtain maximum diffraction efficiency.

b) If the input optical beam is a uniform plane wave, what is the angular divergence (in degrees) between the half-power points of the diffracted beam, and at what angle does it leave the modulator?

c) Make a sketch of the modulator, labeling the angles found in a) and b). (Do not attempt to draw the angles to scale.)

9.4 In a given hybrid OIC, a GaAs waveguide ($n = 3.59$) is butt-coupled to a LiNbO$_3$ Bragg modulator ($n_{Bulk} = 2.2$). (i) If an optical beam of vacuum wavelength $\lambda_0 = 1.15\,\mu m$ is traveling in the GaAs waveguide, incident on the modulator at an angle of 93° relative to the direction of propagation of the acoustic waves, for what acoustic wavelength Λ will the modulator product the maximum output optical intensity? (ii) At what angle will the optical beam be traveling once it has left the modulator (in air)?

9.5 Given a Bragg modulator with $l = 50\,\mu m$, in which the acoustic velocity is 250 m/s, could the modulator be used to impress a 100 MHz bandwidth signal on an optical beam from a He-Ne laser ($\lambda_0 = 0.6328\,\mu m$)?

9.6 An acousto-optic, Bragg-type modulator/beam deflector is fabricated in a waveguide material which has refractive index $n = 2.2$ and an acoustic velocity $v_a = 2800\,m/sec$.

a) If the modulator is to be operated at an acoustic frequency of 100 MHz what should be the spacing between the interdigitated metal fingers of the acoustic transducer? Draw a sketch of the transducer fingers and label the spacing for clarity.

b) What is the angle through which the optical beam will be deflected if the light source is a He-Ne laser with wavelength $\lambda_0 = 6328$ Å?

9.7 A Bragg-type acousto-optic beam deflector is built in a crystalline material in which the acoustic velocity is 2500 m/s. The crystal has a refractive index $n = 1.6$.

a) What is the acoustic wavelength required to deflect a He-Ne laser beam ($\lambda_0 = 6328$Å)) at a Bragg angle of 1°?

b) What is the corresponding acoustic frequency?

c) Sketch an interdigitated metal-finger transducer such as could be deposited on the surface of the crystal to launch a surface acoustic wave with the wavelength calculated in (a). Show how the voltage would be applied to the fingers and label the acoustic wavelength with an arrow between the appropriate fingers.

9.8 An acousto-optic, Bragg-deflection-type RF spectrum analyzer is made in a planar waveguide in LiNbO$_3$ with an index n = 2.0. Signals with modulation frequencies ranging from 50 to 500 MHz are analized by using them to drive the acousto-optic transducer which produces a traveling acoustic wave in the LiNbO$_3$ with a velocity $v_a = 3000\,m/s$. The optical beam has a wavelength in air of $\lambda_0 = 6328$Å and a width $b = 1\,mm$. The interaction length $\ell = 5\,mm$.

a) What is the angular deflection (in degrees) when the modulation frequency is 500 MHz?

b) What is the number of resolvable spots (the ratio of the envelope width to the spot width) at a frequency of 500 MHz?

10. Basic Principles of Light Emission in Semiconductors

This chapter is the first in a sequence of four chapters that describe the light sources used most often in integrated-optic applications. Because of their convenience, gas lasers are frequently used in the laboratory to evaluate wave-guides or other integrated-optic devices; however, semiconductor lasers and light-emitting diodes are the only practical light sources for use in optical integrated circuits, due to their small size and compatibility with monolithic (or hybrid) integration. Also, LED's and laser diodes are widely used in fiberoptic applications, because they can be modulated at high frequencies and can be efficiently coupled to the micrometer-size wave-guiding core of the fiber.

Since its invention in the early nineteen sixties, the semiconductor laser has matured into a highly sophisticated, efficient light source. In order to understand the subtleties of its operation, and how to best use it in a given application, one must be familiar with some of the basic principles of light generation (and absorption) in solids. However, a thorough understanding of the quantum mechanical theory of the interaction of radiation and matter is not necessary. Hence, in this chapter, light generation and absorption in a semiconductor are described in terms of a simplified model that avoids much of the mathematical complexity of the full theory, but nevertheless incorporates all of the essential features of the phenomena that are involved. In this model, both electrons (or holes) and photons of light are described in terms of their energy, mass and momentum within the crystalline material, taking into account the effect of the lattice atoms on these quantities. When this is done, optical processes such as emission and absorption can be conveniently seen to be regulated by the familiar laws of conservation of energy and momentum.

10.1 A Microscopic Model for Light Generation and Absorption in a Crystalline Solid

10.1.1 Basic Definitions

In considering the interaction of light with the electrons of a semiconductor, a convenient model to use is a microscopic (as opposed to macroscopic) *crystal momentum* model [10.1] in which the electron is considered to have a crystal momentum defined by

$$p = \hbar k,$$

(10.1.1)

where k is the wavevector of the electron state (Planck's constant h equals 6.624×10^{-27} ergs and $\hbar = h/2\pi$). The vector p is not the classical momentum of a free electron mv, where m denotes the mass and v the velocity. The crystal momentum p includes the effect on the electron of the atoms of the crystal, and it is in general related to the electron energy and velocity in a rather complicated way. In the equations of motion of an electron in a crystal, p plays the role of momentum and the effective mass m^* plays that of mass,x by analogy with the classical equations of motion of a free electron in a vacuum. Crystal momentum and effective mass can be defined for holes in the valence band just as is done for electrons in the conduction band. However, in general. p and m^* for electrons will be different from those for holes in a given material.

Interacting with the electrons and holes in a semiconductor, there can be photons of light. These discrete quanta, or relatively localized units of optical energy, maintain their identity as units throughout the processes of emission, transmission, reflection, diffraction, absorption, etc. The photon is described in a microscopic model by defining its mass and momentum as follows. First, consider a photon of light in a vacuum. The energy E of the photon is given by

$$E = h\nu,$$

(10.1.2)

where ν is the frequency of the radiation in s^{-1}. If (10.1.2) is combined with the well known relation

$$c = \nu\lambda_0,$$

(10.1.3)

the result is that the energy of the photon is given by

$$E = \frac{hc}{\lambda_0},$$

(10.1.4)

where c is the velocity and λ_0 is the wavelength of light in a vacuum. It is often convenient to remember that (10.1.4) becomes

$$E = \frac{1.24}{\lambda_0},$$

(10.1.5)

in the commonly used unit system in which E is given in electron volts (eV) and wavelength is in micrometers.

From the basic equation giving the energy of a photon (10.1.4), it is possible to designate its *mass* and *momentum*. It is known from relativistic theory that, in general, the energy of a particle and its rest mass are related by [10.2]

$$E = mc^2.$$

(10.1.6)

Combining (10.1.4) and (10.1.6) yields

$$E = mc^2 = \frac{hc}{\lambda_0}.$$ (10.1.7)

Although a photon has no rest mass, one can identify

$$m = \frac{h}{c\lambda_0}.$$ (10.1.8)

By analogy with the classical momentum p, which is equal to the product of mass times velocity, the momentum of a photon is given by

$$p = mcu,$$ (10.1.9)

where u is a unit vector in the direction of travel of the photon. Combining (10.1.9) and (10.1.8), the momentum can also be written as

$$p = \frac{h}{\lambda_0}u = \hbar k,$$ (10.1.10)

where k is the wavevector which is equal to $(2\pi/\lambda_0)u$.

It can thus be seen from (10.1.2–10) that, in this microscopic model, photons may be described equivalently by either their energy, wavelength, frequency, mass or momentum. Note that, in the case of the first four, the direction of travel must also be specified in order to completely describe the photon. The phase of photons is also important in cases involving coherent radiation, as is described in following sections of this chapter. Equations (10.1.2–10) were developed for the case of a photon in a vacuum. However, to use them to describe photons in a solid it is merely necessary to replace c and λ_0 respectively with v and λ, the velocity and wavelength of light in the material, respectively.

The interaction of photons of radiation with electrons and holes in a semiconductor can be considered as *particle* interactions by using the definitions presented in the preceding paragraphs. In that case, the usual rules of conservation of energy and momentum regulate the optical processes.

10.1.2 Conservation of Energy and Momentum

The effect of energy and momentum conservation can be conveniently illustrated with regard to the phenomenon of optical absorption. In Chap. 5, absorption loss was considered from the macroscopic point of view, being characterized by an absorption coefficient α. The microscopic model described in Sect. 10.1.1 can be used to explain further details of the absorption process. The strongest absorption in semiconductors occurs through electronic excitation, when an electron in a given energy level absorbs a photon and makes a transition to a higher energy state. Electronic transitions are subject to certain selection rules – the most basic being that *energy* and *momentum* of the electron and photon must be conserved. (From a theoretical viewpoint, it is a conservation of wavevector k, resulting from periodicity of the crystal structure, which leads to conservation of momentum, defined as $\hbar k$.) Thus, if only photons and electrons are involved,

$$E_i + h\nu_{phot} = E_f,\tag{10.1.11}$$

and

$$p_i + p_{phot} = p_f \quad \text{or} \quad k_i + \frac{2\pi}{\lambda_{phot}}u = k_f,\tag{10.1.12}$$

where the subscripts i and f refer to the initial and final states of the electron. The momentum of a photon in the visible or infrared range of the spectrum is very much smaller than the momentum of a thermally excited electron; hence the momentum selection rule is approximately $k_i = k_f$, and the electron experiences practically no change of momentum. This type of a transition, involving only an electron and photon is called a *direct* transition. It appears on the E versus k diagram as a *vertical* transition, as shown in Fig. 10.1. Note that, in general, the conduction band minimum and the valence band maximum do not have to occur at the same point in k space. If they do, the material is said to have a direct bandgap; if they do not, it is an indirect bandgap material.

Even though a photon cannot transfer significant momentum to an electron, it is possible to have electronic transitions in which k changes. In these processes, a phonon also takes part, or equivalently speaking, momentum is transferred to or absorbed from the lattice atoms. The vibrational energy of the lattice atoms is quantized (as the light energy) into phonons of energy $\hbar\omega$ (ω: angular frequency in rad/s). The phonons are characterized by a wavevector q. The conditions restricting an absorptive transition in which a photon, electron and phonon are involved, are then given by

$$k_i \pm q = k_f,\tag{10.1.13}$$

and

$$E_i + h\nu_{phot} \pm \hbar\omega_{phon} = E_f,\tag{10.1.14}$$

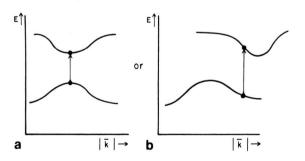

Fig. 10.1a,b. Direct absorptive electronic transitions; **a** in a direct bandgap material; **b** in an indirect bandgap material

where the wavevector $\boldsymbol{k} = (2\pi/\lambda_0)\boldsymbol{u}$ of the photon has been neglected in (10.1.13). An E versus $\boldsymbol{k}$ diagram plot of the process is shown in Fig. 10.2. An electron absorbs a photon and at the same time absorbs or emits a phonon. Such transitions are called *indirect*, and appear as diagonal transitions on the E versus $\boldsymbol{k}$ diagram, as shown in Fig. 10.2. Since both a photon and a phonon are necessary for an indirect transition, while only a photon is required to cause a direct transition, the direct transition is a more probable one. For this reason, semiconductors with direct bandgaps are more optically active than those with indirect bandgaps. This feature is particularly important in the case of light emitters, as is described later in this chapter.

The diagrams of Figs. 10.1 and 2 shown only interband transitions, between valence and conduction bands; however, both direct and indirect transitions also can take place within a band (in*tra*band) or between energy states introduced by dopant atoms and/or defects. In all cases, the principles of conservation of energy and momentum (wavevector) apply. These types of absorption transitions are shown in Fig. 10.3. Intraband absorption can occur by either electrons in the conduction band or by holes in the valence band, and hence is called *free carrier* absorption, as mentioned previously in Chap. 5. Free carrier absorption is usually taken to include transitions of electrons from donor states to the conduction band and transitions of holes from acceptor states to the valence band.

The effects of energy and momentum conservation that have been illustrated by reference to the optical absorption process apply equally well to the generation of photons in a semiconductor. In fact, such considerations are

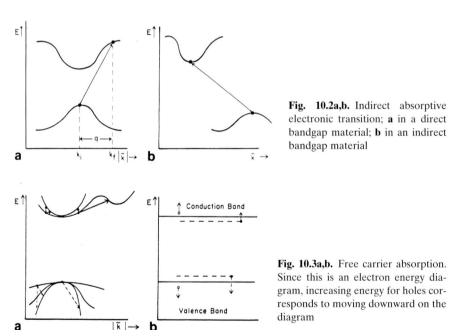

Fig. 10.2a,b. Indirect absorptive electronic transition; **a** in a direct bandgap material; **b** in an indirect bandgap material

Fig. 10.3a,b. Free carrier absorption. Since this is an electron energy diagram, increasing energy for holes corresponds to moving downward on the diagram

even more important in the case of light emission, as is discussed in more detail in Sect. 10.2.

10.2 Light Emission in Semiconductors

Photon generation in a semiconductor usually results from carrier recombination. The generation process can be divided into two types, spontaneous emission, in which a hole and electron randomly recombine, and stimulated emission, in which a hole and electron are stimulated to recombine by an already existing photon.

10.2.1 Spontaneous Emission

In absorption, an electron absorbs a photon and makes a transition to a higher energy state. In emission, just the opposite occurs. An electron makes a transition from a higher energy state to a lower one, and in the process loses energy that is emitted as a photon of light. The significant light emitting transitions in a semiconductor are interband transitions, in that they occur between conduction and valence bands and/or certain states within the bandgap caused by impurity dopants or defects. Since the bandgaps of semiconductors are typically a few tenths eV to a few eV, the emitted wavelengths are generally in the infrared part of the spectrum, and correspond roughly to the absorption edge wavelength of the semiconductor. As in the case of absorption, certain selection rules limit the possible emissive transitions. Conservation of energy and momentum (or wavevector) are required. Thus, for a direct transition

$$E_i - E_f = h\nu_{phot} \tag{10.2.1}$$

and

$$\boldsymbol{k}_i - \boldsymbol{k}_f = \frac{2\pi}{\lambda_{phot}} \boldsymbol{u}, \tag{10.2.2}$$

where $\boldsymbol{u}$ is a unit vector specifying direction only. Note that for a direct transition $2\pi/\lambda_{phot} \ll |\boldsymbol{k}_i|$ and $|\boldsymbol{k}_f|$, so that $|\boldsymbol{k}| \cong |\boldsymbol{k}_f|$, as for a direct absorptive transition. Emission of a photon can also occur in an indirect transition, in which case conservation of momentum requires

$$\boldsymbol{k}_i - \boldsymbol{k}_f \pm \boldsymbol{q} = \frac{2\pi}{\lambda_{phot}} \boldsymbol{u}, \tag{10.2.3}$$

where $\boldsymbol{q}$ is the wavevector of an absorbed or emitted phonon. As in the case of absorption, the necessity of having a phonon present greatly reduces the probability of a transition and hence of the generation of a photon.

A third requirement for an emissive transition is that there must be a filled upper, or initial, energy state and a corresponding empty lower, final, energy state, with a difference in energy between them equal to the energy of the photon to be emitted. This rule seems obvious and trivial, and plays little role in the case of absorption, because the absorbing material is usually near thermal equilibrium, and consequently there are always filled lower states and empty upper states at the proper levels. (Except in the case of less-than-bandgap radiation, for which there are no upper states available, and hence absorption can not occur.) In the case of emission, the preceding rule is much more important. For example, for an intrinsic semiconductor at thermal equilibrium, there are relatively few holes in the valence band and electrons in the conduction band at room temperature and below. In the case of a doped semiconductor, there may be many electrons in the conduction band, as for n-type material, but there are then very few holes in the valence band (because of the constancy of the np product). The opposite is true for a p-type semiconductor. Hence, the radiative recombination that occurs at thermal equilibrium produces very few photons, which are reabsorbed before leaving the crystal. In order to obtain significant light emission from a semiconductor, one must therefore somehow move away from thermal equilibrium to produce more electrons in the conduction band and more holes in the valence band.

The increased concentrations of holes and electrons can be created in several different ways, and the process is generally called *pumping* the material. For example, if high intensity, greater-than-bandgap light is directed onto the semiconductor, interband absorption occurs and results in many hole-electron pairs being created. These holes and electrons are initially *hot* but very rapidly *thermalize*, i.e., electrons settle to the bottom of the conduction band and holes rise to the top of the valence band through interaction with the lattice, as shown in Fig. 10.4. Thermalization is typically very fast, on the order of 10^{-14} s. After thermalization, holes and electrons recombine; this recombination can be radiative, resulting in the emission of a photon of approximately bandgap energy. The lifetime, or time it takes on the average for hole-electron pair to recombine after it is created, is typically about 10^{-11}s in a pumped

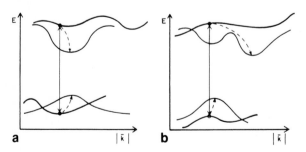

Fig. 10.4a,b. Thermalization of energetic carriers. After being pumped to higher energy levels, both electrons and holes *settle* to their respective bandedge states. **a** Direct bandgap material; **b** indirect bandgap material

semiconductor (with direct bandgap). Note that, in an indirect bandgap semiconductor, thermalization causes holes and electrons to settle into states with
significantly different wavevectors (Fig. 10.4). This means that a phonon is
needed for recombination to be possible, as well as a hole and electron; hence
it is a lower probability process. The typical lifetime for an indirect recombination is as much as 0.25 s; before that time has passed, the hole and electron on
the average will have already recombined through some nonradiative process,
giving up energy to the lattice or to defects, etc. Hence the efficiency of
indirect radiative recombination is very low. The quantum efficiency for a
direct bandgap material can approach 1, but for an indirect gap it is generally
0.001 or lower. However, indirect emissive transitions do occur, and in some
indirect bandgap materials (for example, GaP) measurable and indeed usable
light emission does occur. Nevertheless it is extremely difficult to make lasers
in indirect gap materials.

The optical pumping method that has been described as a means of
producing the required increased concentration of electrons and holes is
somewhat unwieldly in that a high intensity source is required; the process is
inefficient, generating a great deal of heat. Also, the output must be optically
filtered to separate the semiconductor emission from reflected pump light.

An alternative pumping technique, which is both efficient and extremely
simple, employs the properties of a p-n junction. The energy band diagram of
a p-n junction light emitter is shown in Fig. 10.5. Under zero bias conditions,
there are many electrons in the conduction band on the n-side and many holes
in the valence band on the p-side, but very few can get over the barrier and
into the junction region. When a forward bias voltage V_0 is applied, the barrier
is lowered and many holes and electrons are injected into the junction region,
where they recombine, resulting in the generation of photons. The light emitted under stimulation from a battery is called electroluminescence, while that
produced by a pumping with a shorter wavelength optical source is called
photoluminescence. A typical electroluminescent diode, or light-emitting diode (LED), is diagrammed in Fig. 10.6. Light is generated in the p-n junction
and leaves the diode by passing through bulk material outside of the junction

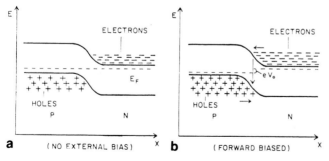

Fig. 10.5a,b. Energy band diagram of a p-n junction light emitter; **a** no external bias voltage
applied; **b** external bias voltage V_0 applied with forward polarity

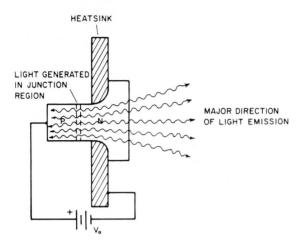

Fig. 10.6. A p-n junction light-emitting diode (LED)

region. Much of the light is reabsorbed before it leaves the diode. Hence one speaks of external quantum efficiency, defined by

$$\eta_{\text{ext}} = \frac{\text{number of photons emitted in desired direction}}{\text{number of hole - electron pairs injected}} \tag{10.2.4}$$

as well as internal quantum efficiency, given by

$$\eta_{\text{int}} = \frac{\text{number of photons generated}}{\text{number of hole - electron pairs injected}} \tag{10.2.5}$$

In designing a light emitter, one would naturally try to minimize reabsorption by making the layer of material between the junction and the surface very thin, and by choosing a material with a very small absorption coefficient if possible. Such diodes have obtained wide usage as panel indicators and display devices because they are small, cheap, more reliable than incandescent devices, and easily incorporated into integrated circuit arrays. LED's are also very useful as light sources for fiber optic communications because they can be modulated at high frequencies (although not as high as for lasers) and can be conveniently coupled to the cores of multimode fibers.

The wavelength spectrum of semiconductor light emitters, regardless of the pumping technique used, is relatively simple when compared, for example, to that of gaseous arc discharge sources, which typically have numerous lines or emission peaks. A good semiconductor light emitter with low defect or extraneous impurity content has a single emission peak at approximately the band-edge wavelength, corresponding to the energy gap. This peak has a halfwidth of typically 200–300 Å, which is considerably wider than gaseous source lines. Extraneous impurities and defect states within the bandgap allow radiative transitions which produce secondary emission peaks at other wavelengths. These secondary peaks indicate that some of the injected hole-electron pairs recombine to produce photons in other than the major band-edge peak. Hence the quantum efficiency at the desired wavelength is reduced. A thorough discussion of the emission spectrum of semiconductor light emitters is beyond the scope of this book. However relatively complete data are available in the literature for most of the commonly used semiconductors [10.3–5].

Even if a semiconductor light emitter is formed in relatively pure and defect free material, so that the light emission is almost entirely at the band-edge wavelength, the peak wavelength is still dependent on temperature and doping, because the bandgap varies with temperature, and also doping changes the effective bandgap by allowing emissive transitions between donor and acceptor states. Thus the emission wavelength of a semiconductor light emitter can be varied over a somewhat limited range by doping it appropriately. For example, consider the case of GaAs doped with the n-type dopants Te, Se or Sn. The center wavelength of the emission peak changes as shown in Fig. 10.7. At relatively low doping levels, recombination occurs between electrons filling donor states 6 meV below conduction band edge and empty states at top of valence band. As doping increases above 5×10^{17} cm^{-3}, the donor band merges with the conduction band and recombination occurs between filled conduction band states near the quasi fermi level and empty valence band states. The p-type dopants Cd and Zn also cause a shift in the emission peak wavelength, as shown in Fig. 10.8. At low p-type dopant

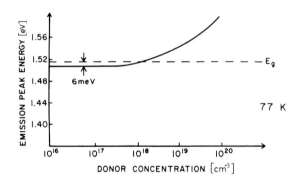

Fig. 10.7. Peak emission wavelength for n-type GaAs as a function of donor concentration [10.6]

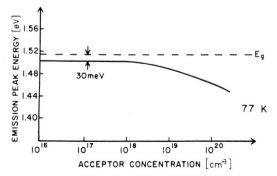

Fig. 10.8. Peak emission wavelength for p-type GaAs as a function of acceptor concentration [10.7]

concentration, the recombination is from filled states at the bottom of the conduction band to empty acceptor states located 30 meV above the valence band edge. As the doping is increased above 10^{18}cm^{-3} the acceptor states broaden into a band, as shown in Fig. 10.9 and recombination occurs between filled states at the bottom of the conduction band and empty states at the top of the acceptor band. Thus the emission peak energy decreases for p-type doping, in contrast to the way it increases for n-type. As the p-type doping is increased above about 2×10^{18}cm^{-3}, the acceptor band merges with the valence band, and hence the radiative recombination may be described as band-band, although it is really from conduction band states to acceptor states. The recombination does not occur only between states at exactly the same energy in the conduction and valence bands, respectively. Instead it occurs between states spread over a range of energies. Hence the emission is not single wavelength but rather has a line-width on the order of hundreds of Angstroms. As the doping is increased for either n- or p-type GaAs, the line-width of emission increases from typically 100 Å, for lightly doped material, to several hundred Angstroms, for heavily doped material. This increase in line-width is due to impurity banding, which makes available pairs of filled and empty states for recombination over a range of energies.

In summary, the principal properties of light emitted by the spontaneous recombination of electrons and holes in a semiconductor are that photon wavelengths are spread over a range of several hundred Angstroms, with the peak wavelength corresponding approximately to the band-edge, but also depending on dopant concentration and type. Photons are emitted more-or-less isotropically, except for the effects of emitter geometry, and there is no fixed phase relationship between photons. The generation of each photon is a spontaneous event, completely unrelated to any other photon. The characteristics of light generated by stimulated emission contrast strongly with those of spontaneously emitted light.

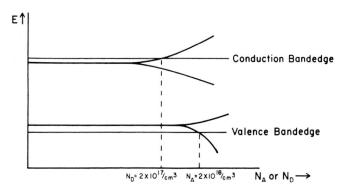

Fig. 10.9. Donor and acceptor energy levels (bands) in GaAs as a function of dopant concentration [10.8]

10.2.2 Stimulated Emission

Spontaneously emitted photons can, of course, be reabsorbed by any of the absorption processes that have been previously described. Alternatively, however, if the photon happens to encounter an electron hole pair of proper energy separation, it can encourage the two to recombine; in this case, a new photon is produced which is an exact duplicate of the one that stimulated the recombination process, i.e., it has exactly the same energy (frequency), direction, phase and polarization. This is stimulated emission, which occurs in addition to spontaneous emission, so that in general the intensity of the radiation produced in the presence of external radiation consists of two parts. One part is independent of the external radiation (spontaneous emission); the other part has intensity proportional to the external radiation, while its frequency, phase, direction of propagation and polarization are the same as those of the external radiation. Note that the *external radiation* can be that generated by other spontaneous and/or stimulated emissions within the sample, and does not have to be truly *external* field.

Except under very special conditions, stimulated emission is less probable than spontaneous emission, and hence a normal light emitter produces mostly spontaneous emission. One has to make a special effort to get a significant amount of stimulated emission. First, the concentrations of holes and electrons in the light emitting region must be so large that most of the states near the bottom of the conduction band are filled with electrons and most of the states near the top of the valence band are empty. This makes it more probable for the photon to cause the transition of an electron from the conduction band to the valence band, producing a duplicate photon, than to be absorbed by causing an electron to move from the valence band up to an empty state in the conduction band. Secondly, in order to have stimulated emission predominate over spontaneous emission, there must be an intense field of photons of the proper energy. The necessity of satisfying these two conditions can be demonstrated mathematically by using a simple two energy level model [10.9], as shown in Fig. 10.10.

At thermal equilibrium (i.e. in the absence of any external *pumping* energy source), most of the electrons in the material will be in the lower energy level. The electron concentrations N_1 and N_2 in the two levels are related by the following expression from Boltzmann statistics:

$$\frac{N_2}{N_1} = e^{-(E_2 - E_1)/kT}, \tag{10.2.6}$$

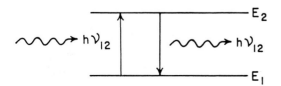

Fig. 10.10. Simplified two energy level model for light emission

where it has been assumed that the two levels contain an equal number of available states. Consider the situation that exists when the material is immersed in a radiation field of photons with energy $h\nu_{12}$, equal to $E_2 - E_1$, such that the total energy in the field per unit volume (at the frequency ν_{12}) is $\rho(\nu_{12})$. The processes that can then occur are absorption, spontaneous emission and stimulated emission. For the steady-state case, these processes are linked by a transition rate equation given by, [10.9]

$$\underbrace{B_{12}N_1\rho(\nu_{12})}_{\text{absorption}} = \underbrace{A_{21}N_2}_{\substack{\text{spontaneous}\\\text{emission}}} + \underbrace{B_{21}N_2\rho(\nu_{12})}_{\substack{\text{stimulated}\\\text{emission}}}, \tag{10.2.7}$$

where B_{12}, B_{21} and A_{21} are proportionality constants known as Einstein coefficients. Note in (10.2.7) that upward or downward electronic transitions are always proportional to the number of electrons in the initial level, N_1 or N_2, respectively, and that both absorption and stimulated emission are proportional to the optical field strength $\rho(\nu_{12})$ as well. The conditions necessary to produce predominance of stimulated emission over both absorption and spontaneous emission can be seen by taking the ratios of the respective transition rates, from (10.2.7) given by

$$\frac{\text{stimulated emission rate}}{\text{absorption rate}} = \frac{B_{21}N_2 p(\nu_{12})}{B_{21}N_2 p(\nu_{12})} = \frac{B_{21}}{B_{12}}\frac{N_2}{N_1}, \tag{10.2.8}$$

$$\frac{\text{stimulated emission rate}}{\text{spontaneous emission rate}} = \frac{B_{21}N_2 p(\nu_{12})}{A_{21}N_2} = \frac{B_{21}}{A_{21}}p(\nu_{12}). \tag{10.2.9}$$

It can be seen from (10.2.6) that N_2 is much less than N_1 at thermal equilibrium, for the usual case of $E_2 - E_1$ being much greater than kT. Hence, for stimulated emission to predominate over absorption, the ratio N_2/N_1 in (10.2.8) must be greatly increased by introducing a pumping source of energy, as described in Sect. 10.2.1. The condition of N_2 being greater than N_1 is called an inverted population. From (10.2.9) it is obvious that an intense flux of photons of the proper energy, i.e. large $\rho(\nu_{12})$, is required in order to make stimulated emission much greater than spontaneous emission. This photon concentration is usually produced by introducing positive optical feedback from mirrors or other reflectors.

The two requirements that there be an inverted population, and that an intense flux of photons of the proper energy be present, are common to all types of lasers. The simplified two-level model that has been used to demonstrate these principles is more appropriate for a gas laser than for semiconductor laser, because the electrons and holes in a semiconductor are distributed over bands of energy rather than occupying two discrete energy levels. Nevertheless the same principles apply. In a semiconductor that is undergoing pumping, electrons fill states at the bottom of the conduction band roughly up to an energy level E_{Fn}, while holes *fill* states at the top of the valence band approximately down to an energy level E_{Fp}, as shown in Fig. 10.11. The energy levels E_{Fn} and E_{Fp} are known as the quasi-Fermi levels for electrons and holes [10.10], since strictly speaking, the Fermi level is defined only for the case of thermal equilibrium. For the nonequilibrium case of external energy input, quasi-Fermi levels must be defined separately for holes and electrons. From Fig. 10.11, it can be seen that a state of population inversion is experienced by any photon of energy $h\nu_{phot}$ when the condition

$$E_{Fn} - E_{Fp} > h\nu_{phot} \tag{10.2.10}$$

is satisfied. The minimum energy condition for population inversion to exist is, of course, given by

$$E_{Fn} - E_{Fp} = E_c - E_v = E_g , \tag{10.2.11}$$

where E_g is the bandgap energy. Thus, in a pumped semiconductor, population inversion exists, and hence stimulated emission dominates over absorption for photons with energy in the range given by

$$E_g \leqq h\nu_{phot} \leqq (E_{Fn} - E_{Fp}). \tag{10.2.12}$$

10.3 Lasing

When the two conditions of population inversion and high photon density are satisfied, a selection process takes place which alters the character of the radiation and results in the generation of significant amounts of stimulated emission. If a suitable optically resonant structure is provided, lasing can occur.

10.3.1 Semiconductor Laser Structures

The conventional structure used for a discrete semiconductor laser is a rectangular parallelepiped such as that shown in Fig. 10.12, which is produced by cleaving the crystal along lattice planes that are perpendicular to the light emitting layer. For example a $\langle 100 \rangle$ wafer of GaAs can be easily cleaved along the four perpendicular $\langle 110 \rangle$ planes to produce such a structure. Typical dimensions for the length L and width W are 300 μm and 50 μm, respectively,

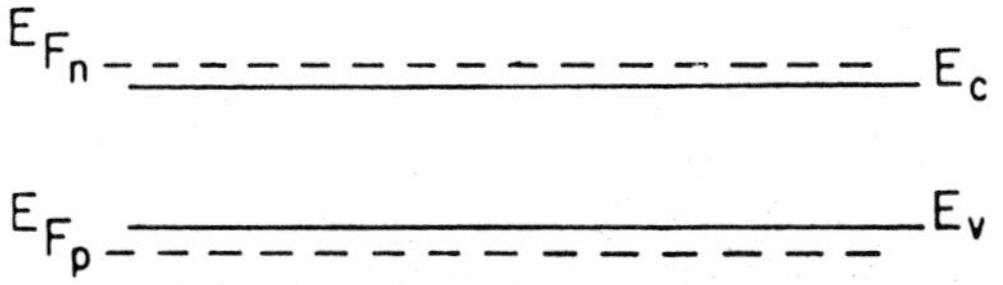

Fig. 10.11. Energy band diagram of a semiconductor with external energy input. E_{Fn} and E_{Fp} are the quasi-Fermi levels for electrons and holes respectively

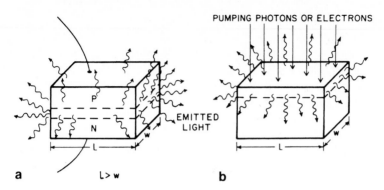

Fig. 10.12 a, b. Semiconductor laser structure; **a** p-n junction pumped (laser diode); **b** electron or photon beam pumped. Light emission is shown as it would be for pumping levels below the threshold for lasing. (See text for explanation). The dashed lines mark the extremes of the pumped region

but great variations from the values are possible, as is discussed in detail in Chap. 11. The two end faces of width W form an optically resonant, reflective structure known as a Fabry-Perot etalon. When such a structure is provided, increasing the pumping power will result in lasing when a certain critical threshold level is exceeded.

10.3.2 Lasing Threshold

As pumping power is increased from zero, an increasing number of photons are emitted by spontaneous emission. Even when pumping power exceeds that required to produce population inversion, the emission, at first, is still composed of predominantly spontaneously generated photons. The spontaneously emitted photons are produced more-or-less equally in all directions, as shown in Fig. 10.12. However, most of these photons start off in directions that very soon carry them out of the inverted population region where predominent stimulated emission can occur, and thus they are unable to reproduce themselves. The few that happen to be traveling in the plane of the inverted population region are able to duplicate themselves by stimulated emission many times before they emerge from the laser. In addition, for any given energy bandgap and distribution of holes and electrons, there will be one particular energy (wavelength) which is preferred (transitions more probable) over others. This wavelength or energy usually corresponds to first order to the peak wavelength at which spontaneous emission takes place in the material. As a result of this preferred energy and direction, when stimulated emission builds up, the emitted radiation narrows both in spatial divergence and spectral linewidth (reduced from several hundred Angstroms to about 25 Å). Note that, as stimulated emission builds up, more hole-electron pairs are consumed per second. Hence, for any given input hole-electron pair generation rate, spontaneous emission is suppressed, because the stimulated emission consumes the generated pairs before they can recombine spontaneously. Such predominant stimulated emission as has been described is known as *super-radiance* radiation. It is not yet the coherent light produced by lasing. Such radiation is due to amplified spontaneous radiation. The photons are not traveling in *exactly* the same direction; they may have slightly different energy, they are not in phase. In other words, the radiation is still incoherent.

When the light emitting region is bounded by parallel reflecting planes as in Fig. 10.12, some reflection will occur at the surfaces and therefore some of the radiation will pass back and forth through the inverted population region and be amplified further. In order to be reflected back and forth in the narrow inverted population layer photons must be *exactly* in the plane of the layer and *exactly* perpendicular to the Fabry-Perot surfaces. The radiation must also be of uniform frequency and phase to avoid destructive interference, because the Fabry-Perot reflectors form a resonant *cavity* in which optical modes are established. Thus, only that small fraction of the super-radiance radiation [10.11] that fulfills all of these requirements continues to survive and becomes the dominant species of radiation which is produced. Such radiation, uniform in wavelength, phase, direction and polarization is coherent radiation. The device is said to be lasing.

The transition between the super-radiance region of output and the lasing region of output (as input power is increased) is taken to be the point at which the amplification of the preferred mode by stimulated emission becomes greater than absorption and all other photon losses, so that the radiation bounces back and forth between Fabry-Perot surfaces, growing in intensity with each bounce (i.e., oscillation sets in). Of course, the radiation does not continue to build up in intensity indefinitely. As the intensity grows, it stimulates more recombinations until concentrations of holes and electrons are being used up at exactly the same rate at which they are being generated by the pumping source, and a dynamic equilibrium is established. In this steady state condition, the laser oscillates at a single peak wavelength (or possibly several discrete wavelenths corresponding to different longitudinal modes) which is highly monochromatic (approximately 1Å linewidth). A standing wave is produced within the crystal, with an integral number of half-wavelengths existing between the Fabry-Perot faces for each mode. Because of the previously described directional selectivity of the reflecting end faces, photons of the lasing mode(s) are emitted predominantly in the direction normal to the end faces, as shown in Fig. 10.13. The light waves (photons) emitted from such a resonant structure are in phase, and in addition, the waves which may be most easily propagated are those with their electric vector perpendicular to the plane of the inverted population layer. Hence the emitted light is normally polarized in that manner.

The transitions between the regions of spontaneous emission, super-radiance emission and laser emission, as the pumping power is increased, are very abrupt. In fact, although the transition between super-radiance and lasing has

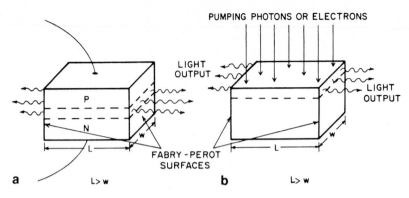

Fig. 10.13a,b. Semiconductor laser operating above the threshold for lasing. **a** p-n junction pumped; **b** electron or photon beam pumped

been described in detail for pedagogic purposes, the actual transition in a normal laser is so abrupt that super-radiance emission cannot be observed. As the pumping power goes above the threshold point, the character of the light emitted abruptly changes to the lasing type, with linewidth decreasing from several hundred Angstroms to one Angstrom, and phase coherence being observable. The intermediate stage of narrowed linewidth due to predominant stimulated emission without phase coherence (i.e. super-radiance) is passed through so quickly that observations cannot be made. Only in laser diode structures with damaged Fabry-Perot reflectors that inhibit lasing, can super-radiance emission be observed.

10.3.3 Efficiency of Light Emission

One important characteristic of semiconductor laser light emission, which has not yet been discussed, is the efficiency of conversion of electrical energy into optical energy. A laser diode operating above threshold is about a hundred times more efficient than a light emitting diode made in the same material. This improved efficiency results from several factors.

As previously mentioned, generation of a photon by stimulated emission has a relatively low probability of occurrence compared to that of spontaneous emission in the case of a noninverted population, such as exists in a light emitting diode. However, when the population is inverted, as in a laser, the probability of stimulated emission becomes very large, even larger than the probabilities of either spontaneous emission or nonradiative recombination of the hole-electron pair. Thus the internal quantum efficiency is about a factor of ten higher in a laser diode than in an LED. In fact, if a laser diode is cooled to 77 K to reduce the thermal spread of electron and hole energies, the internal quantum efficiency can approach 100%.

Once the photons are generated, they also experience less loss in a laser than in an LED. Photons in the lasing mode travel mostly in the inverted population region, in which inter-band absorption is suppressed by a lack of electrons near the valence band edge and empty states at the conduction band edge. Photons in an LED travel mostly in noninverted, bulk material. Thus internal reabsorption of generated photons is much less in a laser. Also, the photons emitted from a laser travel in a much less divergent beam than those emitted from an LED, making it possible to collect (or couple) more of the light output at a desired location. Because of the combined effects of reduced internal reabsorption, greater beam collimation and increased internal quantum efficiency, a semi-conductor laser can be expected to have an external quantum efficiency roughly a hundred times greater than its LED counterpart. Since both the laser and LED operate at approximately the same voltage, an increase in external quantum efficiency translates into a proportional increase in overall power efficiency.

In this chapter, the basic principles of light emission in semiconductor lasers and LED's have been discussed, and a distinction has been made between the characteristics of the light emitted by each. In Chap. 11, the laser diode is considered in more detail, and quantitative relations are developed to evaluate such important characteristics as mode spacing, threshold current density and efficiency. In addition, some other basic types of semiconductor lasers are discussed.

Problems

10.1 a) Why do semiconductor light emitting diodes have a characteristic emission line width on the order of hundreds of Angstroms while semiconductor lasers have line widths ~ 1 Å ?
b) In order to produce a semiconductor laser, what two conditions basic to the gain mechanism must be satisfied?
c) To what extent is the light output of a semiconductor laser collimated? (Give a typical divergence angle.) What can be done to produce greater collimation?

10.2 Given that

$$B_{12}N_1\rho(v_{12}) = A_{21}N_2 + B_{21}N_2\rho(v_{12})$$

show that for a very large photon density $\rho(v_{12}) \to \infty$ it follows that

$$B_{12} = B_{21.}$$

10.3 If the absorption of a photon causes an electron to make a direct transition (i.e., no phonon involved) from a state with wavevector $|k| = 1.00 \times$

$10^7 \mathrm{cm}^{-1}$ in a given direction to a state with wavevector $|k| = 1.01 \times 10^7 \mathrm{cm}^{-1}$ in the same direction, what was the wavelength of the photon?

10.4   Why is a semiconductor laser inherently more efficient than a light-emitting-diode?

10.5   Sketch three different plots of typical diode laser emission spectra which show output light intensity vs. wavelength a) below threshold, b) at threshold, c) above threshold.

10.6   a) Draw the (E vs. x) energy band diagram of a $p - n$ junction laser under forward bias and indicate the inverted population region, as well as E_{Fn}, E_{Fp}, E_{c}, E_{v} and E_{g}.
    b) What energies establish the long and short wavelength limits of the laser emission spectrum?

11. Semiconductor Lasers

In the preceding chapter, we discussed the basic principles of light emission in a semiconductor. Probably the most significant feature of this light emission is that it is possible to design a light source in such a way that stimulated emission of photons will predominate over both spontaneous emission and absorption. If a resonant reflecting structure such as a pair of plane, parallel end faces is provided, a lasing mode can be established and coherent optical emission will result. In this chapter, we consider several basic semiconductor laser structures and develop the quantitative theory necessary to calculate their expected performance characteristics.

11.1 The Laser Diode

Since its inception in 1962 [11.1–4], the semiconductor laser diode has been developed from a laboratory curiosity into a reliable and marketable product, which has gained widespread acceptance as a source of light in many varied applications. In this section, we discuss the basic structure and performance of the discrete laser diode as opposed to the monolithically integrated laser diode of the optical integrated circuit, which will be covered in Chaps. 12 and 13.

11.1.1 Basic Structure

The p-n junction laser diode is an excellent light source for use in optical integrated circuits and in fiber-optic signal transmission applications because of its small size, relatively simple construction and high reliability. To date, most laser diodes have been made in GaAs, $Ga_{(1-x)}Al_xAs$, or $Ga_xIn_{(1-x)}As_{(1-y)}P_y$, but other materials no doubt eventually also be used extensively to obtain emission at different wavelengths, once the fabrication technology has been developed. The basic structure of a p-n junction laser diode is shown in Fig. 11.1. A p-n junction is usually formed by epitaxial growth of a p-type layer on an n-type substrate. Ohmic contacts are made to each region to permit the flow of electrical current which is the *pumping* energy source required to produce the inverted population in the active region adjacent to the junction. Two parallel end faces are fabricated to function as *mirrors* providing the optical

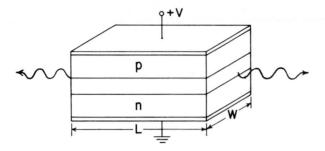

Fig. 11.1. Basic structure
of a p-n junction laser

feedback necessary for the establishment of a lasing optical mode. The device shown in Fig. 11.1 is a discrete diode structure, such as might be used in conjunction with an optical fiber transmission line. In later chapters, we shall consider a much more sophisticated, multi-layered structure with more elaborate optical feed-back schemes that are better suited for monolithic integration into optical integrated circuits. However, the fundamental structure provides a convenient basis for the development of a theoretical description of laser performance that can be easily adapted to more complex devices.

11.1.2 Optical Modes

The partially reflecting end faces of the laser diode provide an optical feedback that leads to the establishment of one or more longitudinal optical modes. Because of their similarity to the plane, parallel mirrors of a Fabry-Perot interferometer [11.5], the laser end faces are often called Fabry-Perot surfaces. When current is passed through the laser diode, light can be generated in the resulting inverted population layer by both spontaneous and stimulated emission of photons as explained in Chap. 10. Because of the reflection that occurs at the Fabry-Perot surfaces, some of the photons will pass back and forth many times through the inverted population region and be preferentially multiplied by stimulated emission. Those photons that are traveling exactly in the plane of the layer and exactly perpendicular to the Fabry-Perot surfaces have the highest probability of remaining in the inverted population layer where they can reproduce themselves by stimulated emission. Hence, they become the photons of the optical mode or modes that are established when steady-state operation is achieved at a given current level. It is possible for other modes to develop so as to correspond to photons following a zig-zag path reflecting off of the side faces of the laser, but in a practical device these modes are usually suppressed by roughening the side faces or by using some equivalent technique to attenuate the undesired mode. The radiation of the lasing mode must also be of uniform frequency and phase to avoid destructive interference. As a result, a standing wave is produced within the laser diode with an integral number of half-wavelengths between the parallel faces.

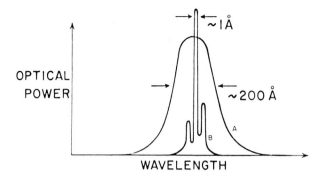

Fig. 11.2. Optical emission spectrum of the diode laser; A spontaneous emission below threshold; B lasing mode structure above threshold

The mode number m is given by the number of half-wavelengths. Thus

$$m = \frac{2Ln}{\lambda_0}, \tag{11.1.1}$$

where L is the length between end faces, n is the index of refraction of the laser material, and λ_0 is the vacuum wavelength of the emitted light. The mode spacing can be determined by taking $dm/d\lambda_0$, keeping in mind that semiconductor lasers are always operated near the bandgap wavelength, were n is a strong function of wavelength. Thus,

$$\frac{dm}{d\lambda_0} = -\frac{2Ln}{\lambda_0^2} + \frac{2L}{\lambda_0} \frac{dn}{d\lambda_0}. \tag{11.1.2}$$

For $dm = -1$, the mode spacing $d\lambda_0$ is given by

$$d\lambda_0 = \frac{\lambda_0}{2L\left(n - \lambda_0 \dfrac{dn}{d\lambda_0}\right)}. \tag{11.1.3}$$

We take dm equal to -1 because a decrease of one in the value of m corresponds to one less half-wavelength between the Fabry-Perot end faces, i.e., an increase in wavelength λ_0.

A typical mode spectrum for a diode laser is shown in Fig. 11.2. Usually several longitudinal modes will coexist, having wavelengths near the peak wavelength for spontaneous emission. The mode spacing for a GaAs laser is typically $d\lambda_0 \cong 3\,\text{Å}$. In order to achieve single mode operation, the laser structure must be modified so as to suppress all but the preferred mode. We will consider such specialized devices in later chapters.

11.1.3 Lasing Threshold Conditions

When a laser diode is forward biased and current begins to flow, the device does not immediately begin lasing. At low current levels, the light that is emitted is mostly due to spontaneous emission and has a characteristic spectral linewidth on the order of hundreds of Angstroms. It is incoherent light. As the pumping current is increased, a greater population inversion is created in the junction region and more photons are emitted. The spontaneously emitted photons are produced going more-or-less equally in all directions. Most of these start off in directions that very soon carry them out of the inverted population region where net stimulated emission can occur, and thus are unable to reproduce themselves. However, those few photons that happen to be traveling exactly in the junction plane and perpendicular to the reflecting end faces are able to duplicate themselves many times before they emerge from the laser. In addition, for any given energy bandgap and distribution of holes and electrons, there is one particular energy (wavelength) that is preferred over others. To first order, this wavelength usually corresponds to the peak wavelength at which spontaneous emission takes place in the material. As a result of this preferred energy and direction, when stimulated emission builds up with increasing current, the emitted radiation narrows substantially both in spectral linewidth and spatial divergence. As stimulated emission builds up, the photon density (intensity) of the optical mode increases, thus leading to a further increase in stimulated emission so that more hole-electron pairs are consumed per second. Hence, spontaneous emission is suppressed for any given input hole-electron pair generation rate, because the stimulated emission uses up the generated pairs before they can recombine spontaneously. Because of the phase condition of (11.1.1), the light produced by stimulated emission in an optically resonant structure like that of the laser diode is coherent, and the device is said to be lasing.

The transition from nonlasing emission to lasing emission occurs abruptly as the current level exceeds the threshold value. As the threshold current is exceeded, the onset of lasing can be experimentally observed by noting the sharp break in the slope of the optical power versus pump current curve (Fig. 11.3), which results because of the higher quantum efficiency inherent in the lasing process (as explained in Sect. 10.3.3). Also, the spectral line-shape of the emitted light abruptly changes from the broad spontaneous emission curve to

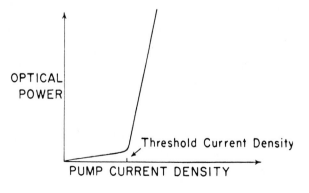

Fig. 11.3. Light output of a diode laser as a function of input pumping current

one consisting of a number of narrow modes as shown in Fig. 11.2. Quantitatively speaking, the lasing threshold corresponds to the point at which the increase in the number of lasing mode photons (per second) due to stimulated emission just equals the number of photons lost (per second) because of scattering, absorption or emission from the laser. In conventional terms used to describe an oscillator one would say that the device has a closed loop gain equal to unity. Using this fact, it is possible to develop an expression for the threshold current as a function of various material and geometric parameters.

We will begin by considering the p-n junction laser structure shown in Fig. 11.1 and by following the approach used by *Wade* et al. [11.6]. The light is emitted from the laser preferentially in the direction perpendicular to the Fabry-Perot surfaces. The transverse spatial energy distribution of the light wave (photon density) is as shown in Fig. 11.4. The photon distribution extends or spreads into the inactive (noninverted) regions on each side of the junction due mainly to diffraction. Thus, there is a light emitting layer of thickness D, which is greater than the thickness d of the active or inverted population layer. For example, in GaAs diodes, $d \cong 1\,\mu m$, $D \cong 10\,\mu m$. It can be seen from the idealized spatial energy diagram that, of the total number of photons existing in the lasing mode at any given time, only a fraction d/D remain in the active region and can generate additional photons by stimulated emission. This effect reduces the gain available from the device.

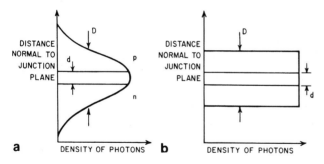

Fig. 11.4a,b. Transverse spatial energy distribution of the light in a diode laser; **a** actual; **b** idealized (area under the idealized curve is adjusted so that d/D remains unchanged)

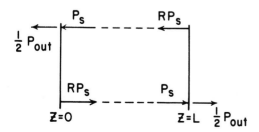

Fig. 11.5. Power flow diagram for a diode laser

To derive a quantitative expression giving the required current density to produce lasing, consider a single pass of a wave of laser light (photon flux) from one Fabry-Perot surface to the other. The power flow diagram looks like that given in Fig. 11.5. P_s is the optical power incident internally on each end face and R is the power reflection coefficient. Under oscillating (lasing) conditions, RP_s grows exponentially with distance during a single pass reaching the value P_s at the opposite Fabry-Perot (F-P) surface. Loss mechanisms present are overcome by the laser gain mechanism of stimulated emission. At each F-P surface, power of $\frac{1}{2}P_{out} = (1 - R)P_s$ is emitted. If α is the loss coefficient [cm^{-1}] for the wave as it travels (including all types of optical loss) and g is the gain coefficient [cm^{-1}], the power as a function of distance is given by

$$P = RP_s \exp\left(g \frac{d}{D} - \alpha \right) Z. \tag{11.1.4}$$

Note that gain occurs only over the inverted population region; hence, g must be multiplied by d/D, while loss occurs wherever the field extends and hence α is not multiplied by anything. For oscillation (the loop gain equals 1), it must be true that

$$P_s = RP_s \exp\left(g \frac{d}{D} - \alpha \right) L, \tag{11.1.5}$$

or

$$\ln\frac{1}{R} = \left(g \frac{d}{D} - \alpha \right) L: \tag{11.1.6}$$

Thus,

$$g \frac{d}{D} = \alpha + \frac{1}{L} \ln\frac{1}{R}. \tag{11.1.7}$$

The gain coefficient g is related to the injected current density of holes and electrons. It can be shown that this relationship is [11.7]

$$g = \frac{\eta_q \lambda_0^2 J}{8\pi e n^2 d \Delta v}, \tag{11.1.8}$$

		for GaAs at 300 K
η_q:	internal quantum efficiency	0.7
λ_0:	vacuum wavelength emitted	9.0×10^{-5} cm
n:	index of refraction at λ_0	3.34
Δv:	linewidth of spontaneous emission	1.5×10^{13} s^{-1}
e:	electronic charge	
d:	thickness of active region	10^{-4} cm
J:	injected current density	

Thus, substituting (11.1.8), which holds both above and below threshold, into the threshold condition (11.1.7), one obtains

$$\underbrace{\frac{\eta_q \lambda_0^2 J_{TH}}{8\pi e n^2 \Delta v D}}_{\substack{\text{effective gain} \\ \text{per pass}}} = \underbrace{\alpha + \frac{1}{L}\ln\frac{1}{R}}_{\text{loss per pass}}, \qquad (11.1.9)$$

or

$$J_{TH} = \frac{8\pi e n^2 \Delta v D}{\eta_q \lambda_0^2}\left(\alpha + \frac{1}{L}\ln\frac{1}{R}\right). \qquad (11.1.10)$$

Thus, the threshold current (density) is that which is necessary to produce just enough gain to overcome the losses.

Note that, from the standpoint of threshold, the light output of the laser at the end faces must be counted as a loss. This is taken account by the term $(1/L)\ln(1/R)$. It may not be immediately obvious that this term represents *loss* due to the power emitted from the end faces but, if we substitute the transmission coefficient, $T = 1 - R$ and do a series expansion of $\ln[1/(1 - T)]$, we get

$$\frac{1}{L}\ln\left(\frac{1}{R}\right) = \frac{1}{L}\ln\left(\frac{1}{1-T}\right) = \frac{1}{L}\left(T - \frac{T^2}{2} + \frac{T^3}{3} - \frac{T^4}{4} + \cdots\right). \qquad (11.1.11)$$

Discarding the higher-order terms in T one obtains

$$\frac{1}{L}\ln\left(\frac{1}{R}\right) \cong \frac{T}{L}. \qquad (11.1.12)$$

T/L represents a loss coefficient [cm^{-1}] obtained by averaging the end *loss* coefficient T over the length. Since T has a value of typically 0.6 (for GaAs), discarding the T^2 and higher-order terms does not produce an accurate result and should not be done in quantitative work. However, the purpose of this example is to qualitatively demonstrate that the nature of the term $(1/L)$ ln $(1/R)$ is that of an *average* loss coefficient per unit length resulting from transmission of photons out of the laser.

From the theoretical development leading to (11.1.10), we see that the spreading of photons (optical fields) out of the inverted population region into surrounding passive regions results in a substantial increase in threshold current density. This fact suggests that the laser diode should be designed so as to make the ratio $D/d = 1$ for optimum performance. In Sect. 11.2 and in Chap. 12 we will discuss ways of making such *confined-field* lasers.

11.1.4 Output Power and Efficiency

Expressions for the overall power efficiency and the power output of a diode laser can be derived as follows: First, consider the losses over a small (incremental) distance ΔZ. To first order, the power loss over this distance is

$$P_{\text{loss}} = P - Pe^{-\alpha\Delta Z} = P(1 - e^{-\alpha\Delta Z})$$
$$\cong P[1 - (1 - \alpha\Delta Z)] = \alpha P \Delta Z, \tag{11.1.13}$$

or

$$dP_{\text{loss}} = aPdZ. \tag{11.1.14}$$

Then the power absorbed over length L is

$$P_{\text{loss}} = \int_{Z=0}^{Z=L} dP_{\text{loss}} = \alpha \int_0^L PdZ. \tag{11.1.15}$$

Similarly, the power generated per pass is given by

$$P_{\text{gen}} = \int_{Z=0}^{Z=L} dP_{\text{gen}} = g\frac{d}{D}\int_0^L PdZ. \tag{11.1.16}$$

Thus, the internal power efficiency η is given by

$$\eta \equiv \frac{P_{\text{gen}} - P_{\text{loss}}}{P_{\text{gen}}} = \frac{g\dfrac{d}{D} - \alpha}{g\dfrac{d}{D}}. \tag{11.1.17}$$

Substituting from (11.1.7)

$$\eta = \frac{\dfrac{1}{L}\ln\dfrac{1}{R}}{\alpha + \dfrac{1}{L}\ln\dfrac{1}{R}}. \tag{11.1.18}$$

Note again that α is the total loss coefficient including both interband and free carrier absorption, as well as any scattering loss. The power out is then given by

$$P_{\text{out}} = \eta P_{\text{in}} = \eta\left[\frac{J}{e}\eta_{\text{q}}(L \times W)hv\right], \tag{11.1.19}$$

where P_{in} is the optical power generated within the laser, and L being the length, W the width and v the lasing frequency. Substituting from (11.1.18) one obtains

$$P_{\text{out}} = \frac{\dfrac{1}{L}\ln\dfrac{1}{R}}{\alpha + \dfrac{1}{L}\ln\dfrac{1}{R}}\frac{J\eta_{\text{q}}}{e}(L \times W)hv. \tag{11.1.20}$$

(Note: P_{out} is power out of both end faces.) The overall power efficiency including the effect of series resistance in the device is thus given by

$$\eta_{\text{tot}} = \frac{P_{\text{out}}}{P_{\text{in, tot}}} = \frac{\dfrac{\dfrac{1}{L}\ln\dfrac{1}{R}}{\alpha + \dfrac{1}{L}\ln\dfrac{1}{R}}\dfrac{J\eta_{\text{q}}}{e}(L \times W)h\nu}{\dfrac{J}{e}(L \times W)h\nu + \underbrace{\left[J(L \times W)\right]^2 R_{\text{series}}}_{I^2 R \text{ loss}}}. \tag{11.1.21}$$

In cases where the series resistance of the diode can be neglected this reduces to

$$\eta_{\text{tot}} = \frac{\dfrac{1}{L}\ln\dfrac{1}{R}}{\alpha + \dfrac{1}{L}\ln\dfrac{1}{R}}\eta_{\text{q}}. \tag{11.1.22}$$

Note that these power and efficiency formulae hold at and above threshold. For current densities below threshold (i.e., before oscillation has set in), the replacement $g(d/D) = \alpha + (1/L)\ln(1/R)$ does not hold, and the expressions would have to be left in terms of g, D, d and α rather than L, R and α.

The efficiency of semiconductor lasers decreases with increasing temperature because of two effects. First, absorption increases [resulting in a larger α in (11.1.18)] and secondly, the quantum efficiency (η_{q}) is reduced. This latter effect occurs because the holes and electrons are *smeared out* in energy over a wider range due to thermal excitation, and hence less hole-electron pairs of the proper energy separation are available for stimulated emission into the lasing mode at any given level of injected input current. These same two effects are responsible for the increase in threshold current density required for lasing, see (11.1.10). Obviously, for a given input current density, the power out is reduced by increasing temperature due to these same two factors. Because of the afore-mentioned bad effects of heat, p-n junction lasers are usually operated on a pulse basis. The *turn-on* times and decay times of the diodes are very short ($\sim 10^{-10}$s or shorter); hence, the diodes can easily be pulsed in good *square wave* 100 ns duration pulses. Overall power efficiencies are generally low (about a few percent), and peak power out is generally about 10 W. Semiconductor p-n junction lasers can be operated at room temperature (or higher) on a cw basis except that enough pumping current must be supplied to achieve threshold, and this much current must be less than that which would destroy the device due to heating. Typical threshold current densities for basic semiconductor p-n junction lasers like that of Fig. 11.1 are on the order of 10^4 A/cm^2, but since areas are about 10^{-3}cm^2, the peak current are typically on the order of 10 A. However, efficiency and power out are strong functions of material properties and device geometry. In later chapters, we will consider advanced geometries that will permit cw operation of laser diodes with threshold currents less than 100 mA.

11.2 The Tunnel-Injection Laser

We include within this chapter on basic types of semiconductor lasers the tunnel-injection laser, proposed by *Wade* et al. in 1964 [11.6, 8]. As the name implies, this device is pumped by a current of electrons or holes which reach the active region by *tunneling* through an energy barrier. The tunnel-injection laser has not achieved much popularity because no one yet has been able to fabricate one with an attainable threshold current density. Nevertheless, it is quite useful to consider the tunnel-injection laser because it was one of the first of the confined-field type lasers to be proposed [11.8–10], and it serves as the basis for a simple model that explains how field-confinement like that used in heterojunction lasers [11.11–13] results in substantially reduced threshold current density and higher efficiency. Modern heterojunction lasers, which are discussed in Chap. 12, have reached a high level of sophistication, providing single mode CW operation at room temperature.

11.2.1 Basic Structure

A semiconductor laser that in principle combines the best features of the p-n junction laser (small size, simplicity, low voltage power supply) within a device structure fabricated in a single type of semiconductor material, so that no junction formation required, is the tunnel-injection laser, shown in Fig. 11.6. In this laser, a single crystal of uniformly doped semiconductor material is used, without a junction. The hole-electron pairs are injected into the semiconductor by tunneling and diffusion. If a p-type semiconductor is used, electrons are injected through the insulator by tunneling while holes diffuse from the source at the contact of metal No. 1 to the semiconductor. If an n-type semiconductor is used, holes are tunneled through the insulator while electrons are injected at the metal No. 1 contact.

The process of tunneling can be better understood by referring to the energy band diagram of Fig. 11.7. The diagram shows the energy band picture

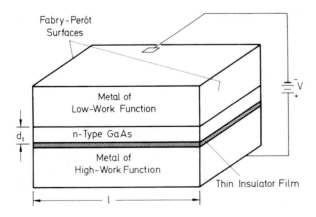

Fig. 11.6. Basic structure of a tunnel-injection laser

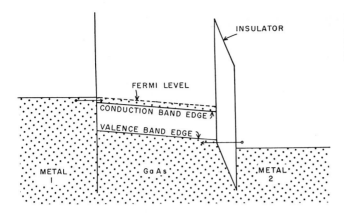

after the required bias voltage has been applied and, as a result, the Fermi levels have shifted to new positions. Note that most of the voltage drops across the insulator; hence, a very high electric field exists in this region. From the diagram, it can be seen that a *hole* in metal No. 2 does not have enough energy (classically speaking) to surmount the insulator barrier. However, quantum mechanical theory predicts that there is a small probability that the hole can pass through the insulator into the semiconductor even if classically it does not have enough energy to do so. In order for a significant number of holes to tunnel through, the insulator must be very thin or the voltage must be very high because the tunneling current density is given by [11.14, 15]

$$J_t = J_0 \left(\frac{E}{E_0} \right)^2 e^{-E_0/E},\tag{11.2.1}$$

where E is the electric field in the insulator and J_0 and E_0 are constants, given by

$$E_0 = \frac{4\varphi^{3/2}(2m^*)^{1/2}}{3hq},\tag{11.2.2}$$

and

$$J_0 = \frac{2q\varphi^2 m^*}{9h\pi^2},\tag{11.2.3}$$

φ being the barrier height, and the other quantities are as defined previously. To obtain the necessary amount of current for laser threshold (about 100 A/cm^2), calculations show that a field strength of about 10^7 V/cm is needed. Thus, when using a 10 V source, one needs an insulator thickness of 10^{-6} cm or 0.01 μm.

So far, all attempts to produce lasing by tunnel-injection have failed because the very thin insulating film is very hard to form without pinholes. Nevertheless, light emitters have been made using the principle of tunnel-injection in both GaAs and CdS. In the case of CdS, tunnel-injection is particularly important because no one has yet been able to form a p-n junction in CdS. Since for light emitters the required current is less than that needed for laser threshold, thicker, more practical films ($\sim 1000\,\text{Å}$) can be used, and these can be made to be reliable.

11.2.2 Lasing Threshold Conditions

The derivation of a threshold current relation for the tunnel-injection laser follows exactly the steps of (11.1.4–10), except that the thickness of the light emitting layer D in the tunnel laser is equal to the thickness of the inverted population region d because the optical fields (or photons) are confined by reflection from the metal layers above and below the active layers. Thus, the value of D in (11.1.10) would be reduced by about a factor of 10 for the case of a GaAs tunnel laser as compared to that of a diode laser. In addition, the loss coefficient α is smaller in the case of the tunnel laser because interband absorption is suppressed in the active region by the inverted population.

As a numerical example, let us compare a conventional GaAs diode laser with $D = 10\,\mu\text{m}$, $d = 1\,\mu\text{m}$, and $\alpha = 35\,\text{cm}^{-1}$ to a GaAs tunnel laser with $d = D = 1\,\mu\text{m}$ and $\alpha = 3\,\text{cm}^{-1}$. For both lasers, we take $R = 0.34$, $L = 1\,\text{mm}$, $\eta_q = 0.7$, $\lambda_0 = 9 \times 10^{-5}\,\text{cm}$, $n = 3.34$ and $\Delta v = 1.5 \times 10^{13}\,\text{s}^{-1}$. Then, from (11.1.10), we calculate

$$J_{\text{th}}(\text{diode}) = 5.43 \times 10^3 \text{ A/cm}^2,$$
$$J_{\text{th}}(\text{tunnel}) = 1.64 \times 10^2 \text{ A/cm}^2,$$

and from (11.1.18) we find

$$\eta(\text{diode}) = 24\%,$$
$$\eta(\text{tunnel}) = 78\%.$$

We have seen that confining the photons of the lasing mode to the active region where they can reproduce themselves by stimulated emission results in a substantial improvement in performance. Although an operational tunnel-injection laser has not yet been fabricated, other lasers employing the field-confinement principle have been made and have indeed yielded the predicted lower threshold current and higher efficiency. In particular, the $Ga_{(1-x)}Al_xAs$ heterojunction laser diode, [11.13], which is discussed at length in the next chapter, uses the index of refraction difference between layers containing different concentrations x of Al to confine the lasing mode to the inverted population region. This device, which has become the most commonly used type of semiconductor laser, also uses the small energy barrier at the heterojunction to confine injected carriers to the active region, thus resulting in further improved performance.

Problems

11.1 We would like to design a p-n junction laser for use as the transmitter in a range finder. The output is to be pulsed with a peak power out of each end of 10 W (only the output from one end will be used) and a pulse duration of 100 ns. Wavelength is to be 9000 Å. Room temperature operation is desired and some of the pertinent parameters have been either measured or established so that:

1) Half-power points of the emission peak for spontaneous emission have been measured for this material at room temperature to be at 9200 and 8800 Å.

2) Index of refraction: 3.3.

3) Thickness of light emitting layer: 10 μm.

4) Thickness of active (inverted pop.) layer: 1 μm.

5) Internal quantum efficiency: 0.7.

6) Average absorption coefficient: 30 cm^{-1}.

7) $W = 300$ μm.

8) Reflectivity of the Fabry-Perot surfaces: 0.4.

a) What must be the separation between Fabry-Perot surfaces if we wish to have a peak pulse current density of $3 \times 10^4 \mathrm{A/cm^2}$?

b) What is the threshold current density?

11.2 In the case of the laser of Problem 11.1, if the heat sink can dissipate 1 W at room temperature, what is the maximum pulse repetition rate that can be used without causing the laser crystal to heat above room temperature (neglect $I^2 R$ loss)?

11.3 For the laser of Problem 11.2, what is the minimum range for which transmitted and target reflected pulses will not overlap in time at the detector, assuming that transmitter and detector are located at essentially the same point and transmission is through air?

11.4 a) Can a light emitter without mirrors or any other optical feedback device produce light by stimulated emission? Is this light coherent?

b) Explain the significance of the *threshold* phenomenon in lasers.

c) Why do confined-field type lasers have lower threshold current and higher efficiency?

d) Why is the divergence angle of the emitted optical beam much greater in a semiconductor laser than it is in a gas laser?

11.5 A semiconductor laser formed in a direct bandgap material is found to have an emission wavelength of 1.2 μm. The external quantum efficiency is 15%.

a) What is the approximate bandgap energy of the material?

b) If the output power is 20 mW, give an approximate estimate of the input current.

11.6 A confined-field laser has been properly designed so that its absorption loss due to interband transitions is negligible ($\alpha_{IB} \cong 0$). The free carrier absorption loss and the scattering loss are $\alpha_{FC} = 5\,cm^{-1}$ and $\alpha_S = 0.5\,cm^{-1}$, respectively. If the length of the laser is increased by a factor of two, by what factor will the threshold current density change? (Assume the end face reflectivity is 65%).

11.7 Derive an expression for the threshold current density J_{th} of an injection laser which has one end face coated with a layer that is totally reflecting ($R = 1$). Assume that the relationship between the input current density J [A/cm²] and the gain coefficient g [cm⁻¹] is given by $J = Kg$, where K is a constant. Also assume that R_0 is the reflectivity of the partially transmitting endface, and that L, α, D, and d are as previously defined.

11.8 The following parameters are known for a semiconductor laser:
Emission wavelength $\lambda_o = 0.850\,\mu m$.
Lasing threshold current $I_{th} = 12$ A
External quantum efficiency = 1% below threshold
External quantum efficiency = 10% above threshold
   a) Sketch the light output power [W] vs. input current [A] curve for this device for currents ranging from 0 to 20 A; label the value of the output power at threshold.
   b) What is the output power [W] for an input current = 18 A?

11.9 A multimode GaAs laser diode is emitting at a wavelength of approximately 0.9 μm. At that wavelength the index of refraction of GaAs is 3.6 and the dispersion is 0.5 μm⁻¹. The laser cavity length of 300 μm. Calculate the mode spacing (in wavelength) at approximately 0.9 μm.

11.10 The external *differential quantum* efficiency of a GaAs laser diode emitting at a wavelength of 0.9 μm is 30%. The applied voltage is 2.5 volts. Calculate the external *power* efficiency of the device.

11.11 A certain semiconductor p-n junction laser when biased as required has a gain coefficient $g = 70\,cm^{-1}$, a loss coefficient $\alpha = 5\,cm^{-1}$, and a length between Fabry-Perot end faces of $L = 300$ μm. It is desired to add anti-reflection coatings to the endfaces. What is the smallest endface reflectivity R that can be used without going below the lasing threshold?
   (Assume that both endfaces have the same reflectivity, and that the confinement factor $d/D = 1$.)

12. Heterostructure, Confined-Field Lasers

In Chap. 11, it was demonstrated that confining the optical field to the region of the laser in which the inverted population exists results in a substantial reduction of threshold current density and a corresponding increase in efficiency. As early as 1963, it was proposed that heterojunctions could be used to produce a waveguiding structure with the desired property of optical confinement [12.1, 2], At about the same time, others proposed using a heterojunction laser structure not for optical field confinement, but to produce higher carrier injection efficiency at the p-n junction, and to confine the carriers to the junction region [12.3, 4]. Actually, all three of these mechanisms are present in a heterostructure laser, and their combined effects result in a device that is vastly superior to the basic p-n homojunction laser.

Because of the technological difficulties involved in the growth of multi-layer heterostructures, several years passed before operational heterostructure lasers were fabricated in 1969 [12.5–7]. These first devices were all fabricated in $Ga_{(1-x)}Al_xAs$ because the close lattice match between GaAs and AlAs, which resulted in minimal strain at the interfaces between the heterostructure layers, made that material preferable to others, such as GaAsP. Even in these early heterostructure lasers, threshold current densities were on the order of $10^3\,A/cm^2$ rather than 10^4 to $10^5\,A/cm^2$, as in a comparable homojunction laser.

Hayashi et al. [12.5], and *Kressel* and *Nelson* [12.6] fabricated single-heterojunction (SH) lasers in 1969, while *Alferov* et al. [12.7] fabricated more effective double-heterojunction (DH) lasers. Since 1969, numerous improvements have been made in the basic heterostructure laser, resulting in devices with threshold current densities on the order of $10^2\,A/cm^2$.

In this chapter, both the basic and the advanced heterojunction laser structures are discussed. The relationships between geometric and materials properties, and the performance characteristics of the laser are described. In Sect. 12.5 of this chapter, we also consider the important question of laser diode reliability. The DH injection laser, while very efficient, is one of the most highly-stressed semiconductor devices in terms of intense optical and electrical fields. Hence a great deal of effort has been expended to produce devices with satisfactory lifetime and limited degradation of performance.

12.1 Basic Heterojunction Laser Structures

12.1.1 Single Heterojunction (SH) Lasers

The simplest heterojunction laser to fabricate is the SH structure shown in Fig. 12.1 [12.5, 6]. In the fabrication of this device, the anomalously fast diffusion of Zn in GaAs [12.8] is utilized to form a diffused p-n junction lying 1 to 2 μm below the $Ga_{(1-x)}Al_xAs$-GaAs heterojunction. If the n- and p-type dopant concentrations are approximately equal on both sides of the p-n junction, the injection current will consist mostly of electrons injected into the p-type layer, because the effective mass of an electron is about 7 times less than that of a hole in $Ga_{(1-x)}Al_xAs$ [12.9]. Thus, the inverted-population region, or active layer, in this type of SH laser is in the p-type GaAs, as shown in Fig. 12.1. This SH laser can be fabricated by using the method of liquid epitaxial growth that is described in Chap. 4, except that a relatively high growth temperature of 900°–1000 °C is used in order to promote the required Zn diffusion into the substrate.

The thickness of the active layer can be selected by controlling the time and temperature of epitaxial growth (and hence Zn diffusion) to produce a thickness of about 1 to 5 μm. However, since the diffusion length of injected electrons is only about 1 μm, increasing the thickness of the p-GaAs layer beyond that value results in decreased efficiency and higher threshold current density, because the inverted population region is still limited to a thickness of about 1 μm by electron recombination [12.10]. Thus, although the optical mode spreads over the entire p-GaAs layer, it can be pumped by stimulated emission over only a 1 μm-thick layer closest to the p-n junction, and reduced efficiency results. In some cases, it might be desirable to increase the p-GaAs layer thickness to greater than 1 μm, even at the expense of increased threshold current density, because reduced diffraction of the optical beam in the thicker layer results in a smaller divergence angle of emitted light in the plane perpendicular to the p-n junction.

In the SH laser, optical confinement occurs only on one side of the light-emitting junction, at the interface between the p-GaAs and p-$Ga_{(1-x)}Al_xAs$ layers. Although there is a waveguiding effect in the depletion

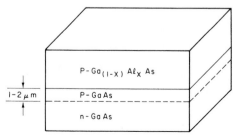

P-Ga$_{(1-x)}$ Al$_x$ As

1–2 μm   P-GaAs

n-GaAs

Fig. 12.1. Single heterostructure laser diode

layer of the p-n junction itself because of carrier concentration reduction [12.11], that phenomenon is generally negligible compared to the substantial confinement that occurs because of the change in index of refraction at the heterojunction. Thus the SH laser structure is only partially effective in producing the desired optical confinement. As a result, SH lasers exhibit higher threshold current densities than do comparable DH lasers. In fact, SH lasers must be operated on a pulsed basis, rather than cw, at room temperature, In many applications, pulsed operation is not detrimental to overall system performance, and may even be beneficial, to improve signal to noise ratio in signal processing. Hence SH lasers have been widely used in the past as sources in a variety of opto-electronic applications. However, when a cw laser diode source operating at room temperature is required, a double heterostructure laser must be used. In recent years the large demand for double heterostructure lasers has resulted in large-volume production and lowered cost, so SH lasers are rarely used.

12.1.2 Double Heterostructure (DH) Lasers

The physical structure of a typical DH laser in GaAlAs is shown in Fig. 12.2, along with a diagram of the index of refraction profile in the direction normal to the p-n junction plane. The basic GaAlAs, three-layer waveguide structure is usually grown on a heavily-doped n+-type substrate, and is capped with a heavily-doped layer of p+-GaAs, to facilitate the formation of electrical contacts. The active region is established on the p-type side of the p-n junction, as explained previously in Sect. 12.1.1. Very often, the active layer will contain a certain concentration of Al in order to shift the optical emission to shorter wavelengths. This topic is discussed in more detail in Sect. 12.3. The thickness of the active layer should be less than 1 μm in order to assure that an inverted population exists throughout the entire layer rather than being limited by the diffusion length for injected electrons. In fact, active layer thickness is often reduced to 0.2 to 0.3 μm in order to produce greater population inversion

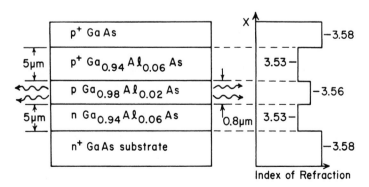

Fig. 12.2. Double heterostructure laser diode

and laser photon density. Typical electrical dopant concentrations are $N_A \cong 2 \times 10^{19} \, \text{cm}^{-3}$ for the p$^+$-layer, $N_A \cong 1 \times 10^{16} \, \text{cm}^{-3}$ for the active layer, $N_D \cong 1 \times 10^{17} \, \text{cm}^{-3}$ in the n-GaAlAs, and $N_D \cong 2 \times 10^{18} \, \text{cm}^{-3}$ in the substrate. The choice of these concentrations is governed by the desire to reduce series resistance in the bulk material, while limiting free-carrier absorption in the light-emitting region. The multilayer $Ga_{(1-x)}Al_xAs$ structure shown in Fig. 12.2 is usually grown by the slidebar method of liquid epitaxial growth that is described in Chap. 4. For a detailed description of the application of this method to DH laser fabrication see *Casey* and *Panish* [12.12]. The newer methods of MOCVD or MBE growth (also described in Chap. 4) offer some advantages for laser diode fabrication. MOCVD permits the processing of relatively large wafers, while quantum well lasers can be made by MBE or MOCVD. However, the high cost of these growth systems has slowed their acceptance for production line use. For a review of the development of heterojunction lasers and the fabrication methods used to produce them see *Kroemer* [12.13], and *Iga* and *Kinoshita* [12.14].

The double heterojunction laser structure provides confinement on both sides of the active region. Because of this confinement of both lasing mode photons and injected carriers to the inverted population region, in which gain due to stimulated emission is possible, the DH laser is highly efficient, and requires minimal threshold current, as compared to other semiconductor lasers.

12.2 Performance Characteristics of the Heterojunction Laser

The superior performance of heterojunction lasers results from the combined effects of optical field confinement and more efficient carrier injection and recombination. In this section these basic features of the phenomena are described, and the differences between the performance of DH, SH and homojunction lasers are explained.

12.2.1 Optical Field Confinement

In Chap. 11, it was shown that the optical field confinement, as characterized by the fraction of lasing mode photons that remain in the inverted population region d/D, strongly affects both threshold current density J_{th} and efficiency η. The quantities J_{th} and η can be calculated using (11.1, 10) and (11.1.17), but one must know the correct value of d/D to use in any given case.

The procedure for determining d/D from first principles is to solve the wave equation, subject to the boundary conditions appropriate for the particular waveguide structure, thereby determining the quantitative expression for the mode shape. The ration d/D is then found by integrating the expression for photon density over the thickness of the inverted population layer and dividing that quantity by the total number of photons in the mode, which is obtained by integrating the photon density expression over the extent of the entire mode. The solution of the wave equation to find the mode shape in a three-layer, symmetric waveguide, as described in Chap. 3, is a lengthy problem, involving extensive computer calculations. However, in many cases, sufficient accuracy can be obtained by using an approximate, but more concise, set of relations developed by *McWhorter* [12.15].

By solving Maxwell's equations, *McWhorter* has obtained relations that give the transverse mode shape in a semiconductor laser for an active region of thickness d and index n_a, bounded by confining regions with indices n_b and n_c, as shown in Fig. 12.3, such that n_a is greater than n_b and n_c. It is assumed that the light confining layers are sufficiently thick so that the tails of the optical mode do not penetrate through to the p+ and n+ contact layers. The relative spatial energy density Φ (photon density) in the active region (a) is given by

$$\Phi = A \cos k_a x + B \sin k_a x \quad \text{for} \quad -d/2 \leq x \leq d/2, \tag{12.2.1}$$

where A and B are constants.

In the confining layers (b) and (c) Φ is given by

$$\Phi = e^{-k_b x} \quad \text{for} \quad x > d/2, \tag{12.2.2}$$

and

$$\Phi = e^{+k_c x} \quad \text{for} \quad x < -d/2. \tag{12.2.3}$$

The extinction coefficients k_b and k_c are determined from

$$k_b = \left(\frac{k_0^2 d}{2}\right)\left(\frac{n_a^2 - \overline{n^2}}{n^2}\right) + \frac{1}{2d}\left(\frac{n_c^2 - n_b^2}{n_a^2 - \overline{n^2}}\right), \tag{12.2.4}$$

and

$$k_c = \left(\frac{k_0^2 d}{2}\right)\left(\frac{n_a^2 - \overline{n^2}}{n^2}\right) + \frac{1}{2d}\left(\frac{n_b^2 - n_c^2}{n_a^2 - \overline{n^2}}\right), \tag{12.2.5}$$

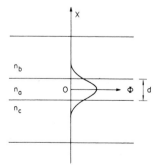

Fig. 12.3. Model for field confinement in a double heterostructure laser

where n is the index of refraction of pure GaAs at the lasing wavelength, and where

$$\overline{n^2} = \frac{n_b^2 + n_c^2}{2}, \tag{12.2.6}$$

and

$$k_0 = 2\pi/\lambda_0. \tag{12.2.7}$$

Once k_b and k_c have been calculated, the constants k_a, A and B can be determined from (12.2.1–3) by matching boundary conditions for continuity at the interfaces in the usual way. Then the confinement factor d/D can be determined from

$$\frac{d}{D} = \frac{\int_{-d/2}^{d/2} \Phi dx}{\int_{-\infty}^{\infty} \Phi dx}, \tag{12.2.8}$$

where Φ is taken from (12.2.1–3).

The calculation of d/D from McWhorter's relations is still a lengthy procedure, even though it is more brief that the direct solution of Maxwell's wave equation. Fortunately, another alternative exists. *Casey* and *Panish* [12.16] have developed an approximate expression for the confinement factor d/D that holds for the special case of a symmetric, three-layer, $Ga_{(1-a)}Al_aAs$-GaAs-$Ga_{(1+a)}Al_aAs$ waveguide with small thickness d. They have shown that, for cases in which

$$d \leqslant \frac{0.07 \lambda_0}{a^{1/2}}, \tag{12.2.9}$$

where a is the atomic fraction of Al in the confining layers, the confinement factor is given by

$$\frac{d}{D} \cong \frac{\int_0^{d/2} E_0^2 \exp(-2\gamma x)dx}{\int_0^{\infty} E_0^2 \exp(-2\gamma x)dx}, \tag{12.2.10}$$

where E_0 is the peak field amplitude and γ is given by

$$\gamma \cong \frac{(n_a^2 - n_c^2)k_0^2 d}{2}, \tag{12.2.11}$$

where n_a and n_c are the indices of refraction in the active and confining layers, respectively. The confinement factor for a $Ga_{(1-a)}Al_aAs$-GaAs three-layer symmetric waveguide is plotted in Fig. 12.4 for various typical values of d and a. These curves are based on the assumption that the laser is operating in the fundamental (TE_0) mode, which is generally the case in a well-designed DH laser. It can be seen from the data of Fig. 12.4 that nearly 100% confinement

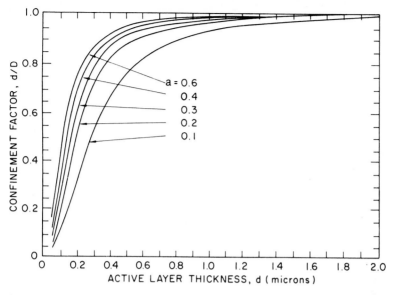

Fig. 12.4. Confinement factor for the fundamental mode in a symmetric, three-layer $Ga_{(1-a)}Al_aAs$-GaAs waveguide [12.16]

can be obtained for active region thicknesses as small as 0.4 µm, without exceeding an atomic fraction of Al equal to 0.6. This is significant because AlAs is hygroscopic, and larger Al concentrations in the $Ga_{(1-a)}Al_aAs$ are therefore unadvisable.

12.2.2 Carrier Confinement

As mentioned previously, some of the researchers who first proposed the heterojunction laser [12.3, 4] did so not becuase of optical field confinement, but rather because the heterojunction offers improved carrier injection efficiency and carrier confinement to the active region. The energy band diagram of a n^+-n-p-p^+ GaAlAs double heterojunction laser is shown in Fig. 12.5. Because the bandgap is larger in the regions with larger Al concentration, discontinuities result in the conduction band at the p-p^+ junction (ΔE_c) and in the valence band at the n-p and n^+-n junctions (ΔE_v). The energy bands are shown in Fig. 12.5 for the case of no applied external bias voltage, and the magnitude of the conduction and valence band discontinuities has been drawn larger than to scale in order to emphasize their presence for pedagogic purposes. When a forward bias voltage is applied to the DH laser, electrons are injected from the n-region into the p-region, forming the desired recombination current. The conduction band discontinuity at the p-p^+ interface ΔE_c provides a barrier to the injected electrons, tending to confine them to the p-region, and to enhance the probability of their recombining with holes by the

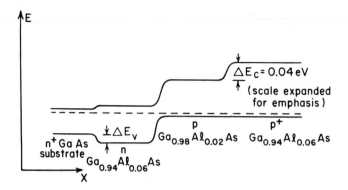

Fig. 12.5. Energy band diagram of a n⁺-n-p-p⁺ double heterojunction laser diode. Zero bias state is assumed, i.e. no externally applied voltage is present

stimulated emission process. The valence band discontinuity at the n-p junction ΔE_v augments the already existing built-in potential barrier, to further prevent hole injection into the n region, thereby improving injection efficiency. Thus the double heterostructure tends to confine both majority and injected minority carriers to the active p-layer. Since the photons of the optical mode are also confined to the active layer by the heterojunctions as described in Sect. 12.2.1, the DH laser provides optimum conditions for establishing both the largest possible inverted population and the greatest photon density in the active layer. It will be recalled from Chap. 10 that these are the two primary requirements for stimulated emission. Thus the DH laser can be expected to significantly outperform the homojunction laser, and indeed it does.

12.2.3 Comparison of Laser Emission Characteristics

Threshold current density for a typical homojunction laser is on the order of 10^4 A/cm², and differential quantum efficiency (above threshold) is about 10%. Even though the SH laser provides photon and carrier confinement at only one interface of the active layer, it is effective enough to reduce the threshold current density to about 5×10^3 A/cm², and to increase differential quantum efficiency to about 40%. It must be remembered, of course, that J_{th} and η_D are functions of temperature, active layer thickness, and dopant concentrations in any given case; so the numbers quoted here are typical values provided for comparison purposes only. Because J_{th} is still relatively large in SH lasers and η_q is only moderate, they cannot be operated on a cw basis. They must be pulsed, usually with 100 ns length pulses at a repetition rate of 100–1000 Hz, to allow time for the junction region to cool between pulses. Nevertheless, peak pulse output power from 10–30 W is possible.

The DH laser, which provides confinement at both interfaces of the active region, has a threshold current density typically from 400 to 800 A/cm², and differential quantum efficiency as high as 55% [12.17]. The DH laser can be operated cw, with output power as high as 390 mW [12.18], although cw power less than 100 mW is more common [12.19]. One disadvantage of the DH laser is that the angular divergence of the beam in the plane perpendicular to the junction can be 20° to 40°, rather than the 15° to 20° commonly exhibited by SH lasers. This divergence, which results from diffraction in the thin active layer, can be reduced by using special large optical cavity (LOC) DH structures, which provide separate confinement of photons and carriers. For broad area homojunction, SH and DH lasers, such as have been described so far, there is no lateral confinement in the plane of the junction, light is free to spread over the full 20 to 80 μm width of the chip. As a result, beam divergence in the junction plane is only about 10° in all cases. Stripe geometry lasers, which are described in Sect. 12.4.1, also confine the light in the lateral direction. This confinement results in a further reduction in J_{th}, but it also causes an increase in beam divergence in the junction plane.

12.3 Control of Emitted Wavelength

One of the more important characteristics of semiconductor lasers is the wavelength at which the peak emission of light occurs. It is often desirable to shift the peak wavelength somewhat from that which is characteristic of the substrate semiconductor material in order to take advantage of waveguide transparency in a given range.

12.3.1 $Ga_{(1-x)}Al_xAs$ Lasers for Fiber-Optic Applications

The peak emission wavelength of a homojunction laser formed in GaAs is about 9200 Å, corresponding to radiative recombination of electrons in donor states 0.005 eV below the conduction band edge and holes in acceptor states 0.030 eV above the valence bandedge. (The bandgap of pure GaAs at room temperature is 1.38 eV [12.20].) While GaAs lasers can be used as light sources with glass optical fiber waveguides, their performance is less than optimum because the fibers can have significant absorption loss at 9200 Å. The spectral attenuation curve for a moderately pure borosilicate glass fiber, typical of those available commercially, is shown in Fig. 12.6. The absorption peak at about 9400 Å is caused by the presence of OH ions in the glass. In order to avoid this absorption, $Ga_{(1-x)}Al_xAs$ heterojunction lasers can be used, with sufficient Al added to the active region to shift the emitted wavelength to 8500 Å. The emission wavelength of $Ga_{(1-x)}Al_xAs$ can be shifted in the range from 9200 Å to about 7000 Å by the addition of Al, as shown in Fig. 12.7. Shorter wavelengths are unattainable because the band gap of the $Ga_{(1-x)}Al_xAs$ becomes indirect for Al concentration greater than about 35%, and as a result internal quantum efficiency is greatly reduced. Because

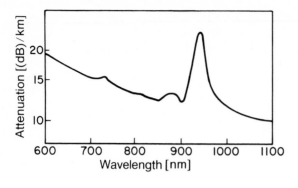

Fig. 12.6. Typical spectral attenuation curve for commercially available multimode glass fiber

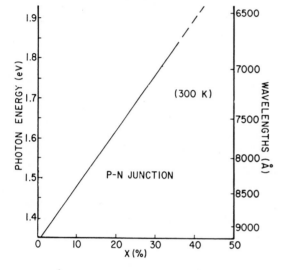

Fig. 12.7. Peak emission wavelength of a $Ga_{(1-x)}Al_xAs$ laser as a function of Al concentration in the active region. X is the atomic fraction of Al

of their compatibility with readily available and inexpensive optical fibers, $Ga_{(1-x)}Al_xAs$ heterostructure lasers, emitting at 8500 Å, are widely used today in local area networks and other short-distance applications.

For long-distance telecommunications the GaAlAs laser has been replaced by GaInAsP laser diodes operating at either 1.3 or 1.55 μm. During the 1980s, the quality of optical fibers continuously improved, due largely to better purification. The OH ions can be removed, along with other contaminants, to produce optical fibers with minimal attenuation, approaching the limit imposed by Rayleigh scattering. The spectral attenuation curve for one such fiber is shown in Fig. 12.8. The minima in atteneuation that occur at 1.3 μm and 1.55 μm are particularly important because material dispersion for a silica-rich core is theoretically equal to zero at a wavelength of 1.27 μm [12.21] and can be shifted to yield zero dispersion at 1.55 μm. Thus, operation at those wavelengths yields not only minimum attenuation, but also minimum dispersion. Of course, in order to obtain minimum dispersion, either a single-mode or a graded index fiber must be used so that modal dispersion is avoided. The

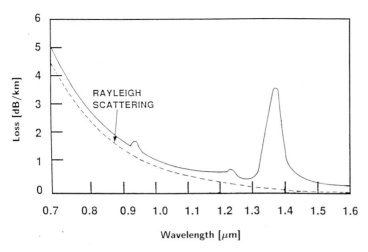

Fig. 12.8. Attenuation in a highly-purified step index single-mode fiber consisting of a borosilicate cladding surrounding a phosphosilicate core

details of optical fiber design are beyond the scope of this book, but are described elsewhere [12.22].

12.3.2 Lasers Made of Quaternary Materials

As described in the previous section, there is a distinct advantage to using a wavelength of about 1.3 μm or 1.55 μm in a fiber-optic system. However, $Ga_{(1-x)}Al_xAs$ lasers cannot emit light at these wavelengths. Because of this limitation, GaInAsP has become the preferred material for fabricating lasers for use in long-distance telecommunications and data transmission. As shown in Fig. 4.12, the bandgap of GaInAsP can be varied from 1.4 eV to 0.8 eV, corresponding to emission wavelengths of 0.886 μm to 1.55 μm. Lattice mismatch at the interfaces between the various layers of the heterostructure, which would result in defect centers that greatly increase absorption and non-radiative recombination, can be avoided by proper choice of the concentrations of the constituent elements. GaInAsP has been demonstrated to be an effective material for producing lasers operating at both 1.3 μm and 1.55 μm wavelengths [12.23]. Double heterostructure lasers of GaInAsP are usually grown by liquid phase epitaxy, and typically have maximum output power in the range of 100 to 200 mW [12.24, 25].

12.3.3 Long-Wavelength Lasers

Semiconductor lasers operating in the 2 to 4 μm wavelength range are of considerable interest for future optical communications systems employing low-loss fluoride glass fibers. Double-heterostructure (DH) lasers in various

III-V materials have been used to achieve emission in this wavelength range. *Mani* etal. [12.26] have reported DH lasers prepared by LPE in the InAsSbP/ InAsSb system, on InAs substrates. These lasers emit at 3.2 µm, with a pulsed threshold current density of 4.5 KA/cm^2 at 78K. *Martinelli* [12.27] has produced DH lasers emitting at 2.55 µm in InGaAsP/InP. Threshold currents of 650 A/cm^2 have been achieved at 80 K; and output powers of several milliwatts were typical.

Rare-earth-doped II-VI materials have also been utilized to produce long-wavelength emitting lasers. For example, Ebe etal. [12.28] have obtained emission at wavelengths between 4.0 and 5.5 µm at 77K in PbEuTe DH lasers. Threshold current was approximately 0.5 kA/cm^2. These devices could also be operated at over 200K ambient temperature with thermoelectric cooling. A lead-salt injection laser has also been reported by Partin [12.29].

12.4 Advanced Heterojunction Laser Structures

12.4.1 Stripe Geometry Lasers

In the discussions of semiconductor lasers thus far, it has been tacitly assumed that the width of the laser was much larger than the thickness of the active layer, so that optical confinement occurred in only one of the transverse directions, i.e. perpendicular to the junction plane. If the width of the laser is limited in some fashion to produce a narrow stripe geometry of typically 5 to 25 µm width, the device will exhibit altered performance characteristics. The most important of these is that the threshold current will be reduced because of the smaller cross sectional area available for current flow. If the width of the stripe is about 10 µm (or less), the lateral confinement will also result in fundamental TE$_{00}$ mode operation, as discussed in more detail in Sect. 12.4.2. Passivation of the longitudinal junction edges by removing them from the side surface of the laser chip reduces leakage current and lengthens the useful life of the laser by reducing degradation.

The lateral edges of the laser may be defined by masked etching of an oxide layer [12.30], as shown in Fig. 12.9 a, or by masked proton bombardment [12.31], to produce high resistivity semi-insulating regions on either side of the stripe, as shown in Fig. 12.9 b. A planar stripe laser can also be formed by masked diffusion [12.32], as shown in Fig. 12.9 c, or by masked etching and epitaxial growth, as shown in Fig. 12.9 d [12.33]. The lasers of Fig. 12.9 a–c are described as being "gain-guided" lasers, because of the change in effective index in the lateral direction is produced by confining the current flow, and hence gain, to a central region. The laser of Fig. 12.9 d is called an "index-guided" laser because the index is directly varied in the lateral direction by changing the chemical composition of the material.

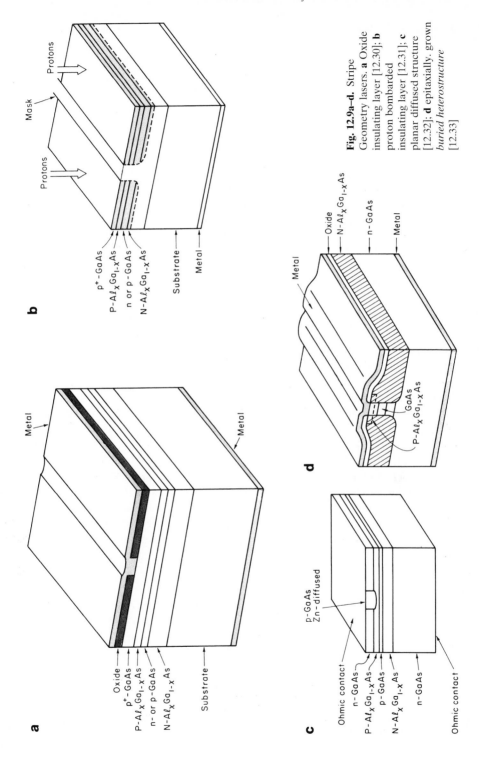

Fig. 12.9a–d. Stripe Geometry lasers. **a** Oxide insulating layer [12.30]; **b** proton bombarded insulating layer [12.31]; **c** planar diffused structure [12.32]; **d** epitaxially. grown *buried heterostructure* [12.33]

12.4.2 Single-Mode Lasers

Most wide-cavity semiconductor lasers oscillate with a multiplicity of transverse and longitudinal modes present, as described in Chap. 11. In many
applications this is not objectionable. However, when a high degree of phase
coherence, or minimal dispersion, is required, as in the case of long-distance
optical fiber communications, single-mode lasers become a necessity. Fortunately, the stripe-geometry gain-guided laser structures described in Sect.
12.4.1 are generally effective in establishing not only a single transverse mode,
but a single longitudinal mode as well, as long as the stripe width is less than
about twice the carrier diffusion length (approximately 3 μm in GaAlAs)
[12.34]. Lasers with broader stripes, up to 29 μm in width, will also oscillate in
a single mode. However mode instability is often observed, accompanied by
kinks in the output curve of optical power versus current density, as shown in
Fig. 12.10. Such kinks are undesirable in themselves, aside from any consideration of mode instability, because they prevent linear modulation of the laser
light output. Mode instability and current kinks are thought to arise in relatively wide stripe-geometry lasers because of localized reductions the carrier
density profile (*hole burning*) caused by carrier consumption at the peak of the
optical mode profile [12.35, 36]. If the width of the stripe is made less than
about two minority carrier diffusion lengths, carriers can diffuse from the
edges of the stripe, where optical intensity is low, to replenish those consumed
in the stimulated emission process. Hence, mode instability and current kinks
are not observed in such lasers [12.34].

Stable single-mode lasers can be produced by gain guiding as described
above or by fabricating a two dimensional structural waveguide, relying on
index of refraction differences confine the optical mode. For example, in the
stripe geometry lasers shown in Fig. 12.9 a–c, optical confinement in the
transverse direction is produced by the variation of the imaginary part of
the dielectric constant, which is related to the gain, rather than by a change in
the real part, which is related to the index of refraction [12.37]. Hence, hole
burning, which causes localized gain saturation, strongly affects optical confinement. In lasers that possess a two-dimensional structural waveguide, such

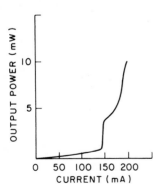

Fig. 12.10. Light output–current characteristic of a stripe-
geometry laser with mode instability

as the buried heterostructure laser of Fig. 12.9 d, gain saturation does not affect optical confinement. Hence stable single-mode oscillation can be obtained for stripe widths in excess of twice the carrier diffusion length [12.38, 39]. In addition to the buried heterostructure, other two-dimensional waveguide structures can be used to produce single-mode lasers [12.40–44]. For a thorough discussion of the various techniques of mode control in semiconductor lasers see *Kaminow* and *Tucker* [12.45]. A survey of various specialized heterostructures and their operating characteristics has been published by *Baets* [12.46].

12.4.3 Integrated Laser Structures

The laser diode structures that have been described in preceding sections are generally suitable for monolithic integration with optical waveguides and other elements of an optical integrated circuit with regard to their dimensions and material composition. However, two significant problems must be dealt with in the design of any specific monolithically integrated laser/waveguide structure. There must be efficient coupling of the light from the laser to the guide, and there must be some means of providing the optical feedback required by the laser. One structure that provides a solution to both of these problems, demonstrated by *Hurwitz* [12.47], is shown in Fig. 12.11. The active region of the laser is doped with Si acceptor atoms, so as to shift the emission wavelength of the GaAs to a 1 μm wavelength that is not strongly absorbed ($\alpha < 1\,\mathrm{cm^{-1}}$) in the n-GaAs waveguide. The rectangular laser mesa, 300 μm long and 45 to 90 μm wide, is produced by masked chemical etching down through the epitaxially grown GaAlAs layers to the substrate. Orientation of the rectangular mask along the $\langle 110 \rangle$ cleavage planes that are perpendicular to the $\langle 100 \rangle$ wafer surface results in reasonably parallel end faces to form the mirrors of the Fabry-Perot cavity. The layers of SiO_2 coating the end faces are necessary because the index difference at the laser-waveguide interface would otherwise be insufficient to provide adequate reflection. After SiO_2 deposition, the GaAs waveguide was grown by vapor phase epitaxy. Room temperature

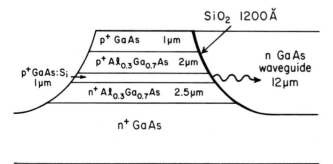

Fig. 12.11. A monolithic laser/waveguide structure in GaAlAs [12.47]

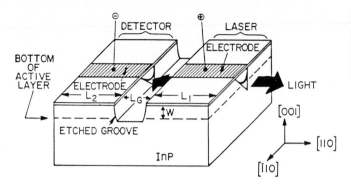

Fig. 12.12. A monolithically integrated GaInAsP laser and monitor photodiode [12.48]

threshold current densities as low as $7.5 \times 10^3 \, \text{A/cm}^2$ were observed for these lasers.

Masked chemical etching of a laser facet has also been used by *Koszi* et al. [12.48] to produce a GaInAsP laser automatically integrated with a monitor photodetector as shown in Fig. 12.12. The laser in that case was a buried-heterostructure, stripe-geometry device emitting at a wavelength of 1.3 μm. The isolation groove between the laser and photodiode was produced by masking with SiO_2, configured by standard photolithography, and etching with a methyl alcohol and bromine solution. The threshold current and external quantum efficiency of the laser were measured to be in the range $I_{th} = 25 - 35$ mA and $\eta_q = 0.06 - 0.09 \, \text{mW/mA}$. Comparable values for similar lasers with cleaved facets were $I_{Th} = 20 \, \text{mA}$ and $\eta_q = 0.19 \, \text{mW/mA}$. The decreased performance of the lasers with etched facets was attributed to a slight tilt in the laser facet. Nevertheless, the I_{th} and η_q of the etched-facet lasers were reasonable for use in optical integrated circuits. A photocurrent of 15 nA per mW of laser optical output power was obtained from the photodiode, and the response was linear.

A technique for producing monolithically integrated laser diodes with cleaved facets has been developed by *Antreasyn* et al. [12.49, 50]. This method, called "stop cleaving", is illustrated in Fig. 12.13, where the thick dark line marks the location of the laser cavity. Holes etched into the substrate are employed to stop the propagation along a cleavage plane, producing cleaved facets which are of comparable quality to conventionally cleaved laser facets. To produce stop-cleaved InGaAsP lasers on an InP substrate, etchants of $4H_2SO_4:1H_2O_2:1H_2O$ and concentrated HCl were used to selectivity etch layers of InGaAsP and InP, respectively. Stripe-geometry, buried-heterostructure lasers fabricated by this method to emit at 1.3 μm had threshold currents as low as 20 mA and differential quantum efficiencies as high as 60% [12.50]. The yield in the cleaving process step was 77%.

There is usually no difficulty in making electrical contact to the p and n sides of the junction of a discrete laser diode, but when the diode is monolithically integrated into an optical integrated circuit it may not be possible to obtain a return path for current through the substrate (for example,

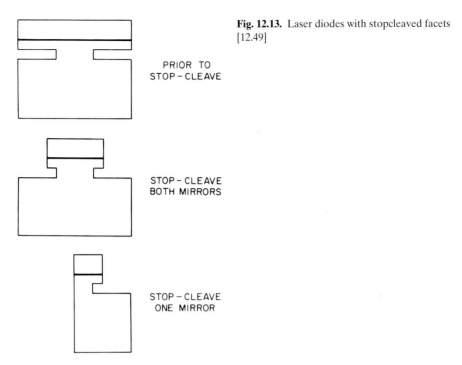

Fig. 12.13. Laser diodes with stopcleaved facets [12.49]

PRIOR TO
STOP – CLEAVE

STOP – CLEAVE
BOTH MIRRORS

STOP – CLEAVE
ONE MIRROR

when a semi insulating substrate is used). In such cases a transverse-junction-stripe (TJS) laser is called for. The TJS laser [12.51, 52], shown in Fig. 12.14, features a transverse junction which has both the p and the n electrical contacts on the top surface. A double heterostructure, stripe-geometry laser cavity is produced in n-type layers grown on a semi-insulating substrate. Then masked Zn diffusion is used to produce a lateral p-n junction. GaAlAs TJS lasers fabricated by MOCVD growth have exhibited threshold current densities as low as 25 mA and external differential quantum efficiencies of 40% [12.52] A TJS laser has been produced in GaInAsP on a semi-insulating InP substrate, by *Oe* et al. [12.53]. Threshold current of 10 mA and maximum CW output power of 10 mW were achieved. Because of the surface configuration of the electrodes TJS lasers are particularly suitable for use at microwave modulation frequencies, with the modulation drive signal being fed to the laser over a metallic strip line deposited directly onto the surface of a semi-insulating GaAs or InP substrate.

A different approach to laser/waveguide integration was demonstrated by *Kawabe* et al. [12.54]. An optically pumped CdS_xSe_{1-x} laser was coupled to a waveguide of the same material by optical tunneling through a CdS confining layer, as shown in Fig. 12.15. Stripe geometry optical pumping by a nitrogen laser above a threshold of 70 kW/cm² produced lasing in the CdS_xSe_{1-x}. A single transverse optical mode was observed in the laser, and this mode was coupled

a

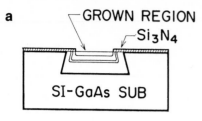

b

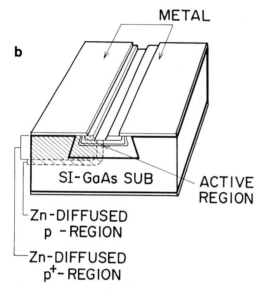

Fig. 12.14. A transverse junction stripe (TJS) laser [12.51]

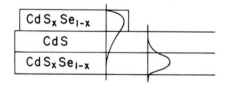

Fig. 12.15. A monolithic laser/waveguide structure in CdSeS [12.54]

into the waveguide as shown in Fig. 12.15 for confining (CdS) layer thickness of 2.5 to 6 μm.

The difficulty of providing Fabry-Perot type feedback can be entirely avoided by employing distributed feedback from a Bragg diffraction grating of the type described in Chap. 9. For example, *Aiki* et al. [12.55] have used that approach to produce a frequency multiplexing light source incorporating six monolithically integrated distributed feedback (DFB) lasers coupled to GaAlAs waveguides. In addition to facilitating monolithic coupling, DFB lasers also provide the advantages of narrower linewidth and better mode stability, as compared to lasers with Fabry-Perot cavities. Because of their importance in optical integrated circuits, DFB lasers are described in detail in Chap. 13.

12.5 Reliability

A key characteristic of laser diodes, which strongly affects system design, is the reliability of the device. Laser diodes are very highly stressed in terms of the current densities and optical field strengths which exist within them, and they generally exhibit some gradual degradation of their operating characteristics as they age, as well as being subject to catastrophic failure under certain conditions.

12.5.1 Catastrophic Failure

Laser diodes are subject to catastrophic failure that occurs when the optical power density at the mirror facets exceeds a certain critical level, generally accepted to be in the range of 2 to 3×10^6 W/cm^2 for GaAlAs lasers and slightly higher for GaInAsP devices. Mechanical damage, in the form of pits and protrusions, usually occurs first near the center of the facet, where the optical field strength is maximum. The threshold power density above which catastrophic failure occurs varies for different lasers, due in part to the fact that the presence of grown-in initial defects in the facet surface enhances the formation of further damage. Catastrophic damage can be reduced by coating the laser facets with a dielectric such as SiO_2, Si_3N_4 or Al_2O_3, with a thickness equal to a half wavelength in the dielectric. Al_2O_3 is closely matched to GaAs in its coefficient of thermal expansion and a has high thermal conductivity. Operating the laser on a pulsed basis rather than CW also results in reduced catastrophic damage.

Catastrophic damage can, of course, be avoided by merely operating the laser well below the damage threshold. However, it must be remembered that transients of only a few microseconds duration are sufficient to destroy the laser. Hence, turn-on transients must be filtered out in cw operation, and ringing must be avoided in pulsed applications. In addition, power supply filtering must be adequate to absorb any random transients that can occur on the line.

12.5.2 Gradual Degradation

The problem of gradual degradation, in which the threshold current density of the laser increases and differential quantum efficiency decreases as the device is operated, was so severe in early laser diodes that device lifetime was only 1 to 100 h, and it was therefore believed that a practical, reliable device would never be made. However, methodical work in a number of different laboratories over a period of years has led to the identification and elimination of most degradation mechanisms, so that lasers with projected lifetimes of 10^6 h are now being made [12.56].

The gradual degradation of laser diodes results from the formation of defects within the active region that act as nonradiative recombination centers. the exact origin of these defects is uncertain. However, a number of possible mechanisms have been proposed, including precipitation of impurity atoms at dislocations [12.57], relaxation of bonding-induced strain by generation of dislocations [12.58], optical generation of defects in strained material [12.59] and optically-enhanced defect diffusion [12.60]. As degradation proceeds, the defects become apparent as *dark line defects* (DLD) [12.61], three-dimensional dislocation networks that grow in size and number during lasing. Defects and contaminant atoms initially present in the laser serve to accelerate DLD growth. Hence, maintenance of the most careful growth procedures is essential if long life lasers are to be produced. $Ga_{1-x}Al_xAs$ diodes with x between 0.05 and 0.1 in the active region have been found to exhibit the least degradation. Since dislocations tend to propagate from the substrate through epitaxial layers, choice of substrates with low dislocation densities is important.

If scrupulous attention is paid to the details of the fabrication process including proper choice of substrate, ultraclean epi-layer growth, carefully controlled contact metalization and bonding, extremely long-lived diodes can result. For example, *Hartman* et al. [12.62] have reported lifetimes longer than two years, and decreases in output power of less than 15% at constant current after one year, in GaAlAs DH, proton-bombarded stripe geometry lasers. From a statistical analysis of accelerated aging tests performed on these devices at 70 °C, they project a mean-time-to-failure at 22 °C of 1.3×10^6 h (longer than 100 years). *Kressel* et al. [12.63] have also obtained similar results for $Al_{0.3}Ga_{0.7}As/Al_{0.08}Ga_{0.92}As$ DH lasers with Al_2O_3 facet coating. Accelerated aging tests on these devices have led to an extrapolated lifetime of about 10^6 h. In GaInAsP, *Oshiba* and *Kawai* [12.64] have reported V-grooved stripe geometry lasers with an output power of 220 mW at 1.3 μm and a median lifetime greater than 10^5 hours.

12.6 Laser Amplifiers

The double heterostructure, p-n junction diode which is used in semiconductor lasers can also function as an optical amplifier. Operating in this mode, the device increases the optical power of light entering one of its facets to produce a higher power output at the other facet. Intensity modulation variations are preserved and amplified, so the diode can be used as a repeater in optical communication systems. The basic mechanism of amplification is that of stimulated emission, just as in a laser. However, in the amplifier the diode is usually biased somewhat below lasing threshold so that oscillation does not occur.

There are three types of semiconductor laser amplifiers: the traveling-wave type, the Fabry-Perot type, and the injection-locking type. In the traveling-wave type of laser amplifier the facets are antireflection coated so that the light waves make only one pass through the inverted population region of the diode. In GaAs and its related ternary and quaternary compounds, the reflectivity of a cleaved endface is approximately 33%. However, it can be reduced to only a percent or two by the application of a properly designed multilayer dielectric antireflection coating. In the Fabry-Perot type of laser amplifier the end-facets are left uncoated. Thus incident light is amplified during successive passes between the end-facets before being emitted at a higher power level. The third type of laser amplifier, the injection-locked amplifier, is different from the other two types in that it is biased above threshold and actually operates as a laser. However, the injection of an intensity modulated light signal at one facet causes the light waves produced by the lasing action to "lock" with the input light waves so as to follow the same intensity and phase variations but at a higher optical power level. In all three types of laser amplifier the source of the added power is, of course, the electrical energy introduced by the bias power supply.

The single-pass gain of the Fabry-Perot laser amplifier and the traveling-wave amplifier are given by [12.65]

$$G_s = \exp[(\Gamma g(N) - \alpha)L], \tag{12.6.1}$$

where Γ is the confinement factor, α is the internal loss coefficient, and L is the length of the cavity. The gain function, $g(N)$, depends linearly on the carrier concentration N so that

$$g(N) = aN - b, \tag{12.6.2}$$

where a and b are constants.

The injection current is given by

$$I = NeV / \tau_s n_i \tag{12.6.3}$$

where e is the electronic charge, V is the active volume, τ_s is the carrier lifetime and n_i is the internal quantum efficiency.

For an ideal traveling wave laser amplifier the total gain is just the gain per single pass G_s as given in (12.6.1). For a Fabry-Perot laser amplifier the gain is enhanced by multiple reflections so that total gain is given by [12.66].

$$G = (1 - R_1)(1 - R_2)G_s / (1 - (R_1 R_2)^{1/2} G_s)^2,$$

where R_1 and R_2 are the endface reflectivities. It might appear that the Fabry-Perot laser amplifier would offer higher gain than the traveling-wave type. However, it must be remembered that the maximum value of G_s (corresponding to the point of lasing threshold) is much larger for a traveling-wave laser amplifier because of the greatly reduced endface reflectivities. The gain of both the traveling-wave and Fabry-Perot laser amplifiers is limited by the fact that they cannot be operated above threshold. However, because of the reduced reflectivity of the mirrors the threshold level is much higher for the traveling-wave device.

Noise also limits the useful gain of these amplifiers, and that of the injection-locked laser amplifier as well. The major source of noise is the background of photons spontaneously emitted in the p-n junction [12.66]. This light is amplified along with the desired signal, reducing the signal-to-noise ratio. In the Fabry-Perot laser amplifier the cavity resonances reinforce and amplify only those frequencies of light which correspond to an integral number of half wavelengths between the two endfaces. The emission spectrum of the amplifier therefore contains a central longitudinal mode, the amplified signal, along with side lobes spaced on the order of 1 nanometer apart corresponding to other longitudinal mode resonances within the cavity. When the amplified output is later detected by a photodiode the various modes combine to generate what is known as "beat" noise. This noise adds to the amplified spontaneous emission noise mentioned previously. The noise spectrum generally spreads over a wider range of wavelengths than does the amplified signal. Thus the signal-to-noise ratio of these devices can be improved by inserting a narrow-bandpass filter at the output.

The Fabry-Perot laser amplifier is suitable for only low gain applications because its maximum gain is limited by the onset of laser oscillation at a relatively low threshold. The frequency of the input signal must be precisely matched to that of the main Fabry-Perot mode for maximum gain and signal-to-noise ratio; thus its bandwidth is limited. The injection-locked laser amplifier also has a limited bandwidth because the frequency of the input lightwaves must be close to that of the laser oscillation frequency if locking is to occur. The traveling-wave laser amplifier has an advantage over the others in that it has high gain, wide spectral bandwidth and low sensitivity to changes in either temperature or drive current level. For example, *Zhang* et al. [12.67] have reported traveling-wave laser amplifiers in GaAlAs with a relatively flat gain of over 20 dB for wavelengths from 886.1 to 868.5 nm wavelength, and *Brenner* et al [12.68] have made InGaAsP/InP traveling wave amplifiers with high gain and integrated them with butt-coupled waveguides for OIC's.

Problems

12.1   Sketch the cross sectional view of a double-heterostructure GaAlAs laser diode. Choose the thickness and aluminum concentration in each layer required to produce light emission at $\lambda_0 = 8500$ Å, along with field confinement to the active layer for the lowest order TE and TM modes only.

12.2   Compare the operating characteristics of homojunction, SH and DH lasers, describing the advantages and disadvantages of each type.

12.3   Derive an expression for the threshold current density J_{th} in a heterojunction laser that has an active layer thickness d, light emitting layer thickness D, average loss coefficient α, and reflecting endfaces with different reflectivities R_1 and R_2. Use the symbols defined in the text for the other parameters, i.e., η_q, λ_0, $\Delta\nu$, etc.

12.4 For small active layer thickness d, the confinement factor d/D may be written as

$$d/D \cong \gamma d.$$

The compositional dependence of the refractive index of $Ga_{1-x}Al_xAs$ at $\lambda_0 = 0.90\,\mu m$ can be represented by

$$n = 3.590 - 0.710x + 0.091x^2.$$

a) Develop an expression for the confinement factor as a function of x, d and λ_0 for a three-layer, symmetric $GaAs-Ga_{(1-x)}Al_xAs$ DH laser with $n_a = 3.590$.

b) For a DH laser with $d = 0.1\,\mu m$, $x = 0.1$ and $\lambda_0 = 0.90\,\mu m$, what is the magnitude of the confinement factor?

c) Compare your answer to Part b, with the data given in Fig. 12.4.

12.5 Sketch the energy-band diagram for the DH laser of Fig. 12.5 when the laser is forward biased to above threshold and show the carrier confinement.

12.6 What are the advantages of a stripe-geometry laser as compared to a broad-area DH laser?

12.7 The external *differential quantum* efficiency of a GaAs laser diode emitting at a wavelength of $0.9\,\mu m$ is 30%. The applied voltage is 2.5 volts. Calculate the external *power* efficiency of the device.

12.8 A certain semiconductor p-n junction laser when biased as required has a gain coefficient $g = 70\,cm^{-1}$, a loss coefficient $\alpha = 5\,cm^{-1}$, and a length between Fabry-Perot endfaces of $L = 300\,\mu m$. It is desired to add antireflection coatings to the endfaces. What is the smallest endface reflectivity R that can be used without going below the lasing threshold?

(Assume that both endfaces have the same reflectivity, and that the confinement factor $d/D = 1$.)

12.9 A double-beterostructure laser diode is used as a Fabry-Perot-type optical amplifier. When biased at the operating point it has a gain coefficient of $40\,cm^{-1}$ a loss coefficient of $5\,cm^{-1}$ and a confinement factor of 0.9. The length between Fabry-Perot faces is $300\,\mu m$ and the reflectivities of the faces are both 33%.

a) What is the overall gain of this optical amplifier?

b) Name the three types of semiconductor-laser amplifiers. (The Fabry-Perot type is one of them.)

13. Distributed-Feedback Lasers

All of the lasers that have been described so far depend on optical feedback from a pair of reflecting surfaces, which form a Fabry-Perot etalon. In an optical integrated circuit, in which the laser diodes are monolithically integrated within the semiconductor wafer, it is usually very difficult to form such reflecting surfaces. They can be formed by etching or cleaving, as described in Chap. 12. However, the planar surface of the wafer is then disrupted, which leads to difficulties in fabricating electrical connections and heat sinks. An alternative approach, which utilizes distributed-feedback (DFB) from a Bragg-type diffraction grating, provides a number of advantages while still utilizing a planar surface geometry.

13.1 Theoretical Considerations

The use of a Bragg-type diffraction grating to deflect an optical beam in a modulator is described in Chap. 9. In that case, the grating structure is usually produced by inducing a periodic change in the index of refraction by means of either the electro-optic or acousto-optic effect. In DFB lasers, the grating is usually produced by corrugating the interface between two of the semiconductor layers that comprise the laser. This corrugation provides 180° reflection at certain specific wavelengths, depending on the grating spacing.

13.1.1 Wavelength Dependence of Bragg Reflections

The basis for selective reflection of certain wavelengths can be understood by referring to Fig. 13.1 which illustrates the reflection of an incident plane wave by a series of reflectors spaced at a distance d. In the original case of x-ray diffraction that was considered by Bragg [13.1], these reflectors were the atomic planes of a crystalline lattice. However, the same effect is observed in the case of reflection from a corrugated grating formed in the junction plane of a semiconductor laser as shown in Fig. 13.2. It is obvious from the diagram shown in Fig. 13.1 that, in order to maintain the phase coherence of the plane wavefront (normal to the rays shown) and thereby avoid destructive interference, the path lengths for reflections from successive reflectors must differ by

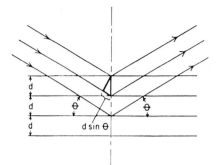

Fig. 13.1. Diagram of Bragg reflection from a periodic structure

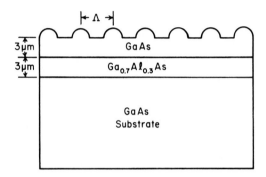

Fig. 13.2. Cross-sectional view of an optically pumped DFB laser

an integral number of full wavelengths. Thus, from geometrical considerations we obtain the Bragg relation [13.1], given by

$$2d \sin\theta = l\lambda, \quad l = 1, 2, 3, \ldots, \tag{13.1.1}$$

where θ is the angle formed by the incident ray and the reflector, and λ is the optical wavelength in the medium. To adapt the relation (13.1.1) to the case of 180° reflection by a grating in a DFB laser, it is only necessary to let d equal the grating spacing Λ, let λ equal λ_0/n_g, where n_g is the effective index in the waveguide for the mode under consideration, and let θ equal 90°. Under these assumptions (13.1.1) becomes

$$2\Lambda = l(\lambda_0/n_g), \quad l = 1, 2, 3, \ldots. \tag{13.1.2}$$

The vacuum wavelength of light that will be reflected through 180° by such a grating is therefore

$$\lambda_0 = \frac{2\Lambda n_g}{l}, \quad l = 1, 2, 3, \ldots. \tag{13.1.3}$$

Although the grating is capable of reflecting many different longitudinal modes, corresponding to the various values of l, usually only one mode will lie within the gain bandwidth of the laser. In fact, because of the difficulty of fabricating a first order ($l = 1$) grating, usually a third-order grating is used, as is discussed in detail in Sect. 13.2.

13.1.2 Coupling Efficiency

The fraction of optical power that is reflected by a grating such as that shown in Fig. 13.2 depends on many factors, including the thickness of the waveguiding layer, the depth of the grating teeth and the length of the grating region. The determination of the fraction of reflected optical power is mathematically quite complex. It is best approached by using the coupled-mode theory, as has been done by *Kogelnik* and *Shank* [13.2], and by *Yariv* [13.3], although the Bloch wave formalism has also been used [13.4]. In the coupled-mode analysis, it is assumed that the grating is only weakly coupled to the optical modes, i.e., the basic mode shape is only slightly perturbed. Using this approach, *Yariv* [13.3] has shown that for a rectangular grating cross section of depth a, as shown in Fig. 13.3, the coupling can be characterized by a coupling coefficient κ, given by

$$\kappa = \frac{2\pi^2}{3l\lambda_0} \frac{(n_2^2 - n_1^2)}{n_2} \left(\frac{a}{t_g}\right)^3 \left[1 + \frac{3(\lambda_0/a)}{2\pi(n_2^2 - n_1^2)^{1/2}} + \frac{3(\lambda_0/a)^2}{4\pi^2(n_2^2 - n_1^2)}\right], \qquad (13.1.4)$$

where t_g is the thickness of the waveguiding layer with index n_2, and l is the order of the harmonic responsible for the coupling, given by

$$l \cong \frac{\beta_m \Lambda}{\pi}, \qquad (13.1.5)$$

where β_m is the phase constant of the particular mode under consideration. The expressions (13.1.4) and (13.1.5) are based on the assumption that β_m is approximately equal to $l\pi/\Lambda$, i.e. the Bragg condition (13.1.3) is satisfied, so that Bragg reflection occurs between the forward and backward traveling modes of order m.

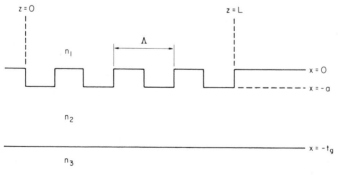

Fig. 13.3. An optical waveguide with a rectangular grating for distributed feedback [13.3]

For the first order ($l = 1$) case, the reflection is essentially limited to coupling between only the forward and backward waves traveling in the $\pm z$ direction. Then the amplitude of the incident wave A_m^+ and that of the reflected wave A_m^- are functions of the distance z that the incident wave travels under the grating, given by the expressions [13.3]

$$A_m^+(z) = A_m^+(O) \frac{\cosh[\kappa(z-L)]}{\cosh(\kappa L)},$$

(13.1.6)

and

$$A_m^-(z) = A_m^+(O) \frac{\kappa}{|\kappa|} \frac{\sinh[\kappa(z-L)]}{\cosh(\kappa L)},$$

(13.1.7)

where L is the length of the grating. The incident and reflected optical powers are, of course, given by $|A_m^+(z)|^2$ and $|A_m^-(z)|^2$, respectively. For sufficiently large arguments of the hyperbolic functions in (13.1.6) and (13.1.7), the incident optical power decreases exponentially with z, as power is reflected into the backward traveling wave. The effective reflection coefficient R_{eff} and corresponding transmissions T_{eff} at the Bragg wavelength are given by

$$R_{eff} = \left| \frac{A_m^-(O)}{A_m^+(O)} \right|^2,$$

(13.1.8)

and

$$T_{eff} = \left| \frac{A_m^+(L)}{A_m^+(O)} \right|^2.$$

(13.1.9)

The relations (13.1.6-9), strictly speaking, are applicable only in the case of first-order Bragg reflection. For higher-order gratings, with periodicity Λ given by

$$\Lambda = \frac{l\lambda_0}{2n_g}, \quad l > 1,$$

(13.1.10)

some optical energy is coupled into waves traveling in directions out of the waveguide plane, as shown in Fig. 13.4. These waves represent a power loss from the lasing mode, and hence produce reductions of both R_{eff} and T_{eff}.

Scifres et al. [13.5] have analyzed the reflections from higher-order gratings as shown in Fig. 13.5. Consider a wave traveling to the right in the corrugated waveguide. In order for the rays scattered from successive teeth to all be in phase, the path lengths must all be integral multiples of the wavelength. Hence a conditions is imposed which requires that

$$b + \Lambda = \frac{l'\lambda_0}{n_g}, \quad l' = 0, 1, 2, \dots,$$

(13.1.11)

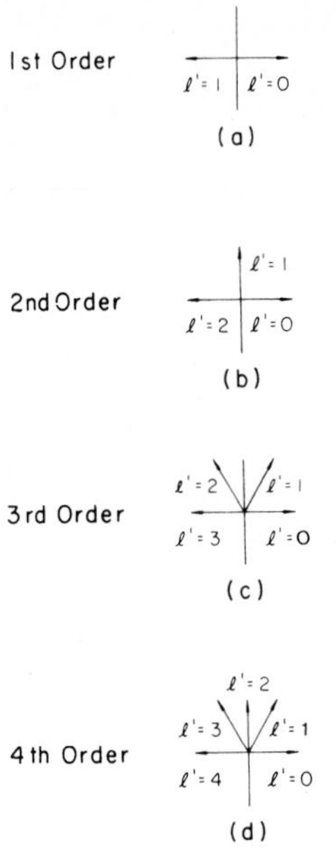

Fig. 13.4. Reflected wave directions for higher-order Bragg diffraction gratings

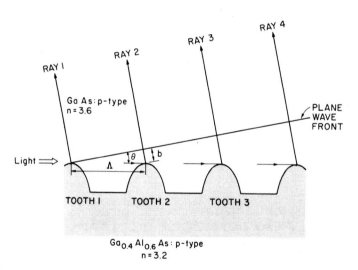

1st Order

$\ell'=1$ | $\ell'=0$

(a)

2nd Order

$\ell'=1$

$\ell'=2$ | $\ell'=0$

(b)

3rd Order

$\ell'=2$ | $\ell'=1$

$\ell'=3$ | $\ell'=0$

(c)

4th Order

$\ell'=2$

$\ell'=3$ | $\ell'=1$

$\ell'=4$ | $\ell'=0$

(d)

Fig. 13.5. Rays scattered from a waveguide with a higher-order Bragg grating [13.5]

where b is the distance shown in Fig. 13.5.

From geometrical considerations, it can be seen that

$$b = \Lambda \sin\theta, \tag{13.1.12}$$

where θ is the angle that the scattered wavefront makes with the plane of the waveguide. Thus, from (13.1.11) and (13.1.12), θ is given by

$$\sin\theta = \frac{l'\lambda_0}{n_g \Lambda} - 1, \quad l' = 0, 1, 2, \ldots. \tag{13.1.13}$$

Consider the case of second-order Bragg diffraction, for which Λ equals λ_0/n_g. In this case (13.1.13) becomes

$$\sin\theta = l' - 1, \quad l' = 0, 1, 2. \tag{13.1.14}$$

The solution for $l' = 0$ corresponds to light scattered in the forward direction, while that for $l' = 2$ gives backward scattering useful for DFB. However, the $l' = 1$ solution yields $\theta = 0$, corresponding to light emitted normal to the waveguide plane. Similar solutions can be obtained for higher-order Bragg reflection gratings. The coupling coefficients have been given by *Streifer* et al. [13.6] for a number of typical cases of scattering from higher-order Bragg gratings.

Generally, the strong coupling in the transverse direction produced by second-order Bragg gratings is undesirable in a DFB laser. Hence, a first-order grating is required to yield optimum performance. If fabrication tolerances prevent the use of a first-order grating, a third-order grating should be used, since coupling to the out-of-plane beams is much weaker than in the second-order case. One exception to this rule is a special, low-divergence-angle laser, demonstrated first by *Scifres* et al. [13.5]. They intentionally used a fourth-order grating to produce a strong transverse coupling in a GaAlAs/GaAs SH laser. The Fabry-Perot end faces of the laser were coated to make them 98% reflective, and the laser output was taken in the transverse direction. Because of the relatively large size of the light emitting surface, diffraction spreading was minimized, and a beam with angular divergence of only 0.35° was produced.

13.1.3 Lasing with Distributed Feedback

It was shown in the previous section that a corrugated grating structure can provide 180° Bragg reflection of an optical wave traveling in close proximity to the grating. If the medium in which the wave is traveling has optical gain, such as that of the inverted population region of a laser diode, this distributed feedback can result in lasing. In a medium with an exponential gain constant γ, the amplitudes of the incident and reflected waves are given by the expressions [13.7]

$$E_i(z) = E_0 \frac{\{(\gamma - i\Delta\beta)\sinh[S(L-z)] - S\cosh[S(L-z)]\}e^{-i\beta_0 z}}{(\gamma - i\Delta\beta)\sinh(SL) - S\cosh(SL)}, \tag{13.1.15}$$

and

$$E_r(z) = E_0 \frac{\kappa e^{i\beta_0 z} \sinh[S(L-z)]}{(\gamma - i\Delta\beta)\sinh(SL) - S\cosh(SL)}, \tag{13.1.16}$$

where

$$S^2 \equiv |\kappa^2| + (\gamma - i\Delta\beta)^2. \tag{13.1.17}$$

The parameter E_0 is the amplitude of a single mode (the one for which net gain is greatest) incident on the grating of length L at $z = 0$, while $\Delta\beta$ is given by

$$\Delta\beta = \beta - \beta_0, \tag{13.1.18}$$

where β_0 is the phase constant at the Bragg wavelength.

The oscillation condition for the DFB laser [13.2] corresponds to the case for which both the transmittance $E_i(L)/E_i(0)$ and the reflectance $E_r(0)/E_i(0)$ become infinite. From (13.1.15-17), it can be shown that this condition is satisfied when

$$(\gamma - i\Delta\beta)\sinh(SL) = S\cosh(SL). \tag{13.1.19}$$

In general, (13.1.19) can be solved to determine the threshold values of $\Delta\beta$ and γ only by numerical solution [13.2]. However, in the special case of high gain ($\gamma \gg |\kappa|, \Delta\beta$) it can be shown [Ref. 13.7, p. 381] that the oscillating mode frequencies are given by

$$(\Delta\beta_m)L \cong -(m + \tfrac{1}{2})\pi. \tag{13.1.20}$$

Since it is known from basic definitions that

$$\Delta\beta \equiv \beta - \beta_0 \cong \frac{(\omega - \omega_0)n_g}{c}, \tag{13.1.21}$$

(13.1.20) can be written in the form

$$\omega_m = \omega_0 - (m + \tfrac{1}{2})\frac{\pi c}{n_g L}, \tag{13.1.22}$$

where ω_0 is the Bragg frequency, and $m = 0, 1, 2, \ldots$. It is interesting to note in (13.1.22) that no oscillation can occur at exactly the Bragg frequency. The mode frequency spacing is given by

$$\omega_{m-1} - \omega_m \cong \frac{\pi c}{n_g L}. \tag{13.1.23}$$

13.2 Fabrication Techniques

The grating structure of a DFB laser is usually formed by masking and then etching the waveguide surface, either chemically or by ion-beam sputtering. Sometimes the two methods are combined in a single process called chemically assisted ion beam etching (CAIBE) [13.8]. The process is generally the same as that described in Sect. 6.4.2 for grating-coupler fabrication. However, some additional factors must be considered in order to produce an efficient laser.

13.2.1 Effects of Lattice Damage

In an integrated DFB laser, such as the one shown in Fig. 13.6, it is necessary to grow one or more epitaxial layers on top of the waveguide layer after the grating has been formed. Lattice damage created during the grating fabrication process will, of course, be detrimental to the optical quality of the subsequently grown epitaxial layers, because point defects and dislocations tend to propagate from the grating region into the epi layer. These defects act as centers for optical absorption and nonradiative recombination. Thus, they act to reduce quantum efficiency and increase threshold current density. The deleterious effects of defects generated during grating fabrication present somewhat of a dilemma to the integrated circuit designer when choosing the fabrication technique, because ion-beam sputter-etching produces the most uniform and controllable grating profile, but chemical etching yields gratings with much less lattice damage.

13.2.2 Grating Location

The approach that is most often used to avoid the effects of lattice damage is to physically separate the grating from the active layer of the laser. For example, *Scifres* et al. [13.9] made the first p-n junction injection DFB laser, by using a diffusion process that separated the p-n junction from the grating as shown in Fig. 13.7. The n-GaAs substrate was first corrugated by ion-beam sputter-etching to produce a third order grating with 3500 Å periodicity. Then the p-GaAlAs layer was grown over the substrate. Zinc was used as the p-type dopant in the GaAlAs because of its anomalously fast diffusion in GaAs. Hence, during the growth of the GaAlAs layer, the zinc diffused into the substrate to produce a p-n junction at about 1 μm below the GaAs-GaAlAs (corrugated) interface. The resulting single heterostructure DFB lasers exhib-

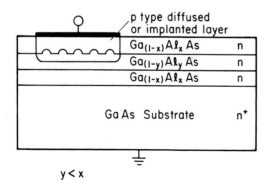

Fig. 13.6. Integrated DFB laser

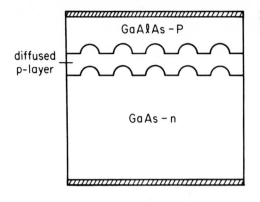

Fig. 13.7. Diffused junction DFB laser [13.9]

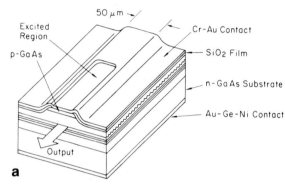

Fig. 13.8a,b. Stripe geometry DFB laser with separate optical and carrier confinement [13.10]. **a** Device structure; **b** close-up section of epilayers

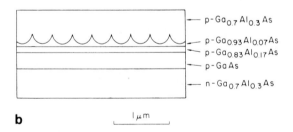

ited threshold densities that were typically equal to those of similar SH lasers with cleaved end faces. Thus, one can conclude that the effects of defects generated during grating fabrication were not severe.

Another approach to separating the grating from the p-n junction was demonstrated by *Nakamura* et al. [13.10], as shown in Fig. 13.8. They used a separate confinement heterojunction (SCH) structure in conjunction with a third-order Bragg grating, in a 50 μm wide stripe geometry laser. The length of the active region was 700 μm. An unpumped waveguide, of 2 – 3 mm length, was continuous with the active region in order to isolate the rear cleaved surface and thereby insure that lasing occurred solely because of distributed feedback rather than mirror reflection. In the structure shown in Fig. 13.8, injected electrons are confined to the 0.2 μm thick active layer by the 0.1 μm thick layer of p-$Ga_{0.83}Al_{0.17}As$, while photons of the optical mode spread to the interface with the p-$Ga_{0.93}Al_{0.07}As$ layer. Thus, the optical mode is perturbed by the grating structure, but the active region (in which recombination occurs) is far enough away from the grating so that nonradiative recombination due to lattice damage is minimized. Devices of this type were operated cw at room temperature, exhibiting threshold current densities of about $3.5 \times 10^3 A/cm^2$. For a grating spacing of 3684 Å, a single longitudinal mode was observed with a wavelength of 8464 Å.

The concept of a separate confinement heterostructure (SCH) laser [13.10] has been further developed and extended to InGaAsP lasers emitting at wavelengths of 1.3 μm and 1.55 μm . For example *Tsuji* et al. [13.11] have produced the SCH buried-heterostructure laser diode shown in Fig. 13.9, which emits at 1.55 μm. This device features a first-order grating of 0.234 μm spacing fabricated by holographic photolithography. A buried-heterojunction stripe geometry was used for current confinement and for transverse mode control. A transparent InP window structure served to isolate one of the end facets to suppress Fabry-Perot oscillation. It should be noted that this type of window structure also suppresses a threshold degeneracy in which two longitudinal modes closely spaced in frequency about the ω_m predicted by (13.1.22) both tend to oscillate [13.12]. The approximate relation (13.1.22) does not reveal this degeneracy, which can be shown to exist only by an exact solution. Lasers fabricated with the structure shown in Fig. 13.9 have been shown to have threshold current as low as 11 mA and to exhibit stable single longitudinal mode CW operation at ambient temperatures up to 106 °C [13.11]. Lifetests on these devices have indicated only minimal degradation over thousands of hours of operation. The average output degradation rate was 0.56% per thousand hours, which is comparable with that of conventional Fabry-Perot buried-heterostructure lasers [13.13].

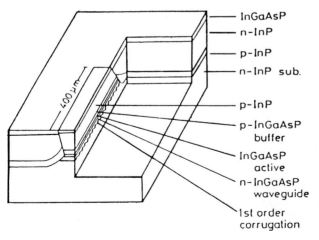

Fig. 13.9. An SCH-DFB-BH laser [13.11]

An alternative approach to separate the grating from the active light emitting region of the laser is to use the laterally-coupled geometry shown in Fig. 13. 10 [13.14]. In fabricating this device photolithograhy was employed to form 2 μm wide photoresist stripes on an MBE grown InGaAs-GaAs-GaAlAs graded-index-separate-confinement-heterostructure (GRINCH) wafer. Chemically assisted ion beam etching (CAIBE) was utilized to etch the stripe ridge to within 0.1 μm of the GRINCH. Then PMMA was applied to the wafer and first-order gratings were exposed by using electron-beam lithography. Following development of the resist, CAIBE was applied to transfer the gratings approximately 700 Å deep into the cap layer and upper cladding beside the ridge.

In the laterally-coupled DFB laser the two-dimensional (transverse and lateral) overlap of the evanescent field to the gratings provides the required

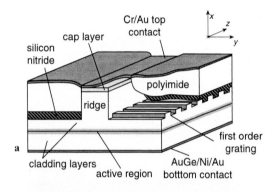

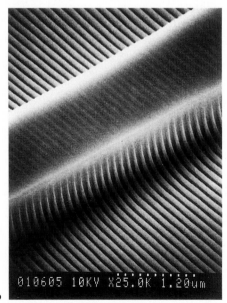

Fig. 13.10. a Diagram of a laterally-coupled distributed feedback (LC-DFB) ridge laser diode [13.14]. **b** Scanning electron micrograph of the first order grating pattern over the 2 μm wide ridge defined in PMMA by electron beam lithography [13.14]

feedback for lasing. The coupling coefficient κ is critically dependent on the ridge etch depth. Too shallow of a ridge depth reduces the transverse overlap of the field with the grating. Etching the ridge too deeply results in more index guiding of the field and reduces the lateral fill factor of the grating.

Devices of the type shown in Fig. 13.10, with cavity length of 250 μm and 1400 Å pitched gratings emitted at a wavelength of 9217 Å. Threshold current of 11 mA, differential slope efficiency of 0.46 mA/mW and single-mode output power of 15 mW were achieved.

13.2.3 DBR Lasers

The active region can also be isolated from the grating region by using the distributed Bragg reflection (DBR) structure [13.15] shown in Fig. 13.11. In such a device, two Bragg gratings are employed, which are located at both ends of the laser and outside of the electrically-pumped active region. In addition to avoiding nonradiative recombination due to lattice damage, placement of two grating *mirrors* outside of the active region permits them to be individually tailored to produce single-ended output from the laser. In order to achieve efficient, single-longitudinal-mode operation, one distributed reflector must have narrow bandwidth, high reflectivity at the lasing wavelength, while the other must have relatively low reflectivity for optimal output coupling.

Stoll [13.16] has theoretically analyzed DBR structures and shown that the transmittivity and reflectivity are given by the expressions

$$T = \frac{\gamma \exp[-i(\beta - \Delta\beta)L]}{(\alpha_p + i\Delta\beta)\sinh(\gamma L) + \gamma \cosh(\gamma L)}, \tag{13.2.1}$$

and

$$R = \frac{-i\kappa \sinh(\gamma L)}{(\alpha_p + i\Delta\beta)\sinh(\gamma L) + \gamma \cosh(\gamma L)}, \tag{13.2.2}$$

where $\gamma^2 \equiv \kappa^2 + (\alpha_p + i\Delta\beta)^2$, and where L is the length of the distributed reflector, and α_p is the distributed loss coefficient in the waveguide passive (unpumped) regions. The coupling coefficient κ and the phase constant deviation from the Bragg condition $\Delta\beta$ are as previously defined. *Stoll* [13.16] has also shown that the longitudinal mode spacing between the mth and $(m \pm 1)$ modes of a DBR laser is given to good approximation by the expression

$$\Delta \equiv |\beta_m - \beta_{m\pm1}| = \frac{\pi}{L_{eff}}, \tag{13.2.3}$$

where the effective cavity length L_{eff} is given by

$$L_{eff} = L_a \left(1 + \frac{L_2}{2L_a(\alpha_p L_1 + 1)} + \frac{1}{2L_a(\alpha_p + \sqrt{\kappa^2 + \alpha_p^2})}\right). \tag{13.2.4}$$

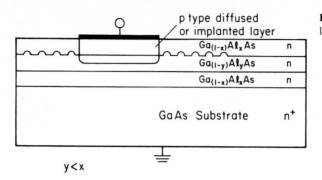

Fig. 13.11. Integrated DBR laser

The parameters L_1 and L_2 are the lengths of the two distributed reflectors, and L_a is the length of the active region. Single-ended optical power emission is assumed to occur from the low-reflectivity DBR of length L_2. Typically, L_1 is on the order of one or two mm, while L_2 is several hundred micrometers. For example, for the case of $\alpha_p = 1\,cm^{-1}$, $L_a = 275\,\mu m$, $k = 20\,cm^{-1}$, $L_1 = 1.47\,mm$, $L_2 = 240\,\mu m$, (13.2.4) yields $L_{eff} = 501\,\mu m$, thus leading to a mode spacing of $\Delta = 63\,cm^{-1}$.

Because they can be conveniently optimized for single-ended output at a preselected wavelength (determined by the grating spacing), DBR lasers are well suited to OIC frequency multiplexing applications, in which a number of sources operating at well-defined wavelengths are desired. For example, *Aiki* et al. [13.17] have monolithically integrated six GaAlAs DBR lasers, coupled into a single waveguide, on a LPE grown wafer. The grating spacing of each laser was chosen to give a wavelength separation of about 20 Å. The lasers exhibited threshold current densities (at room temperature) in the range from 3 to $6 \times 10^3\,A/cm^2$, and had spectral width of 0.3 Å.

DBR lasers have also been fabricated in InGaAsP/InP to produce emission at 1.55 μm wavelength. For example, *Kaiser* et al. [13.18] have integrated a tunable DBR laser, butt-coupled to a directional coupler on an InP substrate by using selective area epitaxy and strip-loaded waveguides. Average coupling efficiencies of 52% and maximum values of 63% were achieved at power levels greater than 1 mW. *Tsang* et al. [13.19] have reported InGaAs/InGaAsP multiquantum-well DBR lasers fabricated by chemical beam epitaxy with CW threshold currents of 10 to 15 mA, slope efficiency as high as 0.35 mW/mA and side mode suppression ratio of 58.5 dB.

13.3 Performance Characteristics

DFB and DBR lasers have unique performance characteristics that give them distinct advantages over conventional reflective-end-face lasers in many discrete-device applications, as well as in OIC's.

13.3.1 Wavelength Selectability

In cleaved-end-face lasers, the light emission occurs at a wavelength deter-mined jointly by the gain curve and the modal characteristics of the laser. Lasing occurs for the mode (or modes) that have the highest gain. When the laser is pumped well above the threshold level, usually a number of longitudi-nal modes lase simultaneously. It is very difficult, if not impossible, to obtain single mode oscillation.

In the case of DFB or DBR lasers, the emission wavelength is, of course, affected by the gain curve of the laser, but it is primarily determined by the grating spacing Λ, as given by (13.1.3). The spacing between the lth and the $l \pm 1$ modes is generally so large compared to the linewidth of the laser gain curve that only one mode has sufficient gain to lase. Thus single-longitudinal-mode operation is obtained relatively easily in distributed feedback lasers. This gives them a distinct advantage over reflective-end-face lasers in many applications. The theoretically expected controllability of emitted wavelength was demonstrated by *Nakamura* et al. [13.20, 21] in the first operational DFB lasers, which were optically pumped devices. The set of laser samples included both SH devices, with a $Ga_{0.7}Al_{0.3}As$ confining layer, and reduced-carrier con-centration waveguide lasers, with a guiding layer dopant concentration of $n = 6 \times 10^{16} cm^{-3}$ on a substrate with $n = 1 \times 10^{18} cm^{-3}$. Both types of devices were cooled to 77 K, and were optically pumped by means of a Rhodamine B dye laser emitting at 6300 Å wavelength. Distributed feedback was provided by a third-order Bragg grating formed at the surface of the semiconductor lasers by ion milling through a photoresist mask to a depth of 500 Å. The dependence of lasing wavelength on grating spacing (in different samples) is shown in Fig. 13.12. Good agreement with the values predicted by (13.1.3) was obtained for a waveguide effective index of refraction $n_g = 3.59$. It was possible to control-lably select laser emission wavelengths over a range of 45 Å by changing the grating period in the range from 3450 to 3476 Å. The controllability of emis-sion wavelength makes DFB lasers particularly useful in wavelength multi-

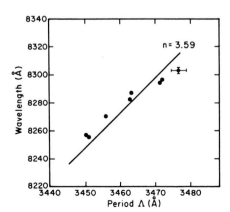

Fig. 13.12. Dependence of laser emission wave-length on grating spacing in a DFB laser [13.21]

plexing applications such as the six laser OIC transmitter for fiber optics produced by *Aiki* [13.17], as described in Chap. 17, or the four laser chip of *Yusuka* [13.22].

As mentioned previously, there is a fundamental oscillation degeneracy in which two longitudinal modes of frequencies close to that predicted by (13.1.22) both have an equal probability of lasing. This degeneracy can be lifted either by providing a window structure in front of one facet as in Figs. 13.8 and 13.9, or by forming a grating with a quarter wavelength phase shift between two equal uniform corrugation sections [13.23, 24].

Fine tuning of the grating-selected wavelength can be accomplished by angling the active stripe at an oblique angle with respect to the axis of the grating bars [13.25]. A shift of approximately $170\,\text{Å}$ was produced by angling the stripe at 7.7 °, while an angle of 3.7° resulted in a shift of a about 60 Å, in lasers operating in the 1.52 to $1.54\,\mu\text{m}$ range.

13.3.2 Optical Emission Linewidth

In addition to providing a means of accurately selecting the peak emission wavelength, grating feedback also results in a narrower linewidth of the optical emission. The spectral width of the emission line is established by a convolution of the laser gain curve with the mode-selective characteristics of the laser cavity. Since the grating is much more wavelength selective than a cleave or polished end-face, the resulting emission linewidth of a DFB or DBR laser is significantly less than that of reflective-end-face laser. The single-mode linewidth of a conventional cleaved-end-face laser is typically about 1 or 2 Å. as described in Chap. 11 (about 50 GHz). However, values reported for modern DFB and DBR lasers with sophisticated grating structures range from about 50 to 100 kHz.

For example, *Okai* et al. [13.26] have demonstrated a linewidth of less than 98 kHz in a three-section corrugation-pitch-modulated multiple-quantum-well DFB laser over a tuning range of 1.3 nm, at a power level of 10 mW. *Mawatari* et al. have reported a linewidth of 80 kHz at a power level of 40 mW in a modulation-doped, strained-multiple-quantum-well DFB laser, and *Okai* et al. have achieved a linewidth of only 56 kHz at a power level of 26 mW in a corrugation-pitch-modulated MQW DFB laser.

The narrower linewidth obtainable with distributed feedback lasers is particularly important in optical communications applications, because the modulation bandwidth is ultimately limited by the linewidth of the laser source.

13.3.3 Stability

The stability of emitted wavelength and threshold current density during operation under varying ambient conditions are of key importance in many applications. Variation of emitted wavelength with junction temperature is a serious problem in conventional cleaved-end-face lasers, with drifts of 3 or 4 Å per degree centigrade typically being observed [13.10]. This drift is extremely detrimental to the operation of optical communication systems in which narrow bandpass optical filters are used to reduce background light intensity reaching the receiver, and thereby improve signal-to-noise ratio. A drift of as little as 10 Å might put the wavelength of the optical beam outside the bandpass of the filter. Not only is the long term drift of wavelength with ambient temperature important, but also transient drift or *chirp* caused by junction heating during pulsed operation or modulation must be considered.

Distributed feedback lasers offer improved wavelength stability as compared to cleaved-end-face lasers, because the grating tends to lock the laser to a given wavelengh. For example, *Nakamura* et al. [13.10] have measured the temperature dependence of emitted wavelength for both cleaved-end-face and DFB lasers fabricated from the same wafer. The data are shown in Fig. 13.13. The length of the cleaved laser was 570 µm, while that of the DFB laser was 730 µm. The grating spacing in the DFB laser was 3814 Å. Both devices were fabricated from the same DH GaAlAs wafer. Note that the cleaved laser exhibited a wavelength drift of 3.7 Å/°C, while the DFB laser drifted only 0.8 Å/°C, over temperature spans of about 100°. However, the DFB laser jumped from the $m = 0$ mode to the $m = 1$ mode at about 280 K. From the

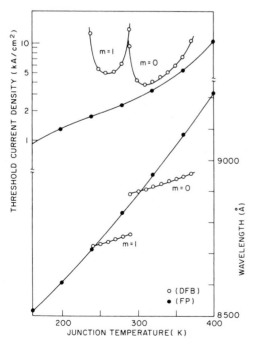

Fig. 13.13. Threshold current density and lasing wavelength as functions of junction temperature [13.10]

threshold current density data shown in Fig. 13.13, it can be seen that the DFB laser required roughly the same current as the cleaved laser when oscillating in the $m = 0$ mode. However, its J_{th} was about a factor of three larger for the $m = 1$ mode. Also the J_{th} of the DFB laser increased sharply at the temperatures where a transition between modes occurred. These rapid increases resulted from mismatch between the laser gain curve and the modal resonances (established by the grating period) at those temperatures. *Martin* et al. [13.14] have made similar measurements of wavelength drift for the laterally-coupled DFB laser of Fig. 13.10, in the temperature range from 5 to 50 °C . A wavelength shift of 0.63 Å/°C was observed for the LC-DFB laser, while a shift of 2.8 Å/°C was measured for a standard Fabry-perot cleaved-endface ridge laser of the same heterostructure and dimensions as the DFB device. Even high-power DFB lasers exhibit low thermal drift when they are properly designed. For example, *Takigawa* [13.29] has reported a 50 mW 780 nm GaAlAs DFB laser which operates in a stable single longitudinal mode with a lasing wavelength temperature coefficient of 0.65 Å/C. In that case a buried twin-ridge heterostructure was used to produce a stable fundamental spatial mode.

The improved temperature stability of distributed feedback lasers results from the fact that the shift of emitted wavelength in the cleaved laser follows the temperature dependence of the energy bandgap, while the shift in wavelength of the DFB laser follows only the temperature dependence of the index of refraction. This improved stability of the DFB laser makes it very useful in applications where wavelength filtering or wavelength division multiplexing are employed. Hence, even in many cases where a discrete laser is being used rather than an OIC, the DFB or DBR laser is to be preferred over a Fabry-Perot end-face laser. DFB and DBR lasers have now become the standards for long-distance transmission in the telecommunications industry. Distance-bandwidth products greater than 2000 GHz·km have been achieved [13.30].

13.3.4 Threshold Current Density and Output Power

The fact that DFB lasers are capable of yielding low threshold current density and relatively high output power is generally recognized today. Some of the data presented in this chapter have illustrated that point. Typical performance characteristics are: CW threshold current density of 10 to 15 mA, output power of 25 to 40 mW, with sidemode suppression ratio 30 to 50 dB [13.14, 19, 28].

Problems

13.1   What is the grating spacing required for a DFB laser to operate at a wavelength $\lambda_0 = 8950\,\text{Å}$ in GaAs if a first-order grating is desired?

13.2   Repeat Problem 13.1 for the case of a laser formed in an active region of $Ga_{0.8}Al_{0.2}As$.

13.3   If the grating of Problem 13.1 is fabricated by using the technique of Fig. 6.10, determine two combinations of prism material and angle (α) that could be used with a He-Cd laser ($\lambda_0 = 3250\,\text{Å}$).

13.4   a) If a surface grating is formed on a waveguide, what is the grating spacing Λ required for fourth-order Bragg diffraction, in terms of λ and n_g?

b) Sketch the reflected wave directions, and indicate the angles between them.

c) In general, can a scattered wave perpendicular to the direction of the waveguide be observed for odd orders, such as the first and third order, of Bragg diffraction?

13.5   A buried heterostructure, multiple contact, distributed feedback (DFB) laser has a grating periodicity of $1200\,\text{Å}$ and a waveguide index of refraction equal to 3.8. The interaction length of the grating is $500\,\mu m$. Assume that only a single mode oscillates and that feedback comes from a first-order Bragg reflection. In this problem you may neglect the slight shift in wavelength which occurs in a DFB laser and assume that emission is at the Bragg wavelength.

a) What is the emitted wavelength of the laser?

b) If current is injected into the modulator section of the diode so as to change the waveguide index by one part in 10^7, what will be the resulting change in laser emitted wavelength (frequency)

14. Direct Modulation of Semiconductor Lasers

In Chaps. 8 and 9 techniques were described for modulating the light of a semiconductor laser by using external electro-optic or acousto-optic modulators. However, it is also possible to internally modulate the output of a semiconductor laser by controlling either the current flow through the device or some internal cavity parameter. Such direct modulation of the laster output has the advantages of simplicity and potential for high frequency operation. The topic of direct modulation of injection lasers is considered in this chapter; this follows the discussions of semiconductor laser fundamental principles and operating characteristics in Chaps. 10–12, so that the reader will be better prepared to appreciate the subtleties of the methods involved.

14.1 Basic Principles of Direct Modulation

The light output of a semiconductor laser can be directly modulated, i.e., made to vary in response to changes within the laser cavity, so as to produce amplitude modulation (AM), optical frequency modulation (FM), or pulse modulation (PM). Most often, the laser output is either amplitude or pulse modulated by controlling the current flow through the device. However, other parameters, such as the dielectric constant or the absorptivity of the laser cavity material can be varied to produce FM, as well as AM and PM of the output. The basic principles of these modulation techniques, most of which were developed during the late 1960's and 1970's, are discussed in this section. More sophisticated improvements, which have led to reliable direct modulation of laser diodes at microwave frequenies in later years, are covered in Sect. 14.2.

14.1.1 Amplitude Modulation

The basic arrangement for amplitude modulation of a laser diode by control of current flow is illustrated in Fig. 14.1. The laser diode must be dc biased to a point above the lasing threshold level in order to avoid the sharp break in the output curve at threshold. Above threshold, the dependence of output power on current is very linear for a well made diode, as discussed in Chap. 12. The ac modulating signal must be isolated from the dc bias supply, and also the dc bias must be prevented from reaching the modulating signal source. At low

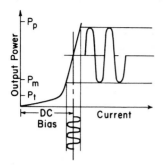

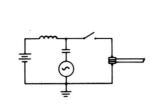

Fig. 14.1. Basic bias circuit and output characteristics for directly AM modulated laser diode

frequencies, this can be accomplished by means of a simple inductor and capacitor, as shown in Fig. 14.1. Above about 50 MHz, more sophisticated high-pass and low-pass filter circuits must be used.

For the direct current modulation method shown in Fig. 14.1, the modulation depth is given by

$$\eta = \frac{P_p - P_m}{P_p},$$ (14.1.1)

where P_p is the peak optical power and P_m is the minimum. The maximum modulation depth over which linear response can be obtained is given by

$$\eta_{max} = \frac{P_p - P_t}{P_p},$$ (14.1.2)

where P_t is the optical output at threshold. Because the power at threshold P_t is usually only about five or ten percent of P_p, the maximum modulation depth can, in principle, be greater than 90%. The expression (14.1.2) is based on the tacit assumption that the dc bias is chosen so that the operating point for zero input signal is at the center of the linear portion of the output curve, i.e., at a power of $(P_p - P_t)/2$.

The amplitude modulation process can be described in terms of a pair of nonlinear rate equations. As given by *Lasher* [14.1], these relations state that

$$\frac{dN_e}{dt} = \frac{I}{eV} - \frac{N_e}{\tau_{sp}} - GN_{ph},$$ (14.1.3)

and

$$\frac{dN_{ph}}{dt} = \left(G - \frac{1}{\tau_{ph}}\right)N_{ph},$$ (14.1.4)

where N_e is the electron inversion number, N_{ph} is the number of photons, I is the current, V is the volume of the active region, τ_{sp} is the spontaneous electron lifetime. τ_{ph} is the photon lifetime, and G is the stimulated emission rate. The relations (14.1.3) and (14.1.4) assume a single lasing mode, into which there is negligible spontaneous emission. For small-signal analysis, a small time vary-ing signal $I(t)$ is imposed on the dc bias current I_{dc}. In the conventional small signal approximation, the resulting small variations in the photon number $n_{ph}(t)$ and the electron inversion number $n_e(t)$ about their average values $\bar{N}_{ph}$ and $\bar{N}_e$ are given by [14.2]

$$\frac{d^2}{dt^2}\left\{\begin{matrix} n_e \\ n_{ph} \end{matrix}\right\} + \gamma\frac{d}{dt}\left\{\begin{matrix} n_e \\ n_{ph} \end{matrix}\right\} + \omega_0^2\left\{\begin{matrix} n_e \\ n_{ph} \end{matrix}\right\} = \left\{\begin{matrix} \dfrac{1}{eV}\dfrac{dI(t)}{dt} \\ \dfrac{g\bar{N}_{ph}I(t)}{eV} \end{matrix}\right\}, \tag{14.1.5}$$

where

$$\omega_0^2 = \frac{1}{\tau_{sp}}\left(gN_0 + \frac{1}{\tau_{ph}}\right)\left(\frac{I_{dc}}{I_{th}} - 1\right), \tag{14.1.6}$$

and

$$\gamma = \frac{1}{\tau_{sp}} + \tau_{ph}\omega_0^2. \tag{14.1.7}$$

In (14.1.5) and (14.1.6), the stimulated emission rate is assumed to be given by

$$G = g(N_e - N_0), \tag{14.1.8}$$

where N_0 is the inversion required to overcome bulk losses, and g is a constant of proportionality. If sinusoidal modulation of the current is used such that

$$I(t) = I_m\cos\omega_m t, \tag{14.1.9}$$

then the modulation depth is given by [14.2]

$$\eta = \frac{n_{ph}(\omega_m)}{\bar{N}_{ph}} = \frac{\dfrac{gI_m}{eV}}{\omega_0^2 - \omega_m^2 + i\omega_m\gamma}. \tag{14.1.10}$$

The expression (14.1.10) exhibits a pronounced peak at a frequency v_{max}, given by

$$v_{max} = \frac{\omega_{max}}{2\pi} = \frac{1}{2\pi}\left(\omega_0^2 - \frac{\gamma^2}{2}\right)^{1/2}, \tag{14.1.11}$$

as shown in Fig. 14.2, in the theoretical results produced by *Ikegami* and *Suematsu* [14.3], who have also observed this peak experimentally. For con-ventional GaAs laser diodes v_{max} is on the order of several gigahertz. The existence of such a peak, followed by a rapid decrease in response to modula-tion of course, implies an upper limit to modulation frequency. However, significant effects that are not included in the simple model presented thus far

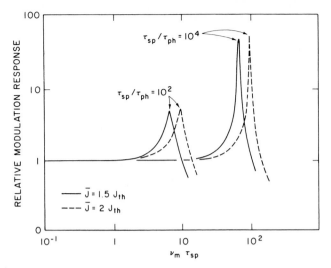

Fig. 14.2. Theoretical dependence of modulation depth on modulation frequency [14.3]. The curve has been calculated based on typical parameters for a GaAs laser and has been normalized to the value of η for ω_m near zero. Curves are plotted for two values of τ_{sp}/τ_{ph} to illustrate the strong effect of changing the Q of the laser cavity

permit specially designed laser diodes to operate at much higher frequencies. These effects are discussed in detail in Sect. 14.2.

14.1.2 Pulse Modulation

Pulse modulation is particularly convenient in semiconductor lasers because of their very short turn-on and turn-off times when the pumping current is pulsed. Pulse rise and fall times on the order of a few hundred picoseconds, typical of DH stripe geometry lasers, permit subnanosecond pulses to be generated, with nanosecond spacing [14.4]. However, the laser must be biased to just below threshold in order to obtain such high speed pulsing.Otherwise, there is an initial delay between the application of the current pulse and the emission of the light pulse, which is given by [14.5]

$$t_d = \tau_{sp} \ln[I_p/(I_p - I_{th})], \tag{14.1.12}$$

where I_p is the peak pulse current. This delay vanishes when the laser is biased to I_{th}.

The maximum permissible duty cycle of a particular laser can also impose a limit to high pulse-repetition-rate operation. In the case of DH stripe geometry lasers, capable of cw operation at room temperature, duty cycle is not a problem. However, in less expensive SH or homojunction lasers, the maximum duty cycle must not be exceeded, or else junction heating will result in wavelength drift and increased threshold current. When duty cycle is a limiting factor, the maximum data rate can be made larger than the pulse repetition rate by using a coding that permits each pulse to carry more than one bit. For example, a 1 ns pulse at an average repetition rate of 30 MHz can transmit a data rate of 150 Mbit/s if pulse-interval modulation is used [14.6].

Laser diodes can also be pulse-width modulated by controlling the width of the drive current pulse. Alternatively, *Fenner* [14.7] has proposed that the initial delay between the application of the current pulse and the onset of light emission can be used to produce pulse-width modulation. From (14.1.12) it can be seen that t_d is a strong function of I_p. Hence, if the driving current pulses are amplitude modulated, the resulting light pulses will be pulse-width modulated. Since the relationship between I_p and t_d is nonlinear, suitable compensation must be provided in the decoding network if this method of pulse-width modulation is used.

Trains of very narrow pulses, at high pulse repetition rates, can be generated in a semiconductor diode laser by taking advantage of the phenomenon of self-pulsing. Note that the form of (14.1.5), for the case of $I(t) = 0$, implies that there exists the possibility of relaxation oscillations at a frequency ω_R, given by

$$\omega_R^2 = \omega_0^2 - \left(\frac{\gamma}{2}\right)^2,$$
(14.1.13)

when the laser is biased above threshold. This *spiking resonance* is usually considered to be detrimental to laser modulation, because it results in the peaking distortion of the AM modulation frequency response curve, as described in Sect. 14.1.1. However, self pulsing due to augmented relaxation oscillations can be used beneficially, in some cases, to produce narrow pulses at GHz rates. For example, *D'Asaro* et al. [14.8] have reported pulse rates between 0.5 and 3.0 GHz. The exact nature of the relationship between the damped relaxation oscillations predicted by (14.1.5) and the self sustained oscillations observed in many laser diodes is not yet thoroughly understood. A number of different models have been proposed, which are discussed in more detail with regard to high frequency modulation in Sect. 14.2.

Self-pulsing junction lasers can be conveniently modulated by analog pulse-position modulation. In this method, the optical pulse rate is frequency-locked to an external signal modulation of the injection current at a frequency near the resonance frequency v_r, or one of its harmonics [14.9]. If the frequency of the locking signal is made to vary as the derivative of the information signal, the deviation of the pulse position from its average value is directly proportional to the information signal itself. Deviations of the pulse rate by as much as 10% of v_r can be obtained with only a few milliwatts of modulating microwave power [14.9]. Alternatively, pulse-position-modulation of self-pulsing lasers can be produced by using the current dependence of the self-induced pulse rate v_r [14.8], or by utilizing an external phase-locked loop to provide regenerative feedback of the self-induced oscillations of injection current that occur simultaneously with the optical oscillations [14.10].

All of the methods of modulating laser diodes discussed thus far rely in some way on the dependence of emitted optical power on injected current. However, it is also possible to directly modulate laser diodes by varying certain parameters of the cavity material. One such modulation scheme can be used to produce a pulse-position modulated output by employing a laser-diode consisting of two sections that are optically coupled, but are electrically isolated by a separation of the upper contact into two parts [14.11]. One section is pumped to produce optical gain, while the other is used as an electrically controllable saturable absorber. The laser-absorber combination is biased just below lasing threshold. Then a small sinusoidal signal applied to the absorber section can trigger laser pulses in synchronism with the sinusoid. These pulses can be pulse-position modulated by frequency modulating the sinusoidal signal applied to the absorber section.

14.1.3 Frequency Modulation

Direct modulation of the optical frequency of a diode laser can be produced by varying the dielectric constant ε of the cavity. The two-section lasers described as pulse modulated devices in Sect. 14.1.2 can also be operated in an FM mode [14.12]. In that case, the current passing through the absorber section of the dual diode is varied in order to produce a change in average cavity ε. Of course, the dual diode must be operated outside of the range of bias conditions for which self-pulsing oscillation can occur; even so, some AM modulation will be present on the output because of variations in average gain produced by the changing current in the absorber section.

An improved version of the two-section laser is the cleaved-coupled-cavity (C^3) laser, in which not only the electrical contacts but also the laser cavity itself are divided in two [14.13]. The C^3 laser is fabricated by first making a standard stripe-geometry Fabry-Perot laser with a cavity length of approximately 250 μm. This device is then recleaved near the middle to produce two coupled laser cavities. These cavities are only separated by a few micrometers and are held in alignment for mounting by a relatively thick electroplated contact on one side. In operation, one section of the laser is pumped above threshold by an injected current I_1, while the dielectric constant of the second section is varied by a modulating current I_2 [14.14]. By controlling I_2 it is possible both to tune the laser to a desired wavelength and to frequency modulate it. For example, a maximum wavelength excursion of 150 Å with a tuning rate of 10 Å/mA has been reported for a GaInAsP C^3 laser emitting at 1.3 μm [14.13].

In addition to permitting frequency modulation of the laser the dual-cavity structure also results in very stable single mode operation. The individual mode spacings of the cleaved cavities are given by

$$\Delta\lambda_1 = \frac{\lambda_0^2}{2n_{\mathrm{eff1}}L_1}$$

(14.1.14)

and

$$\Delta\lambda_2 = \frac{\lambda_0^2}{2n_{\mathrm{eff2}}L_2},$$

(14.1.15)

where λ_0 is the peak emission wavelength, n_{eff1} and n_{eff2} are the effective indices of refraction and L_1 and L_2 are the lengths of the two cavities. Because the two cavities are strongly coupled, only those modes which coincide spectrally will constructively combine to form the resultant modes of the coupled cavity. Thus the mode spacing of the coupled cavity is given by [14.13]

$$\Delta\lambda_c = \frac{\Delta\lambda_1\Delta\lambda_2}{|\Delta\lambda_1 - \Delta\lambda_2|} = \frac{\lambda_0^2}{2|n_{\mathrm{eff1}}L_1 - n_{\mathrm{eff2}}L_2|}.$$

(14.1.16)

Since $\Delta\lambda_c \gg \Delta\lambda_1$ or $\Delta\lambda_2$, only one mode near the peak of the gain curve survives. The laser thus is very stable when tuned to a desired frequency by the DC component of I_2 and frequency modulation can be produced by applying a small-signal AC modulation to I_2. In addition, a stable single mode also is maintained if direct amplitude modulation is produced by varying I_1 in the usual way. For example, it has been shown that a stable, single-frequency mode can be maintained under 2 Gb/s direct modulation with error rates of less than 10^{-10} [14.15].

The stable tuning and modulation characteristics of GaInAsP C³ lasers make them very attractive for use in wavelength division multiplexed transmitters for long-distance telecommunications. The C³ structure can also be used to make highly stable single mode lasers in GaAlAs [14.16].

Frequency modulation, with negligible AM, can be achieved by using acoustic waves to produce a change in cavity ε [14.17, 18]. Longitudinal sound waves passing through the laser diode in the direction perpendicular to the junction cause a shift of the lasing mode to a frequency v given by

$$v = v_0 + A\exp\left(-\frac{\omega_a^2 W^2}{8v_a^2}\right)\cos(\omega_a t),$$

(14.1.17)

where v_0 is the frequency of the mode in the absence of the acoustic waves, A is a constant proportional to the peak acoustic intensity, ω_a is the acoustic frequency, v_a is the acoustic velocity in the material, and W is the width of the optical mode at $\frac{1}{3}$ of its peak amplitude. *Whitney* and *Pratt* [14.19] have experimentally demonstrated modulation rates as high as several hundred megahertz using this acoustic wave direct modulation technique. As in the case of Bragg type acousto-optic modulators, the frequency response of the transducer used to launch the acoustic waves is presently the predominant factor limiting modulation bandwidth. Theoretical calculations predict that the maximum bandwidth of an acoustically modulated GaAs laser, with 400 µm cavity length, can be as high as 43 GHz, being limited ultimately by mode jumping [14.18].

14.2 Microwave Frequency Modulation of Laser Diodes

In the previous section, a glimpse was provided of the microwave frequency modulation capabilities of the injection laser. Because such modulation plays a key role in determining the upper limit to the information carrying capacity of systems employing optical carrier frequencies, it is appropriate to now review in detail the accomplishments and limitations of the various techniques that have been used (or proposed) to directly modulate diode lasers at frequencies in excess of a GHz.

14.2.1 Summary of Early Experimental Results

It was early realized that the inherently short electron and photon lifetimes of a semiconductor laser make it well suitable for applications requiring microwave frequency modulation [14.20, 21]. Well before the advent of laser diodes that could operate at room temperature, modulation at X-band (8–12 GHz) frequencies [14.22], and even as high as 46 GHz [14.23] was demonstrated in cryogenically cooled devices. In the five years immediately following these initial demonstrations, work continued in order to measure lifetimes and to study resonance phenomena [14.24–29]. In all cases, the laser diodes were cryogenically cooled (typically to 77 K), and were mounted in either microwave waveguides [14.22, 29] or in a coaxial line [14.25, 26, 28]. Modulation frequencies in the mm wave range, as high as 46 GHz, were reached, but generally with modulation depths of only a few percent [14.29]. When room-temperature cw lasers became available in 1970 [14.30, 31], further attempts at microwave modulation were directed toward the use of these devices, because cooling to 77 K presented a serious obstacle to any practical application.

During the 1970s work done on the microwave modulation of CW, room-temperature lasers demonstrated that these devices could be directly modulated at frequencies as high as 10 GHz with modulation depth of a few percent [14.32] and in the range from 1–3 GHz with modulation depth as great as 30% [14.33–39]. However, response at frequencies greater than 3 GHz was limited both by the response characteristics of the laser itself, and by the parasitic inductance, capacitance and resistance of the diode package and surrounding microwave cavity and/or transmission line.

14.2.2 Factors Limiting Modulation Frequency

The factors that can set an upper limit to the frequency at which a laser diode can be directly modulated are summarized in Table 14.1. The first four items

Table 14.1. Factors limiting the modulation frequency of a laser diode

Delayed turn-on/turn-off	Parasitic capacitance and inductance
Transient multimode excitation	RF leakage radiation
Relaxation oscillations	RF absorption loss
Self-sustaining oscillations	Impedance mismatch

on the list are inherent in the laser diode chip itself, while the remaining four are associated with the diode package and microwave circuitry. Of course, not all of these factors will be important in every case. However, any one of them acting alone can be sufficient to limit the maximum modulation frequency.

The optical turn-on delay associated with the buildup of the inverted carrier population was discussed in Sect. 14.1.2. It can be essentially eliminated by biasing the laser diode just below threshold. The optical turn-off delay is unlikely to be a limiting factor at frequencies up to 100 GHz since in GaAs and GaAlAs the carrier lifetime in the inverted population region is approximately 10^{-11}s [14.27], and the photon lifetime is even shorter [14.25]. However, the carrier lifetime is about a hundred times longer in noninverted material. Hence, biasing the laser close to threshold also has the effect of eliminating the relatively slowly decaying *tail* of photons that are spontaneously emitted while the laser current is below the threshold level.

Transient multimode excitation produces a broadening of the output spectrum of a pulsed laser at the onset of the current pulse [14.40]. The effect is particularly important at high bit rates because such transient excitation of multiple modes can affect coupling into an optical fiber, producing pulse shape distortion. In such applications a superior quality, low threshold single mode laser is recommended.

The possibility of damped relaxation oscillations of the optical intensity is apparent from the form of (14.1.5) with $I(t)$ equal to zero. When the drive current is suddenly turned on, damped relaxation oscillations can occur at a frequency ω_R given by

$$\omega_R^2 = \omega_0^2 - (\gamma/2)^2, \tag{14.2.1}$$

where ω_0 and γ are as defined in Sect. 14.1.1. In pulsed laser applications, these oscillations take the form of *ringing* of the leading edge of the optical pulse, and can significantly increase the bit error rate. A detailed study of the waveforms of such oscillations and their effect on high data-rate modulation of laser diodes has been performed by *Channin* et al. [14.41, 42].

Lau and *Yariv* [14.43] have reported that distortion (peaking) of the modulation frequency response curve due to relaxation oscillation resonance can be eliminated by reducing the reflectivity of one of the laser endfaces. Such feedback suppression reduces the photon lifetime, and can extend the flat range of the modulation response curve.

By referring to (14.1.6, 7, 10, 11) one can see that v_{max} and the modulation depth both can be increased in three ways – by decreasing the photon lifetime τ_{ph}, by increasing the photon density n_{ph}, and by increasing the optical gain coefficient g. Note that increasing the drive level I_{dc}/I_{th} accomplishes all three of these desired effects. The response curve for a laser designed especially for microwave modulation is shown in Fig. 14.3 [14.44]. The dark solid curve is the laser response when loosely coupled to an external cavity, and the light solid curve is the response when tightly coupled. This laser has a short cavity length (~100 μm) and unequal endface reflectivities to increase the peak photon density and decrease photon lifetime, and it was operated at a high drive level. A window structure was incorporated at the output facet so that a higher photon density could be reached without producing facet damage. The 3 dB bandwidth of this laser was approximately 10 GHz, and response could be peaked at higher frequencies by coupling the laser to an external cavity. The effect of a high drive level on the microwave response of a laser diode can be seen in the data of Fig. 14.4 [14.45]. The device in that case was a 1.3 μm GaInAsP buried-heterostructure laser with a short cavity length of 100 μm. Cooling the device slightly to 20 °C improves the frequency response some-what by increasing the gain coefficient. The observed 18 GHz bandwidth of this laser was determined to be the result of a convolution of an inherent device bandwidth of 22 GHz and a 10 GHz roll-off frequency due to electrical parasitics. To make a very high frequency laser one must make the parasitics small as well as making the resonance frequency high. The constricted-mesa laser [14.46], shown in Fig. 14.5, has a very low capacity because the bonding pad capacitance is typically only 0.35 pF due to the thick polyimide layer under the bonding pad, and the bonding pad itself is only 50 to 100 μm wide. The

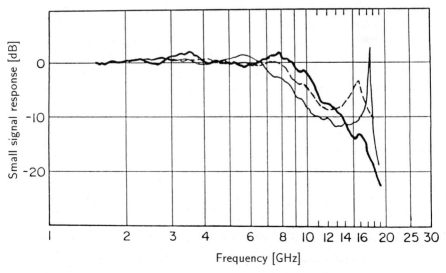

Fig. 14.3. Modulation response of a high frequency laser diode [14.44]

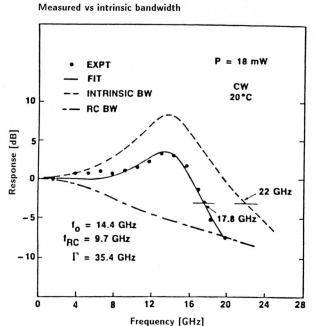

Measured vs intrinsic bandwidth

Fig. 14.4. Frequency response of a GaInAsP BH laser [14.45]

capacitance of the current-confining constricted-mesa structure is also very low (0.18 pF). Once again, the strong effect of drive level can be seen in the response data of Fig. 14.5.

Similar results have been obtained by Chen et al. [14.47] who used an airbridge contact configuration and a narrow mesa to reduce parasitic effects, resulting in a 3 dB bandwidth of 18 GHz in a DFB laser emitting at 1.3 μm. Their device also incorporated a Multi-Quantum-Well (MQW) structure which further enhances laser performance. (The improvements in laser performance resulting from the inclusion of a MQW structure are described in detail in Sect. 16.2). Modulation of a MQW DFB laser at 1.55 μm wavelength with a bandwidth of 20 GHz has been reported by Lipsanen et al. [14.48]. In that case, the intrinsic bandwidth of the laser was estimated to be 35 GHz in the absence of parasitics.

The effect of bias current on maximum modulation frequency, which was observed in the laser of Fig. 14.5, is also apparent in the performance of an InGaAs/GaAs/AlGaAs, MQW ridge laser that has been described by Ralston et al. [14.49]. They observed a modulation bandwidth of 24 GHz at a bias current of 25 mA, which increased to 33 GHz at a bias of 65 mA. Threshold current was approximately 10 mA in these devices. Although the positive effect on modulation bandwidth of increasing bias current is clearly apparent, it should be noted that the current levels in these lasers are still relatively low. This can be attributed to a sophisticated laser geometry employing a short cavity, MQW and ridge-waveguide structure.

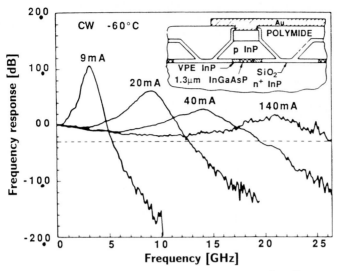

Fig. 14.5. Frequency response of a constricted mesa laser [14.46]

When one considers modulation frequencies in the mm-wave range (above 30 GHz) it is best to use a traveling-wave approach. Tromborg et al. [14.50] have developed a traveling-wave analysis of a general class of semiconductor lasers which includes multisection DFB and DBR lasers as well as gain-coupled DFB lasers. Both intensity modulation and frequency modulation are considered.

14.2.3 Design of Laser Diode Packages for Microwave Modulation

In addition to the four preceding frequency limiting factors inherent in the laser diode chip itself, there are a number of factors associated with the device package and microwave circuit that must be considered as well. Parasitic capacitance and inductance of the laser package are most often the dominant factors that limit modulation frequency when conventional low-frequency laser diodes are used at frequencies above 1 GHz. A diagram of a typical low-frequency laser diode package is shown in Fig. 14.6. It can be seen that the bias lead forms a coaxial capacitor with the threaded stem of the header, which has a capacitance of typically 3 pF. Thus, at a frequency of 10 GHz, a shunt capacitive reactance of only about 5 Ω is presented to the microwave signal when it is applied to the bias lead. Because the lead wire from the feedthrough post to the laser chip has an inductance of about 2 nH, corresponding to a series inductive reactance of approximately 140 Ω at 10 GHz, it is clear that any attempt to apply a 10 GHz modulating signal to the laser chip via the bias lead will result in virtually all of the signal being shunted to ground by the capacitance.

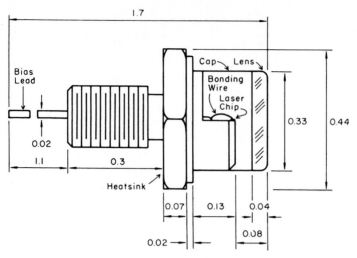

Fig. 14.6. Diagram of a typical commercially available laser diode package. All dimensions are shown in inches

The unshielded bond wire spanning the gap between the feedthrough post and the diode chip also acts as an *antenna* at frequencies on the order of 10 GHz, permitting significant amounts of microwave energy to be radiated by the laser diode. Not only is this energy lost for optical generation, but it can cause serious RFI (radio frequency interference) by coupling to nearby devices and circuits.

Additional loss of microwave modulating energy occurs in the conventional laser diode package because of rf absorption in the dielectric material used to insulate the bias lead and in the metallic surfaces.

The combined action of all of the microwave loss mechanisms that have been described, coupled with the fact that much of the incident microwave energy is reflected because of the poor impedance match between the diode package and the waveguide or transmission line, leads to the inevitable result that little or no microwave energy actually reaches the laser chip. Obviously, a specially-designed laser diode package is needed for microwave frequency applications.

The most effective way of avoiding the deleterious effects of the diode package is to mount the laser diode directly into a microwave strip-line [14.51]. This type of microwave transmission line, which consists of a relatively thin metallic strip on the surface of a metal-backed dielectric substrate, can be conveniently fabricated on semi-insulating GaAS or InP substrates as well as on ceramic substrates. When designed to the proper dimensions for the frequency band in use, such a strip-line has very low parasitics and very low microwave loss. Commercially available laser diodes intended for use at frequencies above 1 GHz are usually mounted on a heat sink with a ribbon electrical lead designed to properly match a strip-line feed. The fact that microwave strip-lines can be fabricated on semi-insulating GaAs or InP substrates makes the monolithic integration of microwave-modulated laser diodes into optical integrated circuits feasible.

14.3 Monolithically Integrated Direct Modulators

The prospect of monolithically integrating the laser diode with its microwave modulator is attractive, not only because of its obvious applicability to optical integrated circuits, but also because it provides the benefit of reducing the parasitic inductance and capacitance of the interconnecting circuit. An example of a possible monolithic combination of a laser diode with a FET modulator is shown in Fig. 14.7. In this device, the lasing junction is formed as an integral part of the drain region of the field effect transistor, with the two being electrically connected in series. The gate of the FET is controlled by the microwave modulating signal, producing a corresponding modulation of the drain current, and hence, the current through the laser diode. Schottky-barrier-gate FET's formed on high resistivity GaAs substrates, as shown in Fig. 14.7, are known to have cut-off frequencies in the range from 10 to 100 GHz, depending mostly on the gate length [14.52, 53]. Because of the monolithic integration of the laser junction with the FET drain, interconnection inductance and capacitance are minimized. Hence, improved high frequency modulation response is to be expected.

The fabrication technology required to produce monolithic integration of laser diodes with microwave modulating devices is quite complex, requiring

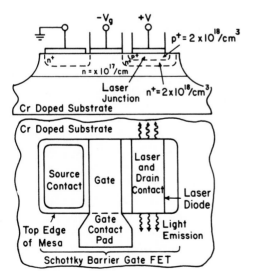

Fig. 14.7. A monolithically integrated laser diode and FET modulator on a GaAs substrate

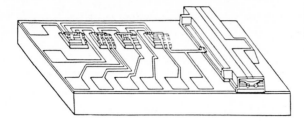

Fig. 14.8. Long wavelength OEIC transmitter in GaInAsP. The substrate is semi-insulating InP. [14. 62]

expertise in two fields that are usually dealt with separately. Nevertheless there are a number of examples of such integration being accomplished successfully. *Margalit* et al. [14.54] produced a monolithically integrated GaAlAs laser and FET on a Cr-doped GaAs substrate that operated at frequencies above 1 GHz. *Ury* [14.55] and *Fukuzawa* [14.56] also produced integrated MESFET and laser chips in GaAs. Following these early demonstrations of laser and FET integration, other researchers have gone on to fabricate monolithically integrated transmitters [14.57–62], receivers [14.63–67], repeaters [14.68, 69] and transceivers [14.70]. These circuits which combine both optical and electronic devices have come to be known as optoelectronic integrated circuits or OEIC's. An example of an OEIC transmitter is shown in Fig. 14.8 [14.62]. This circuit combines a buried-heterostructure laser with three heterojunction bipolar transistor (HBT) drivers and a FET preamplifier. The HBT's are used as drivers because of their high current capability, while the FET is used as a preamplifier because of its high input impedance. This OEIC is capable of operating at a data rate greater than 21 Gb/S.

A monolithic laser/modulator structure such as those shown in Fig. 14.7 and Fig. 14.8 not only increases the efficiency of microwave coupling, but also provides an opportunity for improved optical coupling, since the laser can be monolithically coupled to a waveguide, as discussed in Chap. 12.

14.4 Future Prospects for Microwave Modulation of Laser Diodes

In this chapter, a variety of techniques for directly modulating semiconductor lasers at frequencies into the mm-wave range have been reviewed. Some impressive results have been obtained, but much work remains to be done before the ultimate limits imposed by carrier and photon lifetimes are reached. New OEIC technology offers the promise of driver circuits with negligible parasitics. Bandwidths as high as 10 GHz have been reached [14.60], and devices operating at even higher frequencies are certain to follow.

Problems

14.1 We wish to modulate the light output of a semiconductor laser having the optical power-current characteristic shown below by directly varying the input current. Sketch the intensity (or optical power density) waveform of the light output for a bias current of 300 mA and an applied current signal which is sinusoidal and has a peak to peak current variation of 200 mA.

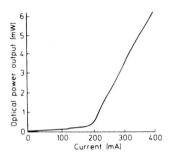

14.2 Repeat problem 14.1 for a bias current of 150 mA, all other conditions remaining the same.

14.3 A cw double heterojunction GaAs-GaAlAs laser diode is to be directly current modulated by a sinusoidal ac signal current of 100 mA peak-to-peak. What must be the minimum dc bias current to the laser to ensure that the laser output linearly follows the input signal, i.e., the output optical signal will be sinusoidal, following the input signal current waveform. Relevant characteristics of the laser diode are as follows:
   1) Emitted wavelength: 9000 Å.
   2) Half power points of the emission peak for spontaneous emission have been measured for this material at room temperature to be at 9200 and 8800 Å.
   3) Index of refraction: 3.3.
   4) Thickness of active layer: 1 μm.
   5) Internal quantum efficiency: 0.8.
   6) Average absorption coefficient: 10 cm^{-1}.
   7) Length between cleaved end faces: 1 mm.
   8) Reflectivity of end faces: 0.4.
   9) Cross-sectional area normal to current flow: 10^{-3} cm^2.

14.4 Given a laser with threshold current $I_t = 100$ mA and peak current $I_p = 300$ mA, what is the maximum modulation depth achievable if the dc bias current is $I_{dc} = 130$ mA? Assume that the laser L-I curve is linear above threshold and that, at threshold, the output power $P_t = 0.05\ P_p$.

14.5   A SH laser diode is to be used for high-speed communications. It must be operated with no dc bias, and in a pulsed mode with no more than a 5% duty cycle, and a pulse width of 0.8 ns. The threshold and peak currents are 100 mA and 7 A, respectively. The spontaneous electron lifetime in the device is 60 ns.

    a) What factor(s) limit the maximum modulation rate?

    b) What is the maximum pulse repetition rate that can be used?

15. Integrated Optical Detectors

Detectors for use in integrated-optic applications must have high sensitivity, short response time, large quantum efficiency and low power consumption [15.1]. In this chapter, a number of different detector structures having these performance characteristics are discussed.

15.1 Depletion Layer Photodiodes

The most common type of semiconductor optical detector, used in both integrated optic and discrete device applications, is the depletion-layer photodiode. The depletion-layer photodiode is essentially a reverse-biased semiconductor diode in which reverse current is modulated by the electron-hole pairs produced in or near the depletion layer by the absorption of photons of light. The diode is generally operated in the *photodiode* mode, with relatively large bias voltage, rather than in the *photovoltaic* mode, in which the diode itself is the electrical generator and no bias voltage is applied [15.2].

15.1.1 Conventional Discrete Photodiodes

The simplest type of depletion layer photodiode is the p-n junction diode. The energy band diagram for such a device, with reverse bias voltage V_a applied is shown in Fig. 15.1. The total current of the depletion layer photodiode consists of two components: a drift component originating from carriers generated in region (b) and a diffusion component originating in regions (a) and (c). Holes and electrons generated in region (b) are separated by the reverse bias field, with holes being swept into the p-region (c) and electrons being swept into the n-region (a). Holes generated in the n-region or electrons generated in the p-region have a certain probability of diffusing to the edge of the depletion region (b), at which point they are swept across by the field. Majority carriers, electrons in (a) or holes in (c) are held in their respective regions by the reverse bias voltage, and are not swept across the depletion layer.

In order to minimize series resistance in a practical photodiode while still maintaining maximum depletion width, usually one region is much more heavily doped than the other. In that case, the depletion layer forms almost

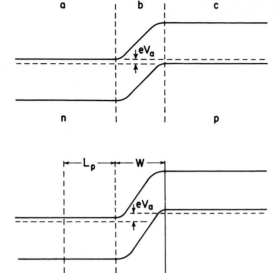

Fig. 15.1. Energy band diagram for a p-n junction diode under application of a reverse bias voltage V_a

Fig. 15.2. Energy band diagram for a p^+-n (high-low) junction diode under application of a reverse bias voltage V_a

entirely on the more lightly doped side of the junction, as shown in Fig. 15.2. Such a device is called a high-low abrupt junction. In GaAs and its ternary and quaternary alloys, electron mobility is generally much larger than hole mobility. Thus, the p-region is usually made thinner and much more heavily doped than the n-region, so that the device will be formed mostly in n-type material, and the p-region then serves essentially just as a contact layer.

For a device with the high-low junction geometry indicated in Fig. 15.2, it can be shown that the total current density J_{tot} is given by [15.3]

$$J_{tot} = q\varphi_0 \left(1 - \frac{e^{-aW}}{(1+\alpha L_p)} \right) + q p_{n0} \frac{D_p}{L_p}, \tag{15.1.1}$$

where φ_0 is the total photon flux in photons/cm^{2}s, W is the width of the depletion layer, q is the magnitude of the electronic charge, α is the optical interband absorption coefficient, L_p is the diffusion length for holes, D_p is the diffusion constant for holes, and P_{n0} is the equilibrium hole density. The last term of (15.1.1) represents the reverse *leakage* current (or *dark* current), which results from thermally generated holes in the n-material. This explains why that term is not proportional to the photon flux φ_0. The first term of (15.1.1) gives the photocurrent, which is proportional to φ_0, and includes current from both the drift of carriers generated within the depletion layer and the diffusion and drift of holes generated within a diffusion length L_p of the depletion layer edge. The quantum efficiency η_q of the detector, or the number of carriers generated per incident photon, is given by

$$\eta_q = 1 - \frac{e^{-\alpha W}}{(1 + \alpha L_p)},$$
(15.1.2)

which can have any value from zero to one. It should be noted that (15.1.1) and (15.1.2) are based on the tacit assumption that scattering loss and free carrier absorption are negligibly small. The effect of these loss mechanisms on the quantum efficiency, when they are not negligible, is discussed in Sect. 15.1.3.

It can be seen from (15.1.2) that, in order to maximize η_q, it is desirable to make the products αW and αL_p as large as possible. When αW and αL_p are large enough so that η_q is approximately equal to one, the diode current is then essentially proportional to φ_0, because the dark current is usually negligibly small.

If the interband absorption coefficient α is too small compared to W and L_p, many of the incident photons will pass completely through the active layers of the diode into the substrate, as shown in Fig. 15.3. Only those photons absorbed within the depletion layer, of thickness W, have maximum effectiveness in carrier generation. Photons absorbed at depths up to a diffusion length L_p from the depletion layer edge are somewhat effective in generating photocarriers, in that holes can diffuse into the depletion layer. Photons that penetrate to as depth greater than $(W + L_p)$ before being absorbed are essentially lost to the photo-generation process because they have such a very low statistical probability of producing a hole that can reach the depletion layer and be swept across. Within the semiconductor, the photon flux $\varphi(x)$ falls off exponentially with increasing depth x from the surface, as shown in Fig. 15.4. Thus,

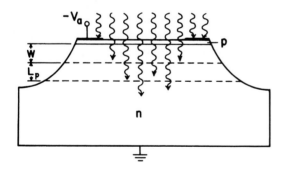

Fig. 15.3. Diagram of a conventional mesa-geometry photodiode with p⁺-n doping profile showing photon penetration

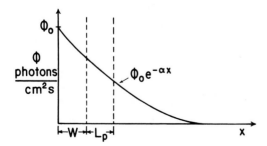

Fig. 15.4. Optical absorption versus depth from the surface in a conventional mesa photodiode

if α is not large enough. many photons will penetrate too deeply before being absorbed, thus producing carriers that (on average) will recombine before diffusing far enough to reach the depletion layer.

Interband absorption is a strong function of wavelength in a semiconductor. The absorption coefficient α usually exhibits a spectral response curve that rises sharply at the absorption-edge (band-edge) wavelength and then saturates at a wavelength that is slightly shorter than the bandgap wavelength, increasing slowly for yet shorter wavelengths. Thus, it is impossible to design a diode with an ideal W for all wavelengths. For wavelengths near the absorption edge, the long-wavelength response of a diode is limited by excess penetration of photons into the substrate, as shown in Fig. 15.3 and 4; its short wavelength response can be limited by too strong an absorption of photons in the p$^+$ layer near the surface, where recombination probability is large.

Aside from the reduction of quantum efficiency that results from poor matching of α, W and L_p, there are some other limitations to depletion layer photodiode performance that are also important. Since W is usually relatively small (in the range from 0.1 to 1.0 μm), junction capacitance can limit high-frequency response through the familiar R-C time constant. Also, the time required for carriers to diffuse from depths between W and $(W + L_p)$ can limit the high frequency response of a conventional photodiode. The waveguide depletion layer photodiode, which is discussed in the next section, significantly mitigates many of these problems of the conventional photodiode.

15.1.2 Waveguide Photodiodes

If the basic depletion layer photodiode is incorporated into a waveguide structure, as shown in Fig. 15.5, a number of improvements in performance are realized. In this case, the light is incident transversely on the active volume of the detector, rather than being normal to the junction plane. The diode photocurrent density is then given by

$$J = q\varphi_0 \left(1 - e^{-\alpha L}\right), \tag{15.1.3}$$

where L is the length of the detector in the direction of light propagation. Since W and L are two independent parameters, the carrier concentration

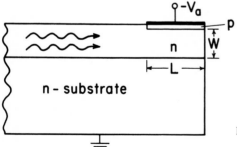

Fig. 15.5. Diagram of a waveguide detector

within the detector volume and the bias voltage V_a can be chosen so that the depletion layer thickness W is equal to the thickness of the waveguide, while L can be made as long as necessary to make $\alpha L \gg 1$. Thus 100% quantum efficiency can be obtained for any value of α, by merely adjusting the length L. For example, for a material with the relatively small value of $\alpha = 30\,\mathrm{cm}^{-1}$, a length of $L = 3\,\mathrm{mm}$ would give $\eta_q = 0.99988$. (Again, it has been tacitly assumed in (15.1.3) that scattering loss and free-carrier absorption are negligible.)

Because a waveguide detector can be formed in a narrow channel waveguide, the capacitance can be very small, even if L is relatively large. For example, for a material with a relative dielectric constant $\varepsilon = 12$, such as GaAs, a 3 mm long detector formed in a 3 µm wide channel waveguide has a capacitance of only 0.32 pF. This capacitance is about a factor of ten less than that of a typical conventional mesa photodiode. Hence, the high frequency response can be expected to be correspondingly improved. Experimentally demonstrated bandwidth of 5 GHz and quantum efficiency of 83% have been obtained with waveguide detectors on GaAs substrate material [15.4], and InGaAs waveguide photodetectors on InP substrates also have exhibited a 5 GHz bandwidth for light in the wavelength range of 1.3–1.6 µm [15.5].

Because all of the incident photons are absorbed directly within the depletion layer of a waveguide photodetector, not only is η_q improved, but also the time delay associated with the diffusion of carriers is eliminated. This result is a further improvement in high frequency response.

Due to the many improvements in performance inherent in the transverse structure of the waveguide detector, as compared to the axial geometry of the conventional mesa photodiode, waveguide detectors should be considered for use in discrete-device applications, as well as in optical integrated circuits. At the present time, waveguide detectors are not commercially available as discrete devices. However, they can be fabricated with relative ease in many laboratories. Hence availability should not long be a problem.

15.1.3 Effects of Scattering and Free-Carrier Absorption

The relations given by (15.1.1–3) neglected the effects of free-carrier absorption and photon scattering on the quantum efficiency of the detector. Because both of these mechanisms result in the loss of photons without the generation of any new carriers, they tend to reduce quantum efficiency. In many cases they can be neglected, and (15.1.1–3) will give accurate predictions. However, when the free carrier absorption coefficient α_{FC} and the scattering loss coefficient α_s are not negligible as compared to the interband absorption coefficient α_{IB}, a more sophisticated expression for η_q is required. Such an expression can be derived as follows. The photon flux at any point located a distance x from the surface of the detector on which the photons are first incident is assumed to have the form given by

$$\varphi(x) = \varphi_0 e^{-\alpha x}, \tag{15.1.4}$$

where in general the loss coefficient α is given by

$$\alpha = \alpha_{IB} + \alpha_{FC} + \alpha_s. \tag{15.1.5}$$

The hole-electron pair generation rate $G(x)$ is given by

$$G(x) = \alpha_{IB} \, \varphi_0 e^{-\alpha x}, \tag{15.1.6}$$

since only α_{IB} results in carrier generation. Thus the photocurrent density is given by

$$J = q \int_0^L G(x) dx \tag{15.1.7}$$

or

$$J = q\varphi_0 \frac{\alpha_{IB}}{\alpha_{IB} + \alpha_{FC} + \alpha_S} (1 - e^{-(\alpha_{IB} + \alpha_{FC} + \alpha_S)L}). \tag{15.1.8}$$

Comparing (15.1.8) with (15.1.3), it is obvious that the effect of additional losses due to scattering and free-carrier absorption is to reduce the quantum efficiency by a factor of α_{IB}/α, even when L is large enough to maximize η_q.

If α_s and α_{FC} are small compared to α_{IB}, as is generally true, (15.1.8) reduces to (15.1.3). However, if the waveguide is inhomogeneous or is unusually rough, or if the detector volume is heavily doped so that α_s and α_{FC} are not negligible, then (15.1.8) must be used.

15.2 Specialized Photodiode Structures

There are two very useful photodiode structures that can be fabricated in either a waveguiding or conventional, nonwaveguiding form. These are the Schottky-barrier photodiode and the avalanche photodiode.

15.2.1 Schottky-Barrier Photodiode

The Schottky-barrier photodiode is simply a depletion layer photodiode in which the p-n junction is replaced by a metal-semiconductor rectifying (blocking) contact. For example, if the p-type layers in the devices of Figs. 15.3 and 15.5 were replaced by a metal that forms a rectifying contact to the semiconductor, Schottky-barrier photodiodes would result. The photocurrent would still be given by (15.1.1) and (15.1.3), and the devices would have essentially the same performance characteristics as their p⁺-n junction counterparts. The energy band diagrams for a Schottky-barrier diode, under zero bias and under reverse bias, are given in Fig. 15.6. It can be seen that the depletion region extends into the n-type material just as in the case of a p⁺-n junction. The

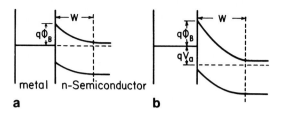

Fig. 15.6 a,b. Energy band diagram for a Schottky-barrier diode; **a** zero bias; **b** reverse biased with voltage V_a

barrier height φ_B depends on the particular metal-semiconductor combination that is used. Typical values for φ_B are about 1 V.

In conventional mesa devices, a thin, optically transparent Schottky-barrier contact is often used (rather than a p^+-n junction) to enhance short-wavelength response, by eliminating the strong absorption of these higher energy photons that occurs in the p^+ layer. In a waveguide photodiode, a Schottky-barrier contact is not needed for improved short-wavelength response because the photons enter the active volume transversely. However, ease of fabrication often makes the Schottky-barrier photodiode the best choice in integrated applications. For example, almost any metal (except for silver) produces a rectifying Schottky-barrier when evaporated onto GaAs or GaAlAs at room temperature. Gold, aluminum or platinum are often used. Transparent conductive oxides such as indium Tin oxide (ITO) and cadmium Tin oxide (CTO) can also be used to eliminate the photon masking effect of the contacts and thereby improve the quantum efficiency. For example, *Parker* [15.6] has reported a Schottky-barrier photodiode based on ITO with a quantum efficiency of 60% and a 3 dB bandwidth of 20 GHz. Photoresist masking is adequate to define the lateral dimensions during evaporation, and no careful control of time and temperature is required, as in the case of diffusion of a shallow p^+ layer.

A detailed discussion of the properties of Schottky-barrier diodes is beyond the scope of this text so that the interested reader should refer to the information available elsewhere [15.7].

15.2.2 Avalanche Photodiodes

The gain of a depletion layer photodiode (i.e. the quantum efficiency), of either the p-n junction or Schottky-barrier type, can be at most equal to unity, under normal conditions of reverse bias. However, if the device is biased precisely at the point of avalanche breakdown, carrier multiplication due to impact ionization can result in substantial gain in terms of increase in the carrier to photon ratio. In fact, avalanche gains as high as 10^4 are not uncommon. Typical current-voltage characteristics for an avalanche photodiode are shown in Fig. 15.7. The upper curve is for darkened conditions, while the lower one shows the effects of illumination. For relatively low reverse bias voltages, the diode exhibits a saturated dark current I_{d0} and a saturated photo-

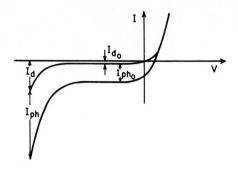

Fig. 15.7. Response curves for an avalanche photodiode

current I_{ph0}. However, when biased at the point of avalanche breakdown, carrier multiplication results in increased dark current I_d, as well as increased photocurrent I_{ph}. It is possible to define a photomultiplication factor M_{ph}, given by

$$M_{ph} \equiv \frac{I_{ph}}{I_{ph0}},$$

(15.2.1)

and a multiplication factor M, given by

$$M \equiv \frac{I_{ph} + I_d}{I_{ph0} + I_{d0}}.$$

(15.2.2)

An exact equation for the current-voltage curve is difficult to obtain in the region of bias in which avalanche breakdown occurs. However, *Miller* [15.8] has represented the functional form of the photomultiplication factor by the expression

$$M_{ph} = \frac{1}{1 - (V_a/V_b)^n},$$

(15.2.3)

where V_b is the breakdown voltage, and n is an empirically determined exponent depending on the wavelength of light, doping concentration, and, of course, the semiconductor material from which the diode is fabricated. For the case of large photocurrent $I_{ph0} \gg I_{d0}$ *Melchior* and *Lynch* [15.9] have shown that the multiplication factor is given by

$$M = \frac{1}{1 - \left(\dfrac{V_a - IR}{V_b}\right)^n},$$

(15.2.4)

where I is the total current, given by

$$I = I_d + I_{ph},$$

(15.2.5)

R being the series resistance of the diode (including space-charge resistance if significant). The derivation of (15.2.4) assumes that $IR \ll V_b$. For the case of I_{d0} and I_d being negligibly small compared to I_{ph0} and I_{ph}, it can be shown that the maximum attainable multiplication factor is given by [15.9]

$$M \cong M_{ph} \cong \sqrt{\frac{V_b}{n I_{ph0} R}}.$$ (15.2.6)

Avalanche photodiodes are very useful detectors, not only because they are capable of high gain, but also because they can be operated at frequencies in excess of 10 GHz [15.10]. However, not every p-n junction or Schottky-barrier diode can be operated in the avalanche multiplication mode, biased near avalanche breakdown. For example, the field required to produce avalanche breakdown in GaAs is approximately 4×10^5 V/cm. Hence, for a typical depletion width of $3\,\mu m$, V_b equals 120 V. Most GaAs diodes will breakdown at much lower voltages due to other mechanisms, such as edge breakdown or microplasma generation at localized defects, thus never reaching the avalanche breakdown condition. In order to fabricate an avalanche photodiode, extreme care must be taken, beginning with a dislocation free substrate wafer of semiconductor material. Generally a guard ring structure [Ref. 15.7, p. 203] must be employed to prevent edge breakdown. *Melchior* et al. [15.11] have given a very detailed description of a process for fabricating avalanche photodiodes in silicon that utilizes both ion-implantation and diffusion doping techniques.

Avalanche photodiodes are highly-stressed devices. Hence, reliability is a question of prime concern. Increasing leakage current due to poor surface passivation or the generation of internal defects during high current pulse operation can lead to degradation of performance as the devices age. Nevertheless, when diodes are carefully fabricated and are hermetically sealed into adequate packages, mean time to failure as high as 10^5 h at $170\,°C$ has been observed [Ref. 15.1, p. 80], which projects to about 10^9 h at room temperature.

15.2.3 p-i-n Photodiodes

In Sect. 15.1.1 it was pointed out that conventional photodiodes must be designed so as to have a large αW product in order to maximize η_q; but one doesn't have complete control over either the depletion width W, which depends on dopant concentrations, or the absorption coefficient, which depends mostly on the bandgap. In the p-i-n photodiode, a very lightly doped "intrinsic" layer is formed between the p and n sides of the diode. This layer generally has a carrier concentration of less than $10^{14}/cm^3$, but it is compensated by a balance of p- and n-type dopants rather than being truly intrinsic. Because of the low carrier concentration, the depletion layer in a p-i-n diode extends completely through the i layer so that the total thickness of the active layer is the sum of the i-layer thickness W_i and the depletion width on the lightly doped (n) side of the junction. Thus the device designer can adjust the total depletion width to produce a large αW product by varying the thickness of the i-layer. The presence of the relatively thick i-layer also reduces the junction capacitance and increases the R-C cutoff frequency of the diode. p-i-n photodiodes are widely used as detectors in optical systems because of their high quantum efficiency (responsivity) and wide bandwidth. For example, *Kato* et al. [15.12] have reported a waveguide p-i-n photodiode operating at 1.55 μm wavelength with a quantum efficiency of 50% and a 3 dB bandwidth of 75 GHZ.

15.2.4 Metal-Semiconductor-Metal Photodiodes

Metal-semiconductor-metal (MSM) photodiodes are surface-oriented devices that feature interdigitated, finger-like, Schottky barrier contacts formed on the surface of a thin semiconducting layer on a semi-insulating substrate. A typical MSM photodiode structure is shown in Fig. 15.8. Carriers generated by the absorption of photons in regions of the semiconducting layer between the contacts are swept by the fringing electric field and collected by the contacts. Holes are collected by the cathode and electrons by the anode. Spacing of the contact fingers must be less than the diffusion length of the carriers in order to produce a high collection efficiency.

Because the contact fingers are very narrow and closely spaced (~1 μm) the capacitance is relatively low and the transit times of carriers are short. Hence, wide-bandwidth operation is possible. *Hsiang* et al. have made Si MSM diodes with 0.2 μm width and spacing with a full-width-half-maximum pulse response of 3.7 ps, corresponding to a 3 dB bandwidth at 110 GHz [15.13]. The Schottky electrodes of the MSM photodiode are essentially identical to the gate metalization of field-effect transistors, which facilies their monolithic integration with FET's. For example, *Mactaggart* et al. [15.14] have made a fully-integrated 400 Mb/s burst-mode data OEIC receiver for application as a phased-array antenna controller. Approximately 350 source-coupled FET logic gates are present on the GaAS chip, along with a 780 nm wavelength MSM photodiode. MSM photodiodes have also been integrated monolithically with High-Electron-Mobility field effect Transistors (HEMT's)

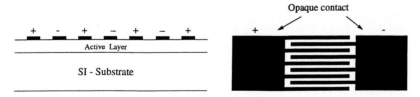

Fig. 15.8. Conventional MSM photodetector

to produce OEIC receivers with bandwidths larger than 14 GHz [15.15, 16]. Molecular beam epitaxy has been used to fabricate InGaAs MSM photodiodes on silicon substrates for monolithic integration [15.17].

The most significant disadvantage of MSM detectors is their inherent low responsivity because the metallization for the surface electrodes shadows the active light-collecting region. However, this problem can be mitigated by using a transparent conducting material for the contact electrodes. *Gao* et al. [15.18] have fabricated InGaAs MSM photodiodes with transparent cadmium tin oxide (CTO) electrodes. The responsivity of these devices to 1.3 μm light was 0.49 A/W, as compared to 0.28 A/W for identical control samples with conventional Ti/Au electrodes. *Seo* et al. [15.19] have reported a similar increase in responsivity from 0.39 A/W for InAℓAS/InGaAs MSM photodiodes when Ti/Au electrodes were replaced by those of indium tin oxide (ITO).

15.3 Techniques for Modifying Spectral Response

The fundamental problem of wavelength incompatibility, which was encountered previously in regard to the design and fabrication of monolithic laser/ waveguide structures in Chap. 12, is also very significant with respect to waveguide detectors. An ideal waveguide should have minimal absorption at the wavelength being used. However a detector depends on interband absorption for carrier generation. Hence, if a detector is monolithically coupled to a waveguide, some means must be provided for increasing the absorption of the photons transmitted by the waveguide within the detector volume. A number of different techniques have proven effective in this regard.

15.3.1 Hybrid Structures

One of the most direct approaches to obtaining wavelength compatibility is to use a hybrid structure, in which a detector diode, formed in a relatively narrow bandgap material, is coupled to a waveguide fabricated in wider bandgap material. The two materials are chosen so that photons of the desired wavelength are transmitted freely by the waveguide, but are strongly absorbed within the detector material. An example of this type of hybrid waveguide/ detector is the glass on silicon structure that was demonstrated by *Ostrowsky* et al. [15.20], as shown in Fig. 15.9. The diode was formed by boron diffusion to a depth of about 1 μm into an n-type, 5 Ω cm silicon substrate. A 1 μm thick layer of thermally grown SiO_2 was used as a diffusion mask. The glass waveguide was then sputter-deposited and silver paint electrodes were added as shown. Total guide loss was measured to be 0.8 dB/cm ± 10% for light of 6328 Å wavelength. The efficiency of coupling between the waveguide and the detector was 80%. However, because the light enters the diode in the direction

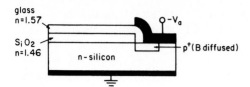

normal to the junction plane rather than parallel to it, this particular waveguide detector geometry does not have many of the advantages described in Sect. 15.1.2. Nevertheless, good high frequency response can be expected. These diffused diodes had a capacitance of only 3×10^{-9} F/cm^2 when reverse biased with V_a equal to 10 V. Thus a detector diode of approximately 10 µm radius, used in conjunction with a 50 Ω load resistance, would have an RC time constant of about 15 ps, implying that modulation of frequencies in excess of 10 GHz could be detected. Another example of hybrid coupling is provided by Koike et al. [15.21], who have coupled a flip–chip Ge photodiode to a dielectric waveguide with a coupling efficiency of 22%.

While hybrid detectors offer the possibility of choosing the waveguide and detector materials for optimum absorption characteristics, better coupling efficiency can be obtained with monolithic fabrication techniques. Monolithically fabricated waveguide detectors also have the advantage that light enters the device in the plane of the junction rather than normal to it.

15.3.2. Heteroepitaxial Growth

The most popular method of monolithically integrating a waveguide and detector is to use heteroepitaxial growth of a relatively narrow bandgap semiconductor at the location where a detector is desired. An example of this approach is given by the InGaAs detector that has been integrated with a GaAs waveguide by *Stillman* et al. [15.22], as shown in Fig. 15.10. In In$_x$Ga$_{(1-x)}$As, the bandgap can be adjusted to produce strong absorption of light at wavelengths in the range from 0.9 to 3.5 µm by changing the atomic fraction x of indium. (Fig. 15.11) The monolithic waveguide detector structure shown in Fig. 15.12 combines an epitaxially grown carrier-concentration-reduction type waveguide with a platinum Schottky barrier detector. A 6000 Å thick layer of pyrolytically deposited SiO$_2$ was used as a mask to etch a 125 µm diameter well into the 5 to 20 µm thick waveguide, and then grow the In$_x$Ga$_{(1-x)}$As detector material. A quantum efficiency of 60% was measured for this detector at a wavelength of 1.06 µm, for low bias voltages. The loss in the waveguide was less than 1 dB/cm. At bias voltage greater than about 40 V, avalanche multiplication was observed, with multiplication factors as high as 50. Optimum performance was obtained at a wavelength of 1.06 µm for an In concentration of $x = 0.2$. The fact that quantum efficiency in this device did not approach 100% more closely was most likely caused by less than optimum

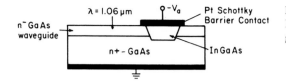

Fig. 15.10. Monolithically integrated InGaAs detector in a GaAs waveguide [15.22]

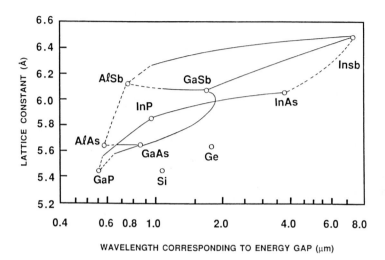

WAVELENGTH CORRESPONDING TO ENERGY GAP (μm)

Fig. 15.11. Dependence of absorption edge wavelength and lattice constant on composition for selected III-V alloys [15.23]

depletion width in the Schottky-barrier diode. The carrier concentration in the waveguide must be very carefully controlled in order to make W equal to the waveguide thickness. Defect centers associated with stress at the GaAs-GaInAs interface may have also played a role in reducing η_q.

In general, the III-V compound semiconductors and their associated ternary (and quaternary) alloys offer the device designer a wide range of bandgaps, and corresponding absorption edge wavelengths. The relationships between bandgaps, absorption edge wavelengths, and lattice constants are shown in Fig. 15.11, as given by *Kimura* and *Daikoku* [15.23]. Dotted portions of the curves correspond to ranges of composition for which the bandgap is indirect. Direct bandgap materials generally have interband absorption coefficients greater than $10^4 \, cm^{-1}$ for wavelengths shorter than the absorption edge, while α may be several orders of magnitude less in indirect gap materials. Nevertheless effective detectors can be made in indirect bandgap materials, especially when the waveguide detector geometry is used, so that the length L can be adjusted to compensate for small α.

The most popular material for the fabrication of detectors in the 1.0 to 1.6μm wavelength range is GaInAsP. Fig. 15.12 shows a planar embedded GaInAs p-i-n photodiode on an InP substrate [15.24]. The device was fabri-

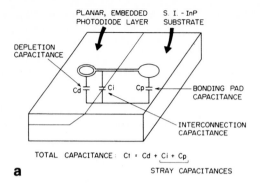

PLANAR, EMBEDDED PHOTODIODE LAYER   S. I. - InP SUBSTRATE

DEPLETION CAPACITANCE

BONDING PAD CAPACITANCE

INTERCONNECTION CAPACITANCE

TOTAL CAPACITANCE: $C_t = C_d + C_i + C_p$

STRAY CAPACITANCES

a

Fig. 15.12.a A planar embedded GaInAs p-i-n photodiode [15.24], **b** cross-section of the planar embedded photodiode [15.24]

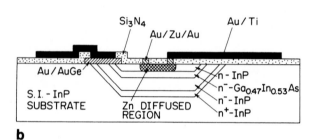

b

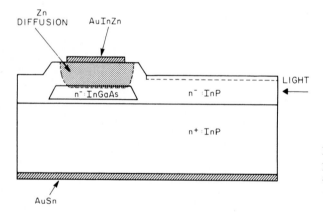

Fig. 15.13. A p-i-n photodiode integrated with a ridge waveguide [15.25]

cated by means of preferential ion beam etching and vapor phase epitaxy. Since the bonding pads lie mostly over semi-insulating substrate material, stray capacitances have been reduced to the point at which total device capacitance is only 0.08 pF for a diameter of 20 µm of photosensitive area. As a result, this device has a cutoff frequency of 14 GHz, limited by the carrier transit time. An integrated waveguide p-i-n photodetector coupled to a ridge waveguide on InP is shown in Fig. 15.13 [15.25]. The device was fabricated by means of a metalorganic vapor phase epitaxial (MOVPE) regrowth technique. The detector can be used in the 1.0 to 1.6 µm wavelength range; and has a 3 dB band-

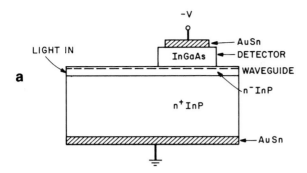

Fig. 15.14. An InGaAs detector grown on top of a ridge waveguide [15.26]

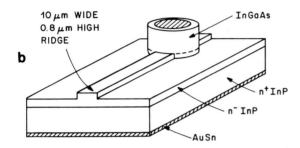

width of 1.5 GHz with a pulse response (full-width-at-half-maximum) of 80 ps at 1.3 μm wavelength. The waveguides had an average propagation loss of 3 dB/cm and 95% of the guided light was coupled into the photodetector.

Another approach to integrating a GaInAs detector and an InP waveguide is shown in Fig. 15.14 [15.26]. In this case an InGaAs detector was grown on top of the waveguide. Since InGaAs has a larger index of refraction than InP, light is coupled out of the waveguide up into the detector. This same approach also has been used by *Erman* et al. [15.27] to couple an InGaAs detector to an inverted rib InP of GaInAsP waveguide on an InP substrate.

15.3.3 Proton Bombardment

In Chap. 4, proton bombardment was described as a method for producing optical waveguides in a semiconductor by generating carrier-trapping defect centers that resulted in reduced carrier concentration and increased index of refraction. In that case, the waveguides were always sufficiently annealed after proton bombardment to remove the optical absorption associated with the trapping centers. However, one of the mechanisms responsible for this absorption is the excitation of carriers out of the traps, freeing them to contribute to the flow of photocurrent. Thus, a photodiode can be fabricated by forming a Schottky-barrier junction over the implanted region, as shown in Fig. 15.15 [15.28]. (A shallow p⁺-n junction could also be used). Photocurrent flows when

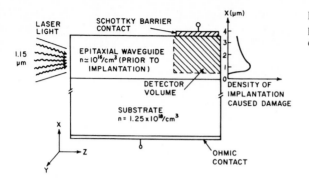

Fig. 15.15. Diagram of a protonimplanted optical detector [15.28]

the junction is reverse biased, because carriers liberated by photo-excitation within the depletion layer of the diode are swept across it by the field. Since a substantial number of the trapping centers have energy levels lying within the forbidden gap, the effective bandgap of the semiconductor is decreased, so that photons having less-than-bandgap energy can be absorbed and take part in the carrier generation process. Thus a proton-bombarded photodiode, made in a given semiconductor, can be responsive to photons that would ordinarily not be absorbed in the material.

For example, *Stoll* et al. [15.28] have made a detector in GaAs that is sensitive to 1.15 μm wavelength radiation. The optical waveguide structure consisted of a 3.5 μm thick n-type epitaxial layer (S-doped, $n \cong 10^{16}$ cm^{-3}) grown on a degenerately doped n-type substrate ($n \cong 1.25 \times 10^{18}$ cm^{-3}). Prior to proton implantation the optical attenuation at 1.15 μm was measured to be 1.3 cm^{-1}, but after implantation with a dose of 2×10^{15} cm^{-2} 300 keV protons in the region where a detector was desired, α increased to over 300 cm^{-1}. A partial annealing of damage at 500 °C for 30 min was performed to reduce α to 15 cm^{-1} in order to allow some optical transmission through the entire length of the implanted region. Then, an Al Schottky-barrier contact was evaporated on top of the implanted region to complete the device. The relative photoresponse of the proton implanted detector as a function of wavelength is shown in Fig. 15.16, along with the corresponding curve for a similar detector formed in unimplanted GaAs. Only negligible response occurs in the unimplanted GaAs at wavelengths longer than 9000 Å, but a substantial absorption *tail* is observed for the proton bombarded detector, extending to wavelengths as long as 1.3 μm.

It must be remembered that, even though α may be relatively small at the longer wavelengths, total absorption over the length of the detector can be quite large. For example, quantum efficiencies as high as 17% have been measured at 1.15 μm wavelength for devices of this type that were only 0.25 mm long. Calorimetric measurements made on proton-bombarded GaAs at a wavelength of 1.06 μm indicate that essentially all of the bombardment-induced optical attenuation can be attributed to absorption rather than diffuse scattering [15.29]. Calculations indicate that interband (carrier producing) absorptive transitions make up about 60% of the total absorption [15.29]. Thus quantum efficiency as high as 60% should be possible with this type of detector.

The principle of operation of the proton-bombarded detector is similar to that of conventional depletion layer photodiodes. Application of a reverse bias to the Schottky barrier produces a depletion layer which, if bias is sufficient and carrier concentration is low enough, extends completely across the high resistivity waveguiding layer to the lower resistivity substrate. Photoexitation of trapped carriers from bombardment-produced defect levels generates free carriers that are swept out of the depletion layer by the field, as shown in Fig. 15.17, thereby causing photocurrent to flow in the external circuit.

Proton bombarded detectors thusfar have been fabricated in GaAs and GaAlAs. However, the same bombardment-generated defect trapping mechanism is known to exist in GaP [15.30], ZnTe [15.31] and CdTe [15.32]. Hence,

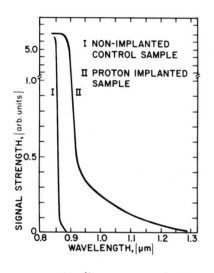

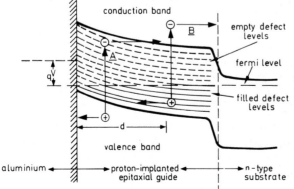

Fig. 15.16. Photoresponse of a proton implanted detector in GaAs

Fig. 15.17. Energy band diagram of a proton-bombarded detector under reverse bias and illumination

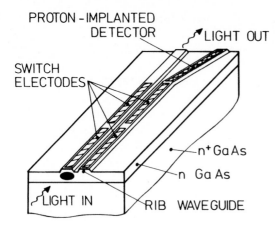

PROTON – IMPLANTED
DETECTOR

Fig. 15.18. An optically bistable device [15.33]

it seems likely that proton bombarded detectors can be made in these materials, and perhaps others as well.

The response time of proton-bombarded detectors is relatively fast for a device which depends on trapping and detrapping. Measurements made in GaAs with a Q-switched Nd:YAG laser at 1.06 μm wavelength indicate response times less than 200 ns, which was the rise-time of the laser pulse.

A proton implanted detector has been employed by *Carenco* and *Menigaux* [15.33] to make an optically bistable device, in GaAs, as shown in Fig. 15.18. The device consists of a rib-waveguide dual-channel directional coupler with a Schottky-barrier proton implanted detector in one of its output arms. The photocurrent output of the detector is amplified by an external electronic circuit and fed back to the control electrodes of the directional coupler switch to provide bistable response to an optical input signal. Operating at a wavelength of 1.06 μm, the bistable device switches in about 1 ns, the optical switching energy required being less than 1 nJ. The quantum efficiency of the proton implanted detector was measured to be 23% at 1.06 μm and 13% at 1.15 μm. The estimated bandwidth of the detector was 1.2 GHz.

15.3.4 Electro-Absorption

One additional method for producing the required shift of the absorption edge to longer wavelength in a monolithic waveguide detector is electro-absorption, or the Franz-Keldysh effect. When a semiconductor diode is reverse biased, a strong electric field is established within the depletion region. This electric field causes the absorption edge to shift to longer wavelengths, as shown in Fig. 15.19. Curve *A* shows the normal unbiased absorption edge for n-type GaAs with a carrier concentration of 3×10^{16} cm^{-3}. Curve *B* is a calculated Franz-Keldysh-shifted absorption edge for an applied field of 1.35×10^5 V/cm, which corresponds to 50 V reverse bias across a resulting depletion width of 3.7 μm.

At a wavelength of 9000 Å this shift corresponds to an increase in α from 25 to $10^4\,\text{cm}^{-1}$ – hardly a negligible effect!

The Franz-Keldysh effect has been well known for many years [15.34]. However it has only been applied to detector design fairly recently. The physical basis for the Franz-Keldysh effect can be understood from the simplified energyband bending model diagrammed in Fig. 15.20. In this diagram x represents distance from the metallurgical junction plane. In the region far from the junction where there is no electric field, photons must have at least the bandgap energy $(E_c - E_v)$ to produce an electronic transition as in (a). However, within the depletion region where field is strong, a transition as in (b) can occur when a photon of less-than bandgap energy lifts an electron partway into the conduction band, followed by tunneling of the electron through the barrier into a conduction band state. The states at the condution band edge are, in effect, broadened into the gap so as to produce a change in effective bandgap ΔE, which is given by [15.34]

$$\Delta E = \tfrac{3}{2}(m^*)^{-1/3}(q\hbar\varepsilon)^{2/3} , \tag{15.3.1}$$

where m^* is the effective mass of the carrier, q is the magnitude of the electronic charge, and ε is the electric field strength.

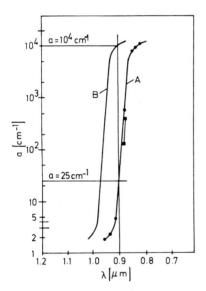

Fig. 15.19. Shift of the absorption edge of GaAs due to the Franz-Keldysh effect. (A) Zero-bias condition; (B) Reverse bias applied to produce a field of $1.35 \times 10^5\,\text{V/cm}$

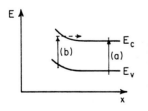

Fig. 15.20. Energy band diagram illustrating the Franz-Keldysh effect. The band bending on the n-side of a p⁺-n junction (or a Schottky barrier junction) is shown for conditions of strong reverse bias

The Franz-Keldysh effect greatly improves the sensitivity of a detector operating at a wavelength near its absorption edge. GaAs waveguide detectors operating at a wavelength of 1.06 μm have been demonstrated by *Nichols* et al. [15.35].

Perhaps the greatest advantage of electro-absorption detectors is that they can be electrically switched from a low absorption state to a high absorption state by merely increasing the reverse bias voltage. This makes it possible to make emitters and detectors in the same semiconductor material that are wavelength compatible. An example of a device making use of this principle is the emitter/detector terminal shown in Fig. 15.21 [15.36]. This device performs the dual function of light emitter, when forward biased, and light detector, when strongly reverse biased. Fabricated in series with a waveguide structure, as shown, it can act as a send/receive tap on an optical transmission line. Because of the large change in α produced by the Franz-Keldysh effect, operation can be very efficient. For example, consider the case of a p+-n junction diode in n-type GaAs with carrier concentration equal to 3×10^{16} cm^{-3}, as before. Application of 50 V reverse bias changes α from 25 to 10^4 cm^{-1} at a wavelength of 9000 Å. Thus, when forward biased, the diode emits 9000 Å light into the waveguide. When reverse biased with $V_a = 50$ V, the diode need have a length of only 10^{-3} μm in order to absorb 99.9% of incident 9000 Å light. When the diode is on *standby* at zero bias, α is just 25 cm^{-1}. Hence, for a typical laser length of 200 μm, the insertion loss is only 2 dB. Such emitter/detector devices may prove to be very useful in systems employing waveguide transmission lines because they greatly simplify coupling problems, as compared to those encountered when using separate emitters and detectors.

Continuing work on light emitting and detecting diode (LEAD) devices has produced a laser/detector in GaAlAs with a differential quantum efficiency of 30.9% in the emission mode, and a responsivity of 0.43 A/W in the detection mode [15.37, 38]. This responsivity at the operating wavelength of 905 nm is notably close to the 0.5 A/W which can be obtained with a silicon p-i-n diode such as is normally used. Laser/detector diodes have been used in a full duplex (simultaneous bidirectional) single optical fiber 16 km transmission system operating at a bit rate of 40 Mb/s [15.39]. In this system the laser/detector diodes function simultaneously as a local oscillator, a mixer and a transmitter. Distributed feedback LEAD devices have been reported by *Sakano* et al. [15.40]. In their InGaAsP OIC, shown in Fig. 15.22, DFB LEAD devices are used alternatively as lasers, detectors or amplifiers, combined with a carrier injection type switch. Thus, this single OIC chip can function in a

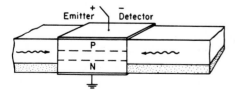

Fig. 15.21. An integrated-optic emitter detector terminal employing the Franz Keldysh effect

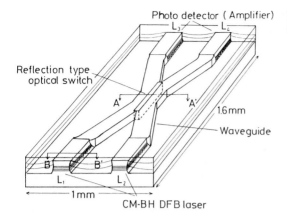

Photo detector (Amplifier)

Fig. 15.22. An InGaAsP OIC featuring DFB LEAD devices [15.40]

Reflection type optical switch

1.6 mm

Waveguide

1 mm

CM-BH DFB laser

variety of ways, such as a modulated or switched optical source, a photodiode-monitored source, or an amplified source. The lasers emit at 1.55 µm and are of the constricted-mesa buried heterostructure type. Typical threshold currents and differential quantum efficiencies are 30 mA and 0.08 mW/mA, respectively. When operated as detectors, the quantum efficiency of the diodes was 0.9% in the photovoltaic (zero-bias) mode. Optical amplification gain of up to 17.7 dB has been realized in these diodes [15.41].

15.4 Factors Limiting Performance of Integrated Detectors

In the design of an integrated optical detector, there are a number of mechanisms that can limit performance in various ways. Not all of these are important in every application. However, the designer (or user) should be aware of the limitations associated with different device types and geometries.

15.4.1 High Frequency Cutoff

A number of the factors that can limit high frequency response have been discussed in Sect. 15.1. These are summarized in Table 15.1, along with some additional frequency-limiting effects. Because of the small area of waveguide photodetectors of the type shown in Fig. 15.5, the RC time constant, which most often limits the response of conventional diodes, can be made small enough to allow frequencies of operation well above 10 GHz, as discussed in Sect. 15.1.2. In this case, other potentially limiting effects must be considered.

The drift time of carriers across the depletion layer can be minimized by designing the device so that the field in the depletion layer is high enough that carriers travel at the scattering limited velocity. For example, in GaAs the scattering limited velocity of 1×10^7 cm/s is reached in electric field strengths

Table 15.1. Factors limiting high frequency response of a depletion layer photodiode

RC time constant due to bulk series resistance and junction capacitance
Carrier diffusion time from regions ouside of the depletion layer
Carrier lifetime and diffusion length
Capacitance and inductance of the package
Carrier drift time across the depletion layer
Carrier trapping in deep levels

greater than about 2×10^4 V/cm. Hence, the transit time across a typical 3 µm wide depletion layer can be made as small as 3×10^{-11} s. However, it is important that the detector be designed so that the depletion layer extends entirely through the waveguide to the substrate, so that all carriers are generated within the depletion layer itself. If depletion is incomplete, carriers generated in undepleted material must diffuse relatively slowly into the depletion region before being collected, a process that occurs over roughly a minority carrier lifetime (about 10^{-8} s in lightly doped GaAs). A short optical pulse will therefore appear to have a long *diffusion tail* when detected.

Carrier trapping in deep levels can also cause a *tail* on the detected waveforms of a short pulse, since detrapping times can be relatively long. Deep level traps are usually associated with defects present in the semiconductor crystal lattice. Hence, special care should be taken in materials selection and device fabrication to minimize the number of defects.

When waveguide detectors are properly designed and are appropriately fed with a microwave transmission line, bandwidths greater than 50 GHz and impulse responses less than 10 ps can be obtained [15.42].

15.4.2 Linearity

A depletion layer photodiode, when reverse biased by more than a couple of volts, operates in the photodiode (or photoconductive) mode. It then functions as a current source, with its current being proportional to the input optical power, up to power levels of typically 1 mW. Hence, in most applications it is a highly linear device. At higher power levels, saturation occurs when the concentration of photo-generated carriers is so large that the field in the depletion layer is reduced by space-charge effects. This field reduction is particularly important in high frequency applications, because it may then result in carriers traveling at less than the saturation-limited velocity.

15.4.3 Noise

The effects of noise in waveguide detectors are essentially the same as they are in conventional photodiodes. The major noise components are thermal noise, arising in bulk resistances of the device, shot noise, associated with nonuniformities of current flow, such as carrier generation and recombination, and background noise, due to photons that are not part of the optical signal entering the detector [15.43].

From a relatively simple model developed by *DiDomenico* and *Svelto* [15.44], it can be shown that the signal-to-noise ratio in a depletion layer photodiode due to the effects of thermal and shot noise is given by [15.45]

$$(S/N)_{\text{power}} = \frac{\eta_q}{4B} M^2 \varphi_0 A \left(1 + \frac{2KT}{q} \frac{(\omega RC)^2}{RI_s} \right)^{-1}, \qquad (15.4.1)$$

where η_q is the quantum efficiency, B is the bandwidth, M is the modulation index (in the case of an intensity modulated light beam), φ_0 is the incident photon flux density, A is the area of the input face, R is the diode bulk resistance, ω is the modulation frequency of the optical signal, I_s is the reverse saturation (dark) current, K is Boltzmann's constant and C is the capacitance. In the case of an avalanche photodiode, there are additional noise souces associated with the statistical nature of the avalanche process that are not considered in (15.4.1) [Ref. 15.1, pp. 72–77].

The waveguide detector has an inherent advantage over conventional detectors in regard to background noise, because the waveguide acts as a filter to eliminate much of the background light. Close matching to the signal wavelength also reduces background noise. For example, if light of 8500 Å wavelength from a GaAlAs emitter is detected by a Si detector, background photons with wavelengths shorter than about 1.2 μm will also be detected. However if, instead, a GaAs detector is used, those photons with wavelengths in the range from 1.2 to 0.9 μm will not contribute to background noise [15.46].

Problems

15.1 We wish to use a photodiode as a detector for a signal of 9000 Å wavelength. Which would be the best choice of material for the photodiode, a semiconductor of bandgap = 0.5 eV, bandgap = 2 eV, or bandgap = 1 eV? Why? (Assume all three are direct gap and are equivalent in impurity content, etc.)

15.2 To improve the signal-to-noise ratio of the diode in Problem 15.1, we wish to use a semiconductor low-pass filter which has the following absorption properties at room temperature:
for 9000 Å radiation $\alpha = 0.2\,\text{cm}^{-1}$
for 7000 Å radiation $\alpha = 10^3\,\text{cm}^{-1}$.

How thick must the filter be to attenuate 7000 Å background noise by a factor of 10^4?

By what factor is the signal (at $9000\,\text{Å}$) attenuated by a filter of this thickness? Neglect reflection at the surfaces.

15.3 If the minimum useful photocurrent of the diode in Problem 15.1 is $1\,\mu A$ (peak pulse value), what is the minimum signal light intensity (peak pulse value) which must fall on the detector? Assume an internal yield or quantum efficiency $\eta_q = 0.8$ and sensitive area $= 10\,\text{mm}^2$.

15.4 Below is a list of semiconductor materials and their bandgaps.

	$E_g[eV]$
Si	1.1
GaAs	1.4
GaSb	0.81
GaP	2.3
InAs	0.36

a) Based solely on bandgap energy, which of these materials could *possibly* be used to make a detector for the light from a GaAs laser?

b) Which of the following ternary compounds could *possibly* be used to make a detector for the light from a GaAs laser?

$$GaAs_{(1-x)}Sb_x \qquad GaAs_{(1-x)}P_x \qquad Ga_{(1-x)}In_xAs$$

c) If a reverse biased Si photodiode with a quantum efficiency of $\eta = 80\%$ and an area of $1\,\text{cm}^2$ is uniformly illuminated with the light from a GaAs laser to an intensity of $10\,\text{mW/cm}^2$ what is the photocurrent which flows?

15.5 a) Determine the maximum value of the energy gap which a semiconductor, used as a photodetector, can have if it is to be sensitive to light of wavelength $\lambda_0 = 0.600\,\mu m$.

b) A photodetector whose area is $5 \times 10^{-6}\,\text{m}^2$ is irradiated with light whose wavelength is $\lambda_0 = 0.600\,\mu m$ and intensity is $20\ \text{W/m}^2$. Assuming each photon generates one electron-hole pair, calculated the number of pairs generated per second.

c) By what factor does the answer to (b) change if the intensity is reduced by a factor of $1/2$?

d) By what factor does the answer to (b) change if the wavelength is reduced by a factor of $1/2$?

15.6 We wish to design a waveguide photodiode in GaAs, with the geometry as shown in Fig. 15.5, for operation at $\lambda_0 = 0.900\,\mu m$ wavelength.

a) If the photodiode has a waveguide thickness and depletion width $W = 3\,\mu m$ at an applied reverse bias voltage of $40.5\,V$, what is the magnitude of the change in effective bandgap due to the electric field? (Assume $m_e^* = 0.067\,m_0$).

b) What length L is required to produce a quantum ifficiency of 0.99? (Assume that scattering loss and free carrier absorption are negligible).

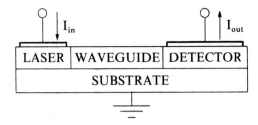

15.7 We wish to design an opto-isolator OIC consisting of a laser, waveguide, and detector monolithically coupled as shown above.
The following assumptions apply:
   1) All coupling of energy from input to output is via photons emitted by the laser and detected by the detector. Electrical leakage currents are negligible.
   2) Neglect optical coupling losses between the laser/waveguide and waveguide/detector.
   3) The current source driving the laser has a maximum current = 5 Amps.
   4) It is required that the current transfer ratio be

$$\frac{I_{out}}{I_{in}} \geq 0.1.$$

Question
Which combination(s) of lasers (a, b, c), waveguides (d, e), and detectors (f, g) described in the following paragraphs would be suitable? For each possible combination calculate I_{out}/I_{in}. *Show all work (calculations) to indicate why certain combinations are possible or not.*

Possible lasers
   a) A DFB GaAs/(GaAl)As laser with grating spacing $\Lambda = 3400$ Å, index in the light emitting layer $n = 3.6$, operating in the third-order reflection mode.
   b) An unconfined field GaAs laser with Fabry-Perot type endface reflectors. The following parameters have been either measured or established.
   1) Half-power points of the emission peak for spontaneous emission have been measured for this material at room temperature to be at 9200 Å and 8800 Å.
   2) Index of refaction = 3.3
   3) Thickness of light emitting layer = 10 microns
   4) Thickness of active (inverted pop.) layer = 1 micron
   5) Internal quantum efficiency = 0.7 (above threshold)
   6) Average absorption coefficient = 30 cm^{-1}
   7) Width = 300 µm
   8) Length = 935 µm
   9) Carrier energy distribution factor $\zeta = 1$
   10) Reflectivity of Fabry-Perot surfaces = 0.4

11) Negligible series $I^2 R$ loss

12) Emission wavelength $\lambda_0 = 9000 \, \text{Å}$

c) A confined field GaAs laser with all parameters the same as in (b) *except for the following*:

3,4) Thickness of the light emitting layer = thickness of active layer = $1 \, \mu\text{m}$

6) Average absorption coefficient = $10 \, \text{cm}^{-1}$

Possible Waveguides

d) $Ga_{0.9}Al_{0.1}As$ on $Ga_{0.8}Al_{0.2}As$ (heteroepitaxial type) thickness of the $Ga_{0.9}Al_{0.1}As$ waveguiding layer = $1.0 \, \mu\text{m}$, length = 5 mm

e) Carrier concentration reduction type in GaAs

substrate $n = 2 \times 10^{18}/\text{cm}^3$

waveguide $n = 1 \times 10^{15}/\text{cm}^3$

thickness of the waveguiding layer = $3 \, \mu\text{m}$

length = 2 mm

Note: Use graphs given in Chap. 4 for determining waveguide index of refration (refraction) and attenuation (absorption loss).

Possible Detectors

f) GaAs schottky barrier avalanche type

depletion width = $3 \, \mu\text{m}$

length = 0.1 mm

photomultiplication factor $M_{ph} = 2$

g) GaAs Franz-Keldysh type

depletion width = $3 \, \mu\text{m}$

length = 0.1 mm

applied electric field in depletion layer = $2 \times 10^7 \, \text{V/m}$

Note: Assume all absorption in the detectors results in carrier generation, i.e. the quantum efficiency equals 1.

15.8 A GaInAs avalanche photodiode is used to detect the pulses of light emitted by a GaInAsP laser at a wavelength of $1.3 \, \mu\text{m}$. The photodiode has a quantum efficiency of 0.85 at this wavelength when operated at low reverse bias voltage, and has a photomultiplication factor of 10^3 when properly biased at avalanche breakdown. The area of the photodiode is $10^{-4} \, \text{cm}^2$.

What photocurrent will be produced when the photodiode is biased in the avalanche regime and a light intensity of $10^{-3} \, \text{W/cm}^2$ from the laser is falling on it?

15.9 The quantum efficiency of an InGaAsP/InP avalanche photodiode is 80% when detecting $1.3 \, \mu\text{m}$ wavelength radiation and biased at low voltage so that no avalanche multiplication is occurring. When biased at higher voltage, for avalanche detection, an incident optical power of $1.0 \, \mu\text{W}$ (at $\lambda = 1.3 \, \mu\text{m}$) produces an output photocurrent of $20 \, \mu\text{a}$.

(a) What is the photomultiplication factor (avalanche gain)?

(b) If the responsivity of the diode falls off rapidly for wavelengths greater than $1.7 \, \mu\text{m}$ what is the energy bandgap in the absorbing region of the diode?

16. Quantum-Well Devices

In all of the devices discussed in previous chapters the dimensions of device structures were large compared to the wavelength of electrons in the device. When the dimensions of the structure are reduced to the point at which they are approaching the same order of magnitude as the electron wavelength some unique properties are observed. This is the case with a class of devices that have come to be known as "quantum well" devices, which feature very thin epitaxial layers of semiconductor material. This chapter will introduce the basic concepts of quantum wells and will describe some of the novel kinds of devices that can be made by using them. Improved lasers, photodiodes, modulators and switches can all be made by employing quantum well structures. Quantum well devices can be monolithically integrated with other optical and electronic devices to produce optical integrated circuits and opto-electronic integrated circuits.

16.1 Quantum Wells and Superlattices

A quantum well structure consists of one or more very thin layers of a relatively narrow bandgap semiconductor interleaved with layers of a wider bandgap semiconductor, as shown in Fig. 16.1. The thickness of the layers is typically 100 Å, or less. An arrangement like that shown in Fig. 16.1, with many layers, is called a "Multiple Quantum Well" (MQW) structure. Single Quantum Well (SQW) structures, with one narrow bandgap layer, are also useful. The required thin layers can be grown by either MBE or MOCVD techniques such as those described in Chap. 4. A typical MQW structure might have about 100 layers. The GaAs-AlAs material system is particularly convenient for the growth of MQW devices because GaAs and AlAs have almost identical lattice constants and thus interfacial strain can be avoided. However, GaInAsP is also a suitable material, as long as the concentrations of the constituent elements are properly chosen to produce lattice matching. (In the case of high power lasers, it is sometimes beneficial to purposely produce a strained-layer structure to reduce losses, as is explained in Sect. 16.2.)

The special properties of these very thin layers result from the confinement of carriers (electrons and holes) in a manner analogous to the well-known quantum mechanical problem of the "particle in a box" [16.1]. In this

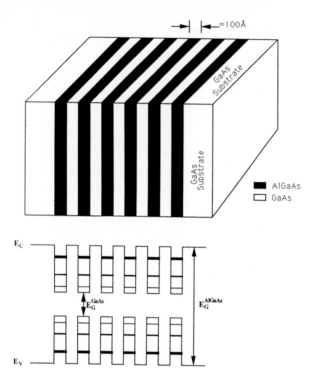

≈100Å

Fig. 16.1. Multiple-quantum-well (MQW) structure

case, the carrier is confined to the narrow bandgap "well" by the larger bandgap "barrier" layers. The magnitude of the wavefunction of the electron (or hole) must approach zero at the barrier wall because the probability of finding the particle within the wall is very small. (If the wall were infinitely high, the wave amplitude would have to equal zero at the barrier.) Hence, the wavefunction must form a standing wave pattern, sinusoidally varying within the well and damping to near zero at the edge of the barrier. The set of wavefunctions which satisfy these boundary conditions corresponds to only certain allowed states for the carrier. The carrier motion is thus quantized, with discrete allowed energies corrsponding to the different wavefunctions.

In the MQW structure described in the previous paragraphs it was assumed that the barrier layers were sufficiently thick that the tail of the wavefunction penetrated only slightly into the barrier layer. Carriers are thus confined to a particular layer and act independently of carriers in the other layers. If a structure is fabricated with many wells and barrier layers, but the barrier layers are made so thin that the electron or hole wavefunction can penetrate the barrier, there is coupling between the allowed states. In that case, the electrons and holes are not clearly confined to a particular state associated with a given well. Instead they exist in a coupled system of states with the wavefunction extending over many wells. Thus the behavior of the carriers is influenced by the long-range periodic modulation superimposed upon the crystalline potential of the host material. The similarity between this situation and that which normally exists for a charge carrier in a crystalline lattice has led to the use of the term "superlattice" to describe such MQW structures with very thin barriers. The formation of a superlattice actually results in the creation of a new material, with a bandgap and electronic and optical properties which may be significantly different from those of the host material.

Both superlattices and quantum well structures can be fabricated in a variety of semiconductor materials. GaAs wells with GaAlAs barriers are generally used for wavelengths near 0.9 μm, while InGaAsP can be used for 1.3 and 1.5 μm wavelengths. Many other III–V and II–VI materials are also suitable. The use of quantum wells and superlattices in opto-electronic devices is a relatively new field of endeavor, and much research is presently being conducted. In the following sections of this chapter some of the more notable examples of quantum well devices ae described, along with the relevant theory of operation.

16.2 Quantum-Well Lasers

In a quantum well laser the quantization of the allowed electron and hole energies into discrete levels within the energy well, as shown in Fig. 16.1, reduces the total number of carriers needed to achieve a given level of population inversion. The free carrier absorption coefficient, which is proportional to the number of carriers, is also reduced. As a consequence, the threshold current density is reduced by approximately a factor of 10 as compared to that of a conventional double heterostructure laser diode.

16.2.1 Single-Quantum-Well Lasers

Only a single-quantum well is required in order to realize the benefits of quantization of the electron and hole states. If it is assumed that the potential well is deep (i.e., that the barrier height is infinite) the solutions of Schrödinger's equation in the nth quantized level is given by [16.2]

$$E_n = \frac{h^2}{2m_e}\left(\frac{n\pi}{L_z}\right)^2 + \frac{h^2}{2m_e}(k_x^2 + k_y^2), \quad n = 1, 2, 3 \ldots . \tag{16.2.1}$$

where the z direction is taken normal to the thin layer which forms the well, m_e is the effective mass, L_z is the thickness of the layer and k_x, k_y and k_z are the magnitudes of the wavevector. A corresponding relation for the energy of a hole in the nth level can be obtained by merely replacing m_e with m_h, the effective mass of a hole, and remembering that increasing energy (hence n) for a hole is in the downward direction in an electron energy diagram such as that shown in Fig. 16.1. Since

$$k_i = \frac{n\pi}{L_i}, \quad i = x, y, z \quad \text{and} \quad L_z \ll L_x, L_y,$$

(16.2.2)

the energy levels in the x and y directions will be closely spaced, and for each allowed value of E_n a two-dimensional energy band will exist in the x–y plane with a density of states that is independent of energy. Thus, for increasing energy, the cumulative density of states function will exhibit a sharp step at each allowed value of E_n. The magnitude of this step $\Delta\rho(E)$ is given by [16.3]

$$\Delta\rho(E) = \frac{m}{\pi\hbar^2},$$

(16.2.3)

where m is the effective mass of the electron or hole, as the case may be.

The assumption of an infinite barrier height in the derivation of (16.2.1) and (16.2.3) is not unduly restrictive. For most device applications, while the wells are finite depth, they are sufficiently deep that only a small perturbation of the wavefunction results as compared to that of the infinite-barrier case. The conduction and valence bands are still quantized into two-dimensional subbands with step-like densities of states. Since photon generation in a quantum well laser occurs through electron transitions from states in one of these subbands of the conduction band to states in one of the subbands of the valence band, the gain curve as a function of photon energy, $\hbar\omega$, has steps which occur at photon energies

$$\hbar\omega = E_g + E_{nc} + E_{nv},$$

(16.2.4)

where E_g is the bandgap energy and E_{nc} and E_{nv} are the energies of the nth levels in the conduction and valence bands, respectively. Plotted curves of gain vs. photon energy calculated by M. Mittelstein have been published by *Yariv* [16.4]. The estimated maximum gain of a single quantum well laser operating in the $n = 1$ transition is $g \simeq 100\,\text{cm}^{-1}$. However, in a practical laser, that gain must be multiplied by a confinement factor to account for the fact that the laser mode does not travel entirely within the active region. The confinement factor Γ_a is given by [16.5]

$$\Gamma_a = \frac{\int_{-L_z/2}^{L_z/2} |E|^2\, dz}{\int_{-\infty}^{\infty} |E|^2\, dz} \simeq \frac{L_z}{W},$$

(16.2.5)

where W is the width of the mode. Despite some loss of the potentially available gain because of modal spreading, the reduction in the number of carriers needed for population inversion, and reduced free carrier absorption, result in about a factor of 10 decrease in the threshold current density of the quantum well laser as compared to that of a conventional laser. The quantiza-

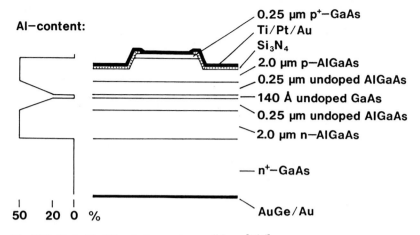

Fig. 16.2. GaAs/GaAlAs single quantum well laser [16.6]

tion of electron and hole allowed energy states also results in a decrease in the linewidth of the emitted radiation.

Quantum wells can be combined with many of the other structures that have been developed over the years to improve laser performance. For example, Fig. 16.2 shows the cross-section of a ridge-waveguide, graded index, separate confinement heterostructure, single quantum well laser [16.6]. This GRIN-SCH-SQW structure yielded a laser with a threshold current density of 20 mA, and a linewidth of only 1.5 MHz at an output power of 5 mW. The sidemode suppression was measured to be greater than 24 dB.

The structure shown in Fig. 16.2 was grown by the MOCVD method, at atmospheric pressure. The SQW was 140 Å thick, and was surrounded by linearly graded GaAlAs layers for efficient carrier collection [16.7] and improved optical confinement [16.8]. The optical confinement factor for this structure was calculated to be 4% [16.9]. Ridge waveguides 5 μm wide and 2 μm deep were fabricated by conventional wet chemical etching, and lasers with cavity lengths ranging from 300 to 800 μm were cleaved and mounted p-side down on heatsinks.

In operation the lasers exhibited a differential quantum efficiency of 66% at a temperature of 20 °C, controlled to within 0.01 °C by a Peltier cooler. The emitted wavelength was 872 nm, and a single longitudinal mode was observed for output power as high as 5 mW. As mentioned previously, the linewidth of the emitted light was unusually narrow. The linewidth-power product for a laser with cavity length $L = 800$ μm was determined to be 6.4 MHz·mW, with the narrowest linewidth of 1.5 MHz being measured at an output power of 5 mW. The linewidth power product was found to follow an L^{-2} dependence on cavity length, L. This laser demonstrates the advantages of combining the GRIN-SCH structure, for enhanced confinement characteristics, and the ridge waveguide, for low threshold current and single mode oscillation, with a long-cavity-length, SQW structure for further reduction of threshold current and linewidth of emission.

Single quantum well lasers with threshold currents less than 1 mA have been produced by adding reflective coatings to the front and rear facets of a GRIN-SCH-SQW structure [16.10]. A minimum threshold current of 0.95 mA was achieved with facet reflectivities of 70%, a cavity length of 250 μm, and an active region stripe width of 1 μm. The SQW in that case was a 100 Å thick layer of GaAs surrounded by 0.2 μm thick, undoped, graded layers of $Ga_{0.8}Al_{0.2}As \rightarrow Ga_{0.5}Al_{0.5}As$. An identical laser, but with uncoated facets, had a threshold current of 3.3 mA. When dielectric coatings ($1/4\lambda Si$, $1/4\lambda Al_2O_3$) were subsequently added to the facets of this control laser, its threshold current was reduced to 0.95 mA. The addition of reflective coatings to the facets of an SQW laser has a much greater effect on its threshold current than the same process would have on that of a conventional DH laser. This is due to the fundamentally lower current density needed to render the thin active layer transparent [16.11].

While the SQW laser diode is a relatively new device, its fabrication has not been confined to just the laboratory. SQW laser became available as commercial products as early as 1987.

16.2.2 Multiple Quantum Well Lasers

Since the gain of a single quantum well laser operating in the $n = 1$ transition is limited to about 100 cm^{-1}, it is often advantageous to use many well and barrier layers to form a multiple quantum well laser. At sufficiently high injection current levels so that the quasi Fermi levels cross over the $n = 1$ levels in both the conduction and valence bands, the gain of an N well structure is just N times that of a single quantum well. However, at lower drive current levels, the gain of the MQW structure is actually less than that of the single quantum well [16.11].

While a higher level of drive current density is required to optimize the gain of an MQW laser, that doesn't mean that a low threshold current cannot be achieved. For example, *Tsang* [16.12] has reported MQW lasers with a threshold current of only 2.5 mA. In that case the laser featured a 4 well structure with 208 Å thick GaAs wells and 76 Å thick barriers of $Ga_{1-x}Al_xAs$. The optimum aluminum concentration in the barriers was shown to be $x \simeq 0.2$. Greater concentration of Al in the barrier layers was found to unduly inhibit the injection of carriers because of the greater barrier height.

MQW lasers exhibit less temperature dependence of the threshold current than do conventional DH lasers. The temperature dependence of threshold current density cay be fitted to the relation [16.13]

$$J_{th}(T) = J_{th}(0) \exp\left(\frac{T}{T_0}\right), \tag{16.2.6}$$

where the constant T_0 gives a measure of the degree of temperature insensitivity. For conventional DH lasers T_0 is usually in the range from 120 to 165 °C, while for MQW lasers T_0 may be over 400 °C. Of course, the value of T_0 depends strongly on the particular MQW structure used and on the specific geometry of the laser diode. However, values of T_0 reported for MQW lasers generally are significantly greater than those for conventional DH lasers. For example, *Chin* et al. [16.13] have measured T_0 for stripe geometry quantum well lasers grown by MOCVD and found $168° \leq T_0 \leq 437$ °C. Diodes which emitted at longer wavelengths were found to exhibit larger values of T_0 (in the wavelength range 8435–8650 Å).

The emission wavelength of a quantum well laser depends strongly on the dimensions of the quantum well structure, since electron transitions occur between the subbands of the conduction band and those of the valence band. *Xu* et al. [16.14] have measured the emission wavelengths of MOCVD-grown MQW lasers in the temperature range from 77 K to 300 K and found a close correspondence to the subband energy levels and to the bandgap energy. In the emission spectra shown in Fig. 16.3 the dominant emission peak corresponds to the $n = 1$ subband and the smaller peak corresponds to the $n = 2$ subband state. The fact that the emission wavelength also depends on the dimensions of the quantum well structure is shown in Fig. 16.4 a, in which the emission spectra for two InGaAsP MQW lasers are displayed [16.15]. The lasers are essentially the same except that sample # 1 has 5 well layers with $L_z = 200$ Å and sample # 2 has 3 well layers with $L_z = 100$ Å. A significant

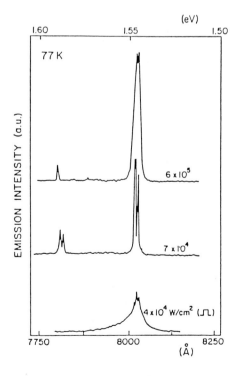

Fig. 16.3. Emission spectra of an MQW laser [16.14]

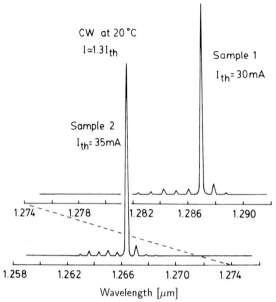

CW at 20°C
$I \simeq 1.3 I_{th}$

Sample 1
$I_{th} = 30 \, mA$

Sample 2
$I_{th} = 35 \, mA$

1.274   1.278   1.282   1.286   1.290

1.258   1.262   1.266   1.270   1.274

Wavelength [μm]

a

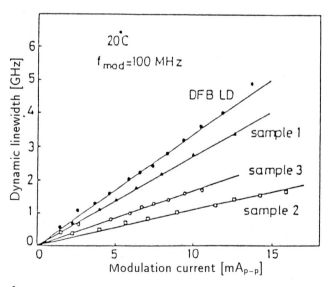

b

Fig. 16.4.a. Emission spectra for two InGaAsP MQW lasers [16.15]. **b** Emission linewidth of InGaAsP MQW lasers [16.15]

difference in the emitted wavelengths is noted due to the quantum-size shift effect [16.16].

The lasers for which the spectra of Fig. 16.4 a were measured also serve as a good illustration of the narrow linewidth obtainable from MQW lasers. These are BH stripe-geometry lasers with a stripe width of about 2.5 μm and a cavity length of 300 μm. Notably, they were fabricated by liquid phase epitaxy (LPE) rather than by MOCVD or MBE. A rapid sliding growth technique was used at a low growth temperature (589 °C) [16.17]. The dynamic spectral linewidths of these lasers were measured for a modulation frequency of 100 MHz. The dynamic linewidth is particularly important in regard to long distance telecommunications because of the problem of frequency shift during pulsing, or "chirping", which was described in Chap. 14. The modulation current dependence of dynamic spectral linewidth is shown in Fig. 16.4b. Samples 1 and 2 are as in Fig. 16.4a; sample # 3 is identical except it has $L_z = 150$ Å. The dynamic linewidth of a conventional DFB laser is also shown in the figure to permit easy comparison. It can be seen that the dynamic linewidth of the MQW lasers decreases as the quantum well width decreases and that it is less than that of the DFB laser for all cases shown. The lasers were DC biased at 1.3 I_{th} for these measurements. The reduced chirping in the MQW lasers can be attributed to reduction of the linewidth enhancement factor α and the confinement factor Γ_a [16.18]. The minimum spectral linewidth of an unmodulated DFB-MQW laser is inherently very narrow, as demonstrated by *Okai* et al. [16.19] who have measured a spectral linewidth of 56 KHz for one at an output power of 25 mW. White noise was found to be the limiting factor.

In the growth of DH lasers and MQW lasers the usual practice is to control the compositions of adjacent layers appropriately so as to produce lattice matching and eliminate interfacial strain. However, recent work has shown that in the case of high power lasers, it may be desirable to purposely produce a controlled amount of strain. Strained InGaAs layers grown on InP have a modified valence band structure [16.20] and an increased conduction band discontinuity [16.21], which it is believed may eliminate Auger recombination and intervalence-band absorption that otherwise cause optical losses [16.22]. A CW output power of 120 mW has been reported for 1.5 μm wavelength modulation-doped InGaAs strained-layer quantum well lasers [16.22]. In these devices, four 30 Å thick $In_{0.8}Ga_{0.2}As$ wells with 1.8% strain separated by three 200 Å thick InGaAsP barrier layers were grown on an InP substrate by low pressure MOCVD. The cavity length was 750 μm. A separate confinement, buried heterostructure was combined with the strained MQW structure. Kink free CW operation was observed at output power levels as high as 120 mW per facet. For pulsed operation with a 400 ns pulse width at a 40 kHz repetition rate, an output power of 360 mW per facet was obtained. From measurements of the differential efficiency of devices with different lengths, a cavity loss of $\alpha = 13$ cm^{-1} and an internal quantum efficiency of $\eta_i \simeq 100\%$ were determined, indicating reduced optical losses.

Strained-layer quantum well heterostructures can also be fabricated in the InGaAs–GaAs–AlGaAs system to produce lasers emitting in the 0.9 to 1.1 µm wavelength range. High power diode lasers operating in that range are attractive as small, efficient, low-cost optical pumping sources for rare earth doped glass lasers. *Coleman* [16.23] has reported stripe geometry, buried heterostructure lasers of this type that exhibit very low threshold currents (7–9mA), stable fundamental optical mode properties to greater than 30 times the threshold current and high power operation (>130mW/uncoated facet). By combining devices of this type in a monolithic parallel array having a cavity length equal to 440µm and a width of 1.6mm, an output power of 2.5 W per uncoated facet was obtained at a wavelength of 1.03 µm. The far field emission pattern of the array indicated that the lasers were phase locked.

The concept of the quantum well laser, in which carriers are narrowly confined in one dimension, can be extended to the two and three-dimensional analogues of the "quantum wire" and "quantum dot" lasers. For example, in a quantum dot laser, the electron would be confined in all three dimensions to a distance on the order of its wavelength in the crystal. One could picture GaAs dots with a GaAlAs host. Theoretical analyses predict that quantum dot or quantum wire lasers would have lower threshold currents, lower noise performance and higher modulation speeds than those of even quantum well lasers [16.24]. While quantum wire and quantum dot lasers are an intriguing concept, it will be very difficult to fabricate them. One approach is to begin with a substrate containing a quantum well structure, and then to define lateral dimensions by electron or ion beam lithography coupled with either etching [16.25] or ion implantation [16.26]. A more sophisticated approach is to produce localized compositional control through growth on patterned substrates [16.27]

The advent of quantum well structures has changed the direction of semiconductor laser research and development over a period of just a few years. Substantial improvements in threshold current density, linewidth and temperature sensitivity have already been demonstrated, even as compared to the properties of the DFB laser. Combining DFB and MQW structures has produced devices in InGaAsP with linewidth-power products of 1.9 to 4.0 with minimum linewidths of 1.8 to 2.2MHz [16.28]. Some quantum well lasers already have found their way to the commercial market. It seems certain that continued development of this relatively new device, perhaps including the development of quantum wire or quantum dot lasers, will lead to further improvements in the operating characteristics of semiconductor laser diodes.

16.3 Quantum-Well Modulators and Switches

The same type of quantum well structures that have proven so useful in laser diodes have also been found to improve the performance of modulators and switches. In these devices the improvement generally results from enhancement of the electro-optic and electro-absorption effects.

16.3.1 Electro-Absorption Modulators

In "bulk"semiconductor crystalline materials the absorption spectrum has a characteristic "absorption edge" at approximately the wavelength corresponding to the bandgap energy. For longer wavelengths interband absorption is absent, but the absorption coefficient α increases rapidly as wavelength is decreased below the bandgap wavelength. For quantum well structures, because of the quantization of both electron and hole energies, the absorption spectrum has a series of steps corresponding to the $n = 1, 2, 3\ldots$ energy levels. These steps are typically 50 to 200 meV in width, and the first step occurs at an energy somewhat above the bandgap energy. In other words, the absorption edge is shifted to a higher energy corresponding to the $n = 1$ transition. In addition, "exciton" peaks are observed at each step. These peaks correspond to absorption of photons which create pairs of electrons and holes that are not completely separated but are orbiting around each other. The pair resembles a hydrogen atom with the hole acting as the proton. Normally, excitons have such a short lifetime that their effects can be observed only at cryogenic temperatures. However, in a quantum well, the hole and electron of the exciton pair are compressed to a closer spacing in one dimension, typically on the order of 100 Å. The spherical shape of the exciton changes to that of a three-dimensional ellipsoid. This greatly increases the coulombic bonding force and makes the exciton more stable, even at room temperature. The presence of excitons is significant because it strengthens the response and sharpens the absorption curve near the bandgap wavelength at which electro-absorption devices work.

In a bulk semiconductor material, application of an electric field induces a small shift in the exciton absorption peak to longer wavelengths because polarization reduces the binding energy. The effect is called the Stark effect by analogy with the Stark shift in the absorption lines of a hydrogen atom in the presence of an electric field. However, in a bulk semiconductor, the small Stark shift is masked by a broadening of the exciton peak due to exciton ionization, so it is not very useful.

If an electric field is applied to a quantum well structure in the direction normal to the planes of the wells, the electrons and holes of the exciton pairs are confined. The hole and electron tend to move to opposite sides of the well and the polarization reduces the binding energy of the pair, but the walls prevent the electron and hole from going too far. Thus, field ionization of the exciton is inhibited, and the sharp exciton absorption peaks are preserved. This "Quantum Confined Stark Effect" QCSE [16.29] is very useful in modulators and switches because even moderate fields ($\sim 10^4$ V/cm) can cause a significant change in absorption [16.30]. The electro-absorption effect is approximately 50 times larger in multiple quantum well structures than it is in bulk semiconductors.

Early QCSE modulators were made in MQW layers of GaAs and GaAlAs in which the light was introduced in the direction normal to the planes of the quantum wells [16.29, 31, 32]. Even though an absorption change $\Delta\alpha$ as much as fifty times larger than that of bulk electro-absorption modulation was observed for drive voltages less than 10 volts, the all-important on-off ratio was limited. The on-off ratio of an electro-absorption intensity modulator is given by [16.33]

$$R = \exp\Delta\alpha L, \tag{16.3.1}$$

where $\Delta\alpha$ is the maximum achievable change in the absorption coefficient and L is the interaction length between the light and the absorptive material. In a transverse QCSE modulator, with light introduced in the direction normal to the planes of the wells, the interaction length L is limited to the thickness of the MQW structure, typically 1 μm at most. Thus, even though $\Delta\alpha$ is large, the on-off ratio will be relatively low. Simply growing more wells in the MQW structure to increase L is not the best approach since it would increase the drive voltage required to achieve a given magnitude of electric field.

Studies of the limits to normal incidence electro-absorption modulation in GaAs/GaAlAs MQW diodes [16.34] have indicated that there is an optimum number of wells for a given bias voltage, which increases roughly linearly with voltage. Performance also depends strongly on the well thickness, and to a lesser extent on background doping. For example, in samples with low back-ground doping ($\sim 10^{14}/cm^3$), optimum performance was $R = 2.1$ for 63 100-Å wells at 5 V operating voltage, and $R = 10$ for 200 100-Å wells at >14 V operating voltage. With a well width of 50 Å, improved performance compared to that of 100-Å well devices could be obtained, but only at operating voltages greater than 28 V. Generally an operating voltage greater than 10 V would be considered excessive.

A more effective way to increase L is to use a waveguide modulator structure such as that shown in Fig. 16.5 [16.35]. In this case the direction of propagation of the light is in the plane of the MQW layers. The light is confined to propagate in this plane by a waveguide structure. Two quantum wells, 94 Å thick, were placed in the center of a p-i-n diode, and light was confined in the plane of the layers by leaky superlattice (SL) waveguides. In this structure L can be made as long as necessary to produce the desired on-off ratio. Also, it is possible to introduce light with either a TE or TM polarization, which permits the fabrication of polarization selective modulators, since absorption is different for the two polarizations. In calculating the on-off ratio for a waveguide structure like that of Fig. 16.5, it is necessary to take into account the fact that the quantum wells make up only a part of the thickness of the waveguide. Thus (16.3.1) becomes [16.36].

$$R = \exp\Gamma\Delta\alpha L, \tag{16.3.2}$$

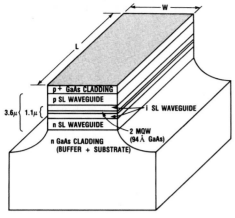

Fig. 16.5. Waveguide MQW electro-absorption modulator [16.35]

where Γ is the overlap integral between the quantum wells and the optical mode of the slab waveguide, $E_{opt}(Z)$:

$$\Gamma \equiv \frac{\int_{MQWs} |E_{opt}|^2 \, dZ}{\int_{-\infty}^{\infty} |E_{opt}|^2 \, dZ}. \qquad (16.3.3)$$

For the device of Fig. 16.5 the calculated value of Γ is 0.010 [16.37]. For a length of $L = 150\,\mu m$, and a width of $W = 40\,\mu m$, the measured on-off ratio was $R = 10$ for a drive voltage of 10 volts. An insertion loss of 7 dB was measured, which was attributed to endface reflections (3 dB), residual absorption in the "on" state (2 dB), and waveguide loss in the leaky waveguide structure (2 dB). Reflection loss, of course, could be greatly reduced by coating the facets with antireflecting dielectric films.

Since external modulators are generally only used at frequencies above 1 GHz, for which direct modulation of the semiconductor laser becomes difficult, the frequency response of electro-absorption modulators is an important parameter. Time domain measurements of the pulse response of the modulator shown in Fig. 16.5 have indicated a full-width-half-maximum (FWHM) pulse resolution of less than 100 ps [16.35]. The frequency response of MQW electro-absorption modulators is generally limited by the *RC* time constant of the modulator and its driver circuit. The series resistance of the MQW diode and its capacitance can both be made very small (resistance of a few ohms and capacitance of a few tenths of a picofarad). Thus operating frequencies greater than tens of GHz can be projected for the device itself. The limiting factors in most cases are the parasitic resistance, inductance and capacitance of the driver circuit, which have limited maximum frequencies. The solution to this problem lies in monolithically integrating the MQW modulator with the laser diode and with the driver circuit on a single semiconductor chip. Monolithically integrated DFB lasers and MQW electroabsorption modulators in GaInAsP have achieved modulation rates as high as 20 Gbit/s [16.38–40]. The work on longer wavelength devices will surely continue, since the major application for QCSE modulators is in long distance telecommunication systems, in which external modulators can be used to avoid high frequency chirping of directly driven laser diodes.

16.3.2 Electro-Optic Effect in Quantum Wells

The strong electro-absorption effect in quantum well devices implies that an electro-optic effect must also be present, because a change in absorption is linked with a change in index of refraction by the Kramers-Kronig relations [16.41]. These relations involve Cauchy integrals which permit a calculation of the index of refraction $n(\lambda)$ when the absorption spectrum $\alpha(\lambda)$ is completely known, or a calculation of the absorption coefficient $\alpha(\lambda)$ when the index spectrum $n(\lambda)$ is completely known. In practical applications $n(\lambda)$ or $\alpha(\lambda)$ are usually measured over only a limited wavelength range and projections are made of their values for wavelengths extending from zero to infinity. *Hiroshima* [16.42] has used the Kramers-Kronig relations to calculate electric field induced refractive index changes in GaAs/GaAlAs quantum well structures, near the excitonic absorption edge. The calculated maximum variation in the refractive index was approximately –0.065 ($\Delta n = 1.8\%$) for an electric field of 80 kV/cm for both 60-Å thick and 100-Å thick GaAs/$Ga_{0.6}Al_{0.4}As$ quantum wells. This value is consistent with the experimental value obtained by *Nagai* et al. [16.43], and is about two orders of magnitude greater than the refractive index variation in conventional III–V semiconductor bulk crystals.

The presence of a significant electro-optic effect could be a problem in MQW electro-absorption intensity modulators if it were strong enough to cause excessive phase modulation of the output light signal, resulting in chirping of the emission frequency. The relative strengths of the electro-absorption and electro-optic effects in MQW devices can be conveniently compared in terms of the "linewidth enhancement factor", given by

$$\text{LEF} \equiv \frac{\Delta n_{\text{real}}}{\Delta n_{\text{imag}}}, \tag{16.3.4}$$

where Δn_{real} and Δn_{imag} are the electric field induced changes in the real and imaginary parts of the index of refraction. Thus, in a purely electro-absorptive material, LEF = 0, while in a purely electro-optic material LEF = ∞. The value of the linewidth enhancement factor has been measured by *Wood* et al. [16.44] for an MQW electro-absorption modulator, and found to be LEF = 1. This measurement was made at the conditions of wavelength, bias and polarization which were optimum for intensity modulation, and hence is representative of typical modulator operating conditions. Since directly modulated lasers have values of LEF in the range of 3 to 6 [16.45], it appears that chirping of electro-absorption modulators due to the electro-optic effect will not be a serious problem. Semi-empirical calculations of the refractive index changes induced by a perpendicular electric field in the quantum wells of a QCSE modulator also have indicated that electro-absorption modulators are capable of very-low-chirp modulation [16.46].

While the electro-optic effect is not large enough to significantly disturb the operation of MQW electro-absorption modulators, it may be possible to use it to make MQW electro-optic phase modulators. Changes of the real part of the index of refraction in the presence of an electric field of as much as 3.7% have been reported for MQW structures [16.44], as compared to typical changes of only a few hundredths of a percent in bulk electro-optic materials. This large index change in MQW devices suggests that the length of electro-optic phase modulators might be significantly shortened by the use of MQWs. Of course, the modulator would have to be designed in such a way as to avoid the possible detrimental effect of electro-absorption intensity modulation.

16.3.3 Multiple Quantum Well Switches

Multiple quantum well structures can also be used advantageously in the fabrication of optical switches. The familiar dual-channel directional coupler switch described in Chap. 8 (Fig. 8.5) can be modified by using an MQW waveguide structure. Wa et al. [16.47] have made such a switch by depositing 20-μm wide gold stripe electrodes onto a waveguide consisting of a 1-μm thick MQW layer on top of a $Ga_{0.7}Al_{0.3}As$ layer, on top of a $Ga_{0.4}Al_{0.6}As$ layer, on a GaAs substrate. The MQW layer was composed of 25 GaAs wells, 100-Å thick, separated by 300-Å thick $Ga_{0.7}Al_{0.3}As$ barrier layers. The ratio of the thickness of the GaAs and GaAlAs layers was chosen so that the MQW layer and the $Ga_{0.7}Al_{0.3}As$ cladding layer could support only the lowest order TE and TM slab modes. The $Ga_{0.4}Al_{0.6}As$ layer was included to prevent any power from the evanescent tail of the guided mode from leaking into the heavily doped ($\approx 10^{18}/cm^3$) substrate. By evaporating two 0.5-μm thick gold stripes on the MQW layer surface, a strain pattern was produced in the semiconductor which resulted in a spatial variation in refractive index via the photoelastic effect, and four channel waveguides were formed underneath the edges of the metal stripes. The two waveguides in the channel regions between the gold stripes (spaced 10 μm apart) formed a dual-channel directional coupler.

This optical switch was tested using an 8500 Å GaAlAs laser diode as the light source. The critical coupling length of the coupler was found to be 1.7 mm, which is shorter than that which would be expected for similarly spaced channel waveguides without an MQW layer. Another interesting feature of this device was that it was possible to produce a phase shift of π radians, and resultant switching, through the nonlinear electro-optic effect, by increasing the optical power output of the laser. The differential phase shift $\Delta\phi$ for an input power P_i is given by [16.47]

$$\Delta\varnothing = \frac{2\pi n_2 p_i}{\lambda_0 A}\left(\frac{1-\exp(-\alpha l)}{a}\right),\qquad(16.3.5)$$

where A is the cross-sectional area of the guided wave, α is the attenuation constant, n_2 is a nonlinear refractive index coefficient, l is the waveguide length and λ_0 is the vacuum wavelength. For the device which has been described, the measured values of the various parameters were: $A = 10\ \mu m^2$, $\alpha = 15\ cm^{-1}$, $l = 2\ mm$, $\lambda_0 = 0.85\ \mu m$, leading to an estimated $n_2 \sim 10^{-7}\ cm^2/W$ based on observed phase shift. Optically controlled phase shift and switching such as was observed in this device are intriguing because they suggest the possibility of all-optical logic circuitry, capable of functioning without any electronic inputs or control signals. However, the required optical power level will have to be greatly reduced from that which was used in this device ($\sim 1\ mW$) before an optical computer of this type would be feasible.

The electro-absorption effect, which was described previously in regard to intensity modulators, can also be used to make optical switches. A 2×2 optical gate matrix switch, monolithically integrating MQW electro-absorption switches and waveguide circuits is shown in Fig. 16.6 [16.48]. Arrays of such switches could be used for constructing large scale optical switching systems. In the switch shown in Fig. 16.6, a small chip size has been obtained by monolithically integrating MQW electro-absorption switches with miniaturized optical splitters and combiners having corner mirrors fabricated by the reactive-ion-beam-etching (RIBE) method. Electrical isolation between the four MQW gates was achieved by forming grooves around the individual gates. The MQW structure consisted of 28 GaAs quantum well layers each 80 Å thick, separated by 80-Å thick GaAlAs barrier layers, incorporated into a p-i-n diode structure.

In operation near the exciton absorption peak wavelength (approximately 838 nm) the MQW gates had an extinction ratio (crosstalk) of 20 dB for an applied voltage of 12 V (for the TE mode). The gate switching response time was measured to be 28 ps. However, the insertion loss of the switch was relatively large – a total of 24 dB, including 9 dB propagation loss, 8 dB mirror loss, 4 dB scattering loss and 3 dB splitting loss.

The MQW electro-absorption switch is attractive because it can provide a high extinction ratio at a lower applied voltage than that required by a conventional DH electro-absorption switch, because of the QCSE. A high switching speed can be achieved because of the smaller device size (and capacitance). Also, MQWs have relatively low absorption saturation. *Wood* et al. have shown that on-off ratios of greater than 10 dB can still be obtained at intensities as high as 500 kW/cm^2, which corresponds to 10.5 mW coupled into a device of 1 μm $\times$ 3 μm cross-sectional area [16.49]. However, the optical losses due to absorption and scattering are relatively high.

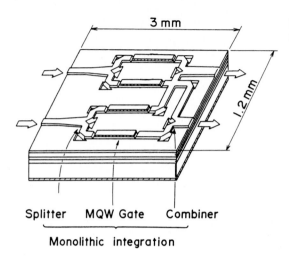

Fig. 16.6. 2×2 MQW optical gate matrix switch [16.48]

3 mm

1.2 mm

Splitter   MQW Gate   Combiner

Monolithic integration

16.4 Quantum-Well Detectors

The unique properties of MQW structures have been shown to be useful in several different types of photodetectors. In some of these devices photocurrent gains as high as 2×10^4 have been observed.

16.4.1 Photoconductive Detectors

An extremely large photocurrent amplification phenomenon has been observed at very low voltages in superlattices composed of 35-Å thick wells of $Ga_{0.47}In_{0.53}As$ interleaved with 35-Å thick barriers of $Al_{0.48}In_{0.52}As$ grown by molecular beam epitaxy [16.50]. Responsivities as high as 4×10^3 A/W were measured at 1.3 μm wavelength with 1.4 V bias, corresponding to a current gain of 2×10^4. Current gain of approximately 50 was observed for a bias voltage as low as 20 mV. This large gain arises from the large difference in tunneling rates of electrons and holes through the barriers because of their different effective masses in the MQW structure. In these semiconductors the electron effective mass is only about $0.05 m_o$; thus, they have a long quantum wavelength and can easily tunnel through the 35 Å thick barriers. The hole effective mass, however, is about $0.5 m_o$, so they have a much shorter wavelength and a much lower probability of tunneling. The substantial difference in the hole and electron tunneling rates results in a long electron lifetime and a large photocurrent gain. Because of the selective tunneling characteristics of the MQW structure, favoring the motion of electrons, these devices have been called effective mass filters (EMF). The EMF has inherently lower dark current and noise than a conventional photodiode because of the low voltage at which it can operate. Also, the EMF can be tuned for maximum gain by changing the thickness of the layers. If the thickness of the layers is adjusted to make it an integral multiple of the electron wavelength, resonant tunneling occurs and the gain is greatly increased. (The reported gain of 2×10^4 was for this resonant tunneling case).

Quantum photoconductivity has also been observed in a forward-biased p^+-n junction with an $Al_{0.48}In_{0.52}As$ (23 Å)/$Ga_{0.47}In_{0.53}As$(49 Å) superlattice in the n layer [16.51]. In that case a photocurrent gain of 2×10^3 was observed. Two other interesting features of the photoresponse were a large blue shift in the spectral response and a reversal of the direction of flow of the photocurrent when the magnitude of the forward bias voltage exceeded the built-in potential of the junction.

16.4.2 MQW Avalanche Photodiodes

$Al_{0.48}In_{0.52}As$/$Ga_{0.47}In_{0.53}As$ multiquantum well structures have also been used to make 1.3 μm wavelength avalanche photodiodes [16.52]. A p^+-i-n^+ structure was grown by MBE, consisting of a 35-period $Al_{0.48}In_{0.52}As$/$Ga_{0.47}In_{0.53}As$, 139-Å thick, MQW region sandwiched between p^+ and n^+ $Al_{0.48}In_{0.52}As$ transparent layers. DC current gain of 62 was measured. These devices were found to have good high frequency response, exhibiting an FWHM pulse response of 220 ps at a photocurrent gain of 12. The dark current at unity gain was 70 nA.

Low dark current InP/$Ga_{0.47}In_{0.53}As$ superlattice avalanche photodiodes, operating at wavelengths in the range of 1.3 μm to 1.6 μm have also been reported [16.53]. These devices had a superlattice consisting of 20 periods with 300-Å wells and 300-Å barriers. A photocurrent gain of 20 was measured at 0.9 V reverse bias, with a dark current of less than 10 nA.

16.5 Self-Electro-Optic Effect Devices

In Sect. 16.3.3, an optically switched MQW switch was described. That device employed the nonlinear electro-optic effect to produce a phase shift and resultant switching in a dual-channel coupler. However, it required an excssively large optical power ($\sim$1 mW) to produce switching. The dream of an optical computer requires low energy switching if it is to be realized. A device which offers promise of meeting that requirement is the self-electro-optic device (SEED) [16.54]. The fundamental concept of the SEED involves the combination of low power quantum well modulators with photodetectors to produce devices with both optcal inputs and outputs. Actually, a single diode can function simultaneously as both modulator and detector. In this case a p-i-n MQW diode is reverse biased through a bias resistor and light of the proper wavelength is introduced in the direction normal to the quantum well layers. The MQW structure is fabricated so as to place the excitonic absorption peak at the operating wavelength with zero bias applied to the diode. Then the bias voltage is chosen so that the excitonic absorption peak is shifted away from the desired operating wavelength when the full bias voltage is applied. Thus with no light falling on the input face of the diode, no photocurrent flows through the resistor and the full bias voltage is applied to the diode. However, when light is present at the input, photocurrent is generated and part of the bias voltage is dropped across the bias resistor. The reduced voltage across the diode results in increased absorption as the excitonic peak shifts toward the operating wavelength. This positive feedback results in a "snowballing" effect which switches the diode to a high absorption, low voltage state when the input optical power exceeds a certain critical threshold level. If the input light level is reduced, the same positive feedback mechanism causes the diode to switch back to a low absorption, high voltage state when a certain lower sustaining level is passed. Thus the SEED is an optically switchable bistable device. An optical crossover switching network in which MQW SEED arrays are used as electrically addressed four-function interchange nodes has been produced by *Cao* et al. [16.55]. This network features a 64 × 64 matrix of high-speed interchange nodes.

By replacing the bias (load) resistor in the previously described circuit with a second identical MQW diode it is possible to make a symmetric SEED (S-SEED) [16.56]. The key feature of the S-SEED is that it has gain, which permits switching at low levels of optical power. The S-SEED has two optical inputs, $P_{in\,1}$ and $P_{in\,2}$, and two optical outputs, $P_{out\,1}$ and $P_{out\,2}$, corresponding to light transmitted through each of the MQW diodes. If $P_{in\,1}$ and $P_{in\,2}$ are maintained equal the S-SEED will be in one state or another, with $P_{out\,1}$ high and $P_{out\,2}$ low, or vice versa, depending on the initial status of the diodes. If $P_{in\,1}$ and $P_{in\,2}$ are then both reduced by an identical amount the S-SEED will not switch. The switching of the device only occurs when the photocurrent in one diode starts to exceed that in the other one. In other words, the switching of the S-SEED is controlled by the ratio of input optical powers, not by their absolute levels. Thus the power in both input beams can be temporarily reduced, during which time an additional small amount of power added to one input beam can cause switching. Then the power of both beams can be increased. In this way a "time-sequential gain" is produced in which a very low power optical input signal can control a much larger output power. Switching energies of a few picojoules have been observed in S-SEEDs, and switching speeds of less than 1 ns have been demonstrated (at higher switching energies). Since these are relatively new devices, the prospects for further improvements in performance are good.

S-SEED arrays as large as 64 × 32 have been produced [16.57]. This array is only 1.3 mm square, with each S-SEED consisting of two MQW mesa diodes, 10 μm × 10 μm in size, connected in series electrically. Only two electrical connections are required for the bias supply. In terms of the number of logic gates available, the 16 × 32 S-SEED array corresponds to a chip with 6,144 logical pinouts, with two inputs and one output for each gate, although the chip actually has 16,384 physical "pinouts" because each gate has two optical input beams and two output beams. (A logical "1" state is one beam more powerful than the other; the "0" state is the reverse. Either $P_1 > P_2$ or $P_2 > P_1$ can be defined as the "1" state.)

S-SEEDs have been proposed for use in other applications, such as dynamic memories and oscillator circuits, as well as for digital logic. It is too soon to say whether the S-SEED will evolve into the basic element of future optical computers, but it is definitely a potential candidate because of its small size and relatively low switching energy.

16.6 Quantum-Well Devices in OEIC's

Quantum well devices have been incorporated into a number of different OEIC's for a variety of applications. The fact that quantum well devices can be fabricated in the same III–V semiconductor materials that are preferred for other opto-electronic devices facilitates their monolithic integration. One example of such integration has already been presented in Sect. 16.3.3, in which a 2×2 MQW gate matrix switch was described. In the remaining paragraphs of this chapter additional examples will be given.

16.6.1 Integrated Laser/Modulators

Since one of the major applications envisioned for MQW devices is the modulation of semiconductor diode lasers, the monolithic integration of laser diodes with MQW modulators is highly desirable. Such integration could reduce the optical coupling losses between laser and modulator, as well as facilitating the introduction of the modulator electrical driving signal if a microwave stripline were deposited on the surface of the chip. Of course, monolithic integration should also ultimately reduce the cost and improve the reliability of the laser/modulator pair.

A monolithically integrated quantum well modulator and laser diode have been made in GaAs/GaAlAs by *Tarucha* and *Okamoto* [16.58]. Both the laser and the modulator had the same MQW structure, as shown in Fig. 16.7. The MQW structure consisted of 16 periods of 80-Å thick GaAs wells and 50-Å thick $Ga_{0.75}Al_{0.25}As$ barriers, sandwiched between 1.5 μm thick $Ga_{0.57}Al_{0.43}As$ cladding layers, in a 20-μm wide stripe waveguide mesa. The laser diode and modulator were separated by a 2.6-μm wide gap formed by reactive ion beam etching. The waveguide length was 110 μm for the laser and 190 μm for the modulator. The capacitance of the modulator was 7 pF for zero bias voltage, and the reverse breakdown voltage was 9 V. The laser threshold current was approximately 30 mA. With the laser diode biased at $1.4I_{th}$, a low frequency modulation depth of 7 dB was measured. The high-frequency modulation response curve was flat out to about 0.3 GHz then fell off to a – 3 dB degrada-

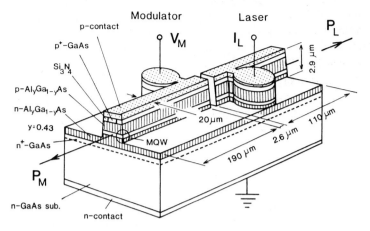

Fig. 16.7. Integrated MQW laser and modulator [16.58]

tion at about 0.9 GHz, limited by the *RC* time constant of the modulator. Reducing the waveguide width should improve not only the modulator frequency response but also the laser threshold current. For 1 3-μm wide, singlemode waveguide the predicted capacitance is 0.4 pF, leading to a projected 10 GHz bandwidth.

An InGaAsP/InP DFB laser and an InGaAs/InAlAs MQW modulator have been monolithically integrated on an InP substrate by *Kawamura* et al. [16.59, 60]. The emission wavelength was 1.55 μm. Notably the DFB laser was grown by liquid phase epitaxy, while the MQW modulator was prroduced by molecular beam epitaxy. First the DFB structure, consisting of a 0.15-μm InGaAsP active layer and a 0.15-μm waveguiding layer (with DFB corrugations), a 2.4-μm InP cladding layer and a 0.5-μm InGaAsP cap, was grown by LPE on a Si-doped InP substrate. This DFB structure was then removed from the portion of the substrate on which the modulator was to be grown by using masked etching. The InGaAs/InAlAs MQW p-i-n modulator structure was then grown by MBE using SiO$_2$ masking. The MQW structure consisted of 40 periods of undoped InGaAs/InAlAs (65-Å wells, 70-Å barriers) sandwiched with a Si-doped n-InAlAs layer and Be-doped p-InAlAs layer and capped with a 0.5-μm InGaAs layer. Finally, the DFB laser was masked and etched to produce a 10-μm wide stripe geometry laser (cavity length = 300 μm), and the MQW p-i-n layers were formed into a mesa-type optical modulator (mesa width 100 μm, cavity length 50 μm). The laser/modulator operated at 1.55 μm with a single longitudinal mode. An off-on ratio as high as 3 dB was observed, with a pulse response time of 300 ps.

Another example of an integrated MQW laser/modulator, of a different type than those described previously, is the monolithically integrated laser transmitter of *Nakano* et al. [16.61]. In that case, an MQW GaAlAs laser was integrated on a semi-insulating GaAs substrate with a driver circuit containing 10 field effect transistors. The FETs were fabricated by selective ion implantation of dopant atoms into the semi-insulating substrate. The laser had a threshold current density of 40 mA, and could be modulated at a data rate of 2 Gbit/s using non-return-to-zero (NRZ) coding.

An example of an integrated MQW device intended for OEIC receiver circuits is the monolithically integrated InGaAs/AlGaAs pseudomorphic single-quantum-well MODFET and avalanche photodiode reported by *Zebda* et al. [16.62]. In this circuit a $1\,\mu m$ gate length FET is combined with a $30\,\mu m \times 50\,\mu m$ APD which contains a 13-period GaAs/AlGaAs ($400/400\,\text{Å}$) MQW structure in its avalanching region.

Clearly MQW lasers, modulators and detectors can be monolithically integrated with other opto-electronic devices and with conventional electronic devices as well. Most of the integration so far has involved only one laser and a modulator. However, more sophisticated circuits will surely evolve. The four-channel transmitter array described in the next section is an example of one such circuit.

16.6.2 A Four-Channel Transmitter Array with MQW Lasers

A monolithically integrated four-channel opto-electronic transmitter array has been fabricated by *Wada* et al. [16.63], as shown in Fig. 16.8. Each transmitter channel consists of a GaAlAs MQW laser diode, a monitor photodiode, three FET's and a resistor. All components are fabricated monolithically on a single semi-insulating GaAs substrate. The mesa-etched, n-channel FET's were designed to have a transconductance of greater than 17 millimhos, and the resistors were made to be $50\,\Omega$ to match the input to an external microstrip line. The laser/photodiode pairs were aligned on the edge of the chip and an embedded heterostructure was used for both the lasers and monitor photodiodes. Etched wells were formed in the substrate to isolate the laser, photodiode and FETs from one another. The laser facets were cleaved after a graded index, separate confinement heterostructure, single quantum well (GRIN-SCH-SQW) laser structure was grown on the substrate by MBE. A ridge waveguide, 3 μm wide, was formed by etching the laser surface prior to metallization. The final chip had dimensions of $4 \times 2\,mm$, with 1-mm separation between the laser stripes. The laser cavity length was $40\,\mu m$, and the separation between the lasers and photodiodes was $60\,\mu m$.

Threshold currents for the lasers were measured to be in the range of $17.5 \pm 3.5\,mA$, and the differential quantum efficiencies were $54.5 \pm 9.5\%$. The *L-I* curve for the lasers was linear up to an output level of 4 mW/facet. Light emission was essentially single mode, at a wavelength of $8340\,\text{Å}$. The photodiode response curve was also linear, with a dark current of less than 100 nA. The power dissipation of the array chip was estimated to be approximately 1.2 W. Modulation of the laser diodes of all four channels at a data rate of 1.5 Gbit/s (NRZ format) was possible. Crosstalk between channels was

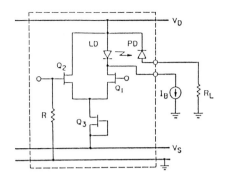

Fig. 16.8. Integrated four-channel transmitter [16.63]

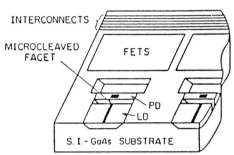

INTERCONNECTS

MICROCLEAVED FACET

FETS

PD

LD

S. I.- GaAs SUBSTRATE

measured to be – 28 dB at a frequency of 500 MHz, rising to a value of – 14 dB at 1 GHz.

Preliminary aging tests on these lasers have determined an output power degradation rate of 8%/kh, after an initial "burn-in" period of 200 h. The overall electrical-optical conversion efficiency was measured to be 6 mW/V, which makes the transmitter suitable for operation with ECL inputs signals. Overall, considering the relatively high levels of integration in this OEIC, the performance characteristics of the various components (and of the complete chip) are quite good. They clearly demonstrate the benefits obtainable by the monolithic integration of quantum well devices. Further improvements in technology will surely lead to even more impressive results.

Problems

16.1 Explain the difference between a multiple quantum well structure and a superlattice.

16.2 For a deep quantum well of GaAs with a thickness $L_z = 70 \text{Å}$, $L_x = L_y = 100 \mu\text{m}$, calculate the magnitude of the steps in the conduction band cumulative density of states function and the energies at which they occur relative to the bottom of the well for the $n = 1, 2$ and 3 levels. (Assume that the well has infinitely high sidewalls.) For $In_{0.53} Ga_{0.47}$ As the effective masses of electrons and holes are $0.045m_0$ and $0.465m_0$, respectively.

16.3   If the quantum well of Problem 16.2 were incorporated into a diode laser, what would be the emission wavelength for electron transitions between the $n = 1$ level in the conduction band and the $n = 1$ level in the valence band? (Assume 300 °K operation.)

16.4   a) Explain how a single p-i-n MQW diode can function as a self-electro-optic-effect device (SEED).

b) Why are SEEDs envisioned as being possible digital logic elements in a computer?

16.5   A single-quantum well of GaAs with a thickness of 60Å is incorporated into a GaAlAs heterojunction laser. The laser has a stripe geometry, being 200μm long and 10μm wide.

a.) What is the emission wavelength if the energy bandgap is $E_g = 1.52\,eV$ in the emitting region and electron transitions occur between the $n = 1$ level in the conduction band and the $n = 1$ level in the valence band?

b). If the gain coefficient of the laser in part (a) is $g = 80\,cm^{-1}$, what would be the gain coefficient of a multiple quantum well laser with 20 identical wells?

17. Applications of Integrated Optics and Current Trends

In the preceding chapters, the theory and technology of optical integrated circuits have been described. Although this a relatively new field of endeavor, numerous applications of OIC's to the solution of current engineering problems have already been implemented and some OIC's are now available as "off-the-shelf" commercial products. Of course, optical fiber waveguides, the companion element of OIC's in an integrated-optic system, are already well recognized as being very useful consumer products. In this chapter, some of the more recent applications of both fibers and OIC's are reviewed, and current trends are evaluated. In this review of representative integrated optic applications, specific systems and companies are named in order to illustrate the international character of the field and the types of organizations that are involved in it. Recommendation of any particular company or its products is not intended or implied. Also, the performance data that are quoted have generally been obtained from news articles and other secondary sources. Hence, they should be interpreted as being illustrative rather than definite.

17.1 Applications of Optical Integrated Circuits

17.1.1 RF Spectrum Analyzer

Probably the earliest demonstration of a multi-element OIC that was performed was the hybrid implementation of the real-time rf spectrum analyzer, which was originally proposed by *Hamilton* et al. [17.1]. The purpose of this spectrum analyzer is to enable the pilot of a military aircraft to obtain an instantaneous spectral analysis of an incoming radar beam, in order to determine if his plane is being tracked by a ground station, air-to-air missile, etc. Obviously, such information is required if he is to be able to quickly take effective evasive action. Of course, the frequency content, or *signature*, of all enemy radar signals that are likely to be encountered would have to be available for comparison, probably stored in the memory of the plane's onboard computer.

A diagram of the integrated-optic spectrum analyzer is shown in Fig. 17.1. Light from a laser source is coupled into a planar waveguide, in which it passes first through a collimating lens, then through a Bragg-type acousto-optic

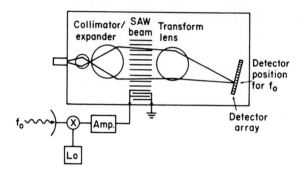

Fig. 17.1. Diagram of an integrated-optic rf spectrum analyzer

modulator. The rf signal to be spectrally analyzed is applied to the acoustic transducer that generates the sound waves, causing them to have a time varying period. Thus the deflection angle of the optical beam at the output of the modulator is a function of the rf signal. A second lens is used to focus the optical beam onto an array of photodetectors. If more than one frequency component is present in the rf signal, the light beam is divided into corresponding components that are focused onto different detector elements. Each detector element represents a particular frequency channel, and, since photodiodes generally have square law response, the output signal from any channel is proportional to the rf power at that frequency. The advantage of an integrated-optic spectrum analyzer, as compared to an electronic one, is that only a few optical elements are needed to perform a function that would otherwise require thousands of electronic elements.

The development of working models of the integrated optic rf spectrum analyzer took place at several different laboratories and extended over a number of years. The first working model was produced by the Westinghouse Advanced Technology Laboratories in 1980 [17.2]. It was fabricated on an X-cut $LiNbO_3$ substrate, approximately $7 \times 2.5 \times 0.3 \, cm^3$, in which a planar waveguide had been produced by indiffusion of titanium at $1000 °C$. The lenses used were of the geodesic variety, formed by machining *dimples* into the surface of the substrate prior to waveguide diffusion. In this type of lens, light waves are still confined by the waveguide, but they follow the longer curved path through the lens region. Since waves traveling near the center of the lens go over a greater path length than waves traveling near its edges, the wavefront is modified, so that focusing can occur. Such lenses can be made with surprising accuracy. The two aspherically-corrected geodesic lenses in the Westinghouse spectrum analyzer had essentially diffraction-limited spot sizes. The silicon diode detector array contained 140 elements, and was butt coupled to the waveguide. Design parameters for the Westinghouse integrated-optic spectrum analyzer are shown in Table 17.1. The input lens focal length was chosen to expand a 6μm GaAlAs laser spot to 2mm by diffraction.

The spectrum analyzer was tested first with a He-Ne laser source of 6328 Å wavelength and found to have a bandwidth of 400 MHz with a resolution of 5.3 MHz. Later results obtained by using a butt-coupled GaAlAs laser

Table 17.1. Westinghouse spectrum analyzer, design parameters [17.2]

Substrate size	$7.0 \times 2.5\,\text{cm}^2$
Front face to collimating lens	2.45 cm
Collimating lens diameter	0.80 cm
Collimating lens focal length	2.45 cm
Spacing between lenses	1.80 cm
Transform lens offset angle	3.79°
Transform lens offset	0.06 cm
Transform lens focal length	2.72 cm
Detector array pitch	12 μm
Laser beam width	6 μm
Number of detector elements	140
SAW transducer type	2 element, tilted

Table 17.2. Westinghouse spectrum analyzer, performance characteristics [17.2]

Center frequency	600 MHz
Frequency bandwidth	400 MHz
Frequency resolution	
with He–Ne 6328 Å source	5.3 MHz
with GaAlAs 8300 Å source	4 MHz
Detector integration time	2 μs
Detector element spacing	12 μm (with no dead space between elements)
Full width at half power	3.4 μm (1.02 × diffraction limited size)
of focused spot in detector focal plane	
Bragg diffraction efficiency	50 to 100%/w

diode emitting at 8300 Å as the source showed an improved resolution of 4 MHz. Other performance characteristics are given in Table 17.2. The 400 MHz bandwidth limitation is mostly caused by the acoustic transducer, and may be improved by using a more sophisticated transducer, as described in Chap. 9. In any case, the spectrum analyzer could be used over a wider frequency range by using a local oscillator and mixer at the input to the transducer, as shown in Fig. 17.1. Thus, heterodyning could be used to electronically shift the 400 MHz bandpass to various center frequencies as desired.

Shortly after a working model of an rf spectrum analyzer was demonstrated by Westinghouse, an alternate embodiment of essentially the same design was demonstrated by Hughes Aircraft Company. The Hughes OIC also followed the basic pattern proposed by *Hamilton* et al. [17.1], as shown in Fig. 17.1. However, it differed from the initial Westinghouse OIC in that it featured a butt-coupled GaAlAs laser diode rather than a He-Ne laser source, and the detector array was composed of silicon charge-coupled-devices (CCD) [17.3, 4] rather than photodiodes. The Hughes spectrum analyzer exhibited a 3 dB bandwidth of 380 MHz with a diffraction efficiency of 5% (at 500 mW rf power) [17.5]. Operating at a wavelength of 8200 Å, the OIC had a resolution of 8 MHz and a linear dynamic range greater than 25 dB. Losses in the two geodesic lenses were measured to be less than 2 dB each.

The two embodiments of the rf spectrum analyzer described above are excellent examples of hybrid optical integrated circuit technology. By fabricating the laser diode in GaAlAs, the detector array in silicon, and the Bragg modulator in LiNbO₃, one can use the best features of all three materials to advantage. The major disadvantage of the hybrid approach is that all of these substrate materials must be carefully aligned and permanently bonded with micrometer-tolerance precision. Thermal expansion and vibration must somehow be prevented from destroying the alignment. Despite these difficulties, hybrid OIC's have been demonstrated to be viable structures, and will continue to be used in many applications even after monolithic technology has been fully developed.

17.1.2 Monolithic Wavelength-Multiplexed Optical Source

One of the applications for which optical integrated circuits were proposed early in the history of the field is an optical-frequency-multiplexed transmitter, such as that shown previously in Fig. 1.1, in which a number of DFB lasers, operating at different wavelengths, are coupled into a single fiber transmission line. An OIC of this type has, in face, been fabricated by *Aiki* et al. [17.6, 7], using GaAlAs monolithic technology. Six DFB lasers, operating at wavelengths separated by 20 Å, were fabricated on a 5 mm square GaAs substrate by a two-step LPE growth process. The lasers had a separate confinement heterostructure (SCH) [17.8]. Third-order gratings were made on the surface by chemical etching, by using a mask made by holographic lithography. The lasers were coupled to undoped $Ga_{0.9}Al_{0.1}As$ waveguides by direct transmission, as shown in Fig. 17.2. The lateral dimensions of the lasers and waveguides

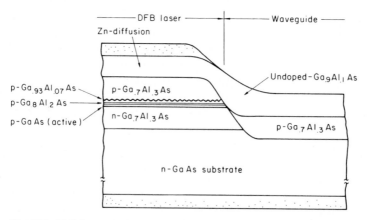

Fig. 17.2. DFB laser coupled to a GaAlAs waveguide by direct transmission [17.8]

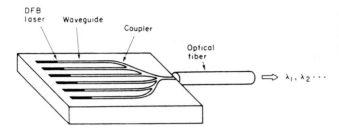

DFB
laser   Waveguide
                    Coupler
                                    Optical
                                    fiber

$\Rightarrow \lambda_1, \lambda_2 \cdots$

Fig. 17.3. Schematic drawing of a wavelength-multiplexed light source [17.6]

were defined by mesa etching down to the GaAs substrate to produce stripes that were 20 μm wide and 3 μm thick. The separation of the lasers was 300 μm, and the waveguides were curved through bends of minimum radius equal to 4 mm, in order to bring them together in a confluent coupler, as shown in Fig. 17.3. The output of the coupler was obtained via a single waveguide that was butt coupled to an optical fiber.

The lasers were operated by applying 100 ns current pluses at a repetition rate of 1 kHz. The differential quantum efficiency of the lasers was measured to be 7%, and the waveguide loss coefficient was about 5 cm^{-1}. The threshold current densities of the lasers were in the range from 3 to 6 kA/cm^2 at room temperature. The wavelength separation between lasers was measured to be 20 ± 5 Å, with the slight variation from the 20 Å nominal value being attributed to nonuniformity in the epitaxially grown layers on the chip. The spectral width of the laser beams was typically 0.5 Å. No difficulty was encountered in separately modulating the six lasers, and the overall differential quantum efficiency, measured at the launching output terminal, was about 30%. Thus, this monolithic chip represents a usable OIC, even though further refinements are certain to yield better efficiency.

Work on monolithic wavelength-multiplexed optical sources employing DFB lasers has continued over the years, with more recent work being directed toward 1.3 μm and 1.55 μm wavelength lasers for optical-fiber telecommunication systems. *Zah* et al. [17.9] have reported a study of wavelength-division multiplexed lightwave systems which has led to the conclusion that, in order to be cost effective, it is necessary to fabricate multi-wavelength laser transmitters by monolithic integration on one chip to reduce the cost of packaging and control circuitry by sharing them among all of the wagelengths. An example of such monolithic integration is provided by the frequency-division multiplexed ten-channel tunable DFB laser array of *Sato* et al. [17.10]. The lasers are tunable, multi-section, quarter-wave-shifted, strained InGaAsP MQW devices. The lasing frequencies of channels are spaced within a 10 GHz range. The linewidth of each channel is less than 2.3 MHz.

17.1.3 Analog-to-Digital Converter (ADC)

An analog-to-digital conversion method, proposed by *Taylor* [17.11, 12], has been implemented by *Yamada* et al. [17.13] in an optical integrated circuit that

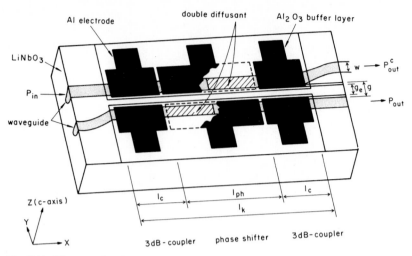

Fig. 17.4. Diagram of an integrated-optic analog to digital converter [17.13]

is capable of one-bit electro-optical AD conversion at a 100 MHz rate. The OIC incorporates two 3-dB couplers and a phase shifter, formed in a pair of straight waveguides, as shown in Fig. 17.4. The waveguides were fabricated by Ti diffusion of a $LiNbO_3$ substrate. The phase shifter was formed by a Ti double-diffusion, as shown in Fig. 17.4. An Al_2O_3 1100 Å thick buffer layer was used, separated by a 4 μm, gap between the waveguides to suppress dc drift. Waveguide spacing was 5.4 μm and device length was about 2 cm.

The configuration of two electro-optic couplers and a phase shifter forms a balanced-bridge modulator, with two complementary outputs which are equally affected by fluctuation of the light source. Hence a serious source of conversion error is inherently eliminated in this OIC. The integrated ADC was operated with a 1.15 μm-wavelength He-Ne laser source at bit rates up to 100 MHz. This initial success of high speed analog-to-digital conversion points the way towards more sophisticated, multi-bit, and monolithic OIC's. However, much remains to be done, especially in regard to development of a monolithic, high-speed, electronic or optical comparator to be incorporated into a fully monolithic ADC system.

17.1.4 Integrated-Optic Doppler Velocimeter

An integrated-optic Doppler velocimeter which employs both an optical fiber link and an OIC to measure velocity has been demonstrated by *Toda* et al. [17.14], a shown in Fig. 17.5. The optical integrated circuit was fabricated in a z-propagation $LiNbO_3$ substrate with Ti diffused waveguides. Laser beam lithography, with 0.2 μm accuracy, was used for waveguide patterning. The light source was a linearly polarized He-Ne laser. TE polarized light was focused into the input waveguide by a 20 × lens and then split by a Y-branch

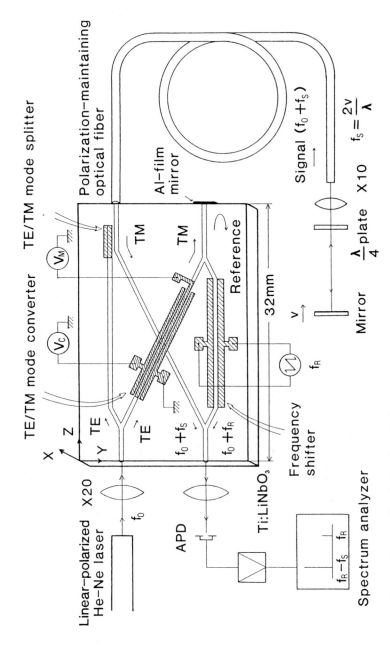

Fig. 17.5. An integrated-optic Doppler velocimeter

coupler into a signal beam and a reference beam. In order to keep the outgoing beams separate from the reflected return beams, TE polarization was maintained on outgoing light while TM polarization was used for reflected light. Of course, polarization-maintaining optical fiber was used. In the case of the reference beam the TE/TM conversion was accomplished by means of an electro-optic mode converter. For the signal beam a quarterwave plate was used to produce TE/TM mode conversion. An absorptive TE/TM mode splitter was used to route the return signal beam to the avalanche photodiode (APD) which served as a mixer and detector. An electro-optic modulator was used to impress the reference modulation frequency f_R onto the reference beam. The Doppler effect produced a shift in the signal beam frequency from f_o to $f_o + f_s$, where f_s is given by

$$f_s = \frac{2v}{\lambda_0},$$
(17.1.1)

where v is the velocity and λ_o is the vacuum wavelength. After being recombined by a Y-branch coupler the reflected signal beam and reference beam were mixed in the avalanche diode. Because of the nonlinear response characteristic of the APD (it is a square-law device, i.e., response is proportional to the square of the electric field strength) the output photocurrent contained a beat frequency component $f_R - f_s$. Thus f_s, and hence the velocity, was determined. For a velocity of 8 mm/s the measured f_s was 25 kHz, and the signal to noise ratio was 25 dB.

This Doppler velocimeter demonstrates that integrated-optic technology can provide the compact and rugged heterodyne optics need for highly accurate measurement of velocity and displacement. By adding a balanced-bridge waveguide optical switch to an interferometric circuit of this type *Toda* et al. [17.15] have produced a time-division-multiplexed Doppler velocimeter which can measure two-dimensional velocity components v_x and v_y.

17.1.5 An IO Optical Disk Readhead

Optical disk information storage has found widespread use for computer data, as well as for video and audio reproductions. High data density and low background noise are key advantages of this method. However, relatively sophisticated optics must be used to insure good resolution and tracking of the light beam that is used to read information off the disk. For example, the optical readheads used in commercially available audio compact disk (CD) players often have eight or nine discrete optical elements, all of which have to be held in exact alignment in the face of much shock and vibration.

As an alternative, an integrated-optic optical disk pickup device capable of detecting readout and focus/tracking error signals has been designed and fabricated by *Ura* et al. [17.16], as shown in Fig. 17.6. The OIC was formed by depositing a planar # 7059 glass waveguide on a SiO_2 buffer layer on a silicon substrate. The light source was a butt-coupled GaAlAs laser diode. A chirped and curved focusing grating pattern coupler [17.17] fabricated by electron beam direct writing lithography [17.18] was used to focus the beam onto the disk, as well as to refocus the reflected beam back into the waveguide. A twin-grating focusing beam splitter served to divide the reflected beam into two beams which were focused onto two pairs of photodiodes formed in the Si substrate.

In operation, the pickup head provides not only a readout signal but also focus and tracking error signals. When the light beam is focused, the return beams hit both diodes of each pair equally. If the readhead is too close to the disk, the beams fall more on the outer diodes, while if it is too far away they fall more on the inner diodes. Tracking error is detected when the total intensity of the return beam reaching the left pair of diodes is not equal to that reaching the right pair. Thus conventional electronic comparators, sensing the photo-currents from the diodes, can be used to develop error signals to drive position correcting actuators.

The OIC readhead of Fig. 17.6, which has dimensions of only $5 \times 12\,\text{mm}$, obviously has the advantage of being relatively insensitive to shock and vibration, as compared to a readhead fabricated from discrete optical components. While this OIC was first proposed as an optical disk pickup device, the same basic arrangement can be used more generally as a fully integrated interferometer position/displacement sensor with direction discrimination [17.19, 20]. Such an interferometric sensor would be useful in a variety of high-precision positioning applications in which submicrometer accuracy is required.

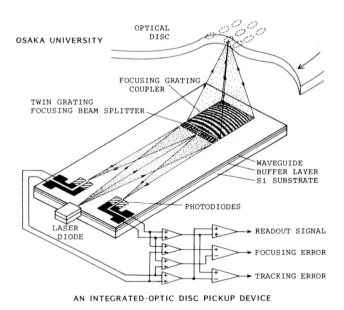

Fig. 17.6. An integrated-optic disc pickup device [17.16]

17.1.6 OIC Temperature Sensor

The integrated-optic temperature sensor [17.21] shown in Fig. 17.7, requires no electrical connection, making it particularly useful in explosive or flammable environments in which an electrical sensor might be dangerous. The OIC is fabricated in a LiNbO$_3$ substrate with Ti diffused waveguides. It features a parallel array of three unequal arm-length Mach-Zehnder interferometers. The optical transmission of each interferometer varies sinusoidally with temperature, as shown in Fig. 17.7b, with a period which is inversely proportional to the optical path length difference between the two arms.

The optical transmission P_{out}/P_{in} at the wavelength λ depends on both the effective index n_{eff} and the path length difference ΔL and is given by [17.21]

$$\frac{P_{out}}{P_{in}} = \frac{\gamma}{z}\left[1 + m\cos\left(\frac{2\pi}{\lambda}b\Delta L T + \Delta\phi_0\right)\right], \tag{17.1.2}$$

where the constant of proportionality b is given by

$$b = \frac{dn_{eff}}{dT} + \frac{n_{eff}}{\Delta L}\frac{d(\Delta L)}{dT}. \tag{17.1.3}$$

Both n_{eff} and ΔL are functions of temperature T. The quantities γ and m are related to the insertion loss and depth of modulation of the interferometer, respectively. (For an ideal device $\gamma = m = 1.0$.) $\Delta\phi_0$ is a constant for a given device.

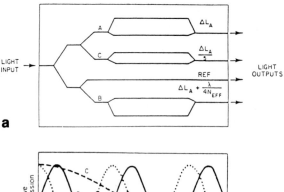

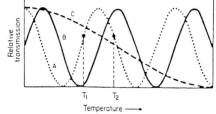

Fig. 17.7a, b. OIC temperature sensor **a** device structure **b** optical transmission characteristics [17.21]

By measuring the transmission of all three inteferometers one can determine the temperature. Two of the interferometers (A and B) have arm-length differences that are almost the same. Thus their transmission curves track close to one another, providing a high resolution in the temperature measurement. The third interferometer (C) has an arm-length difference which is only approximately one-fifth those of A and B. Thus one can determine which peak of the A and B transmission curves is being measured and temperature measurements can be made over a wide range. It has been reported that this temperature sensing OIC can measure with an accuracy of $2 \times 10^{-3} °C$ over a $700 °C$ range, when used with a 6328 Å He-Ne laser as the light source. The sensor, of course, would be mounted at the point at which temperature measurement was desired and the optical input and output would be via optical fiber. Since the OIC chip is about 1 cm on a side, temperature measurements can be made on relatively small objects. Since the measurement signal is entirely optical, this device is relatively immune to electrical noise.

17.1.7 IO High Voltage Sensor

Integrated-optic Mach-Zehnder interferometers can be used to sense high voltage as well as temperature. A diagram of such a device is shown in Fig. 17.8 [17.22]. The waveguides are formed by Ti diffusion into a LiNbO$_3$ substrate. In this circuit the two branches of the interferometer are covered by metal electrodes which form a capacitive voltage divider. The electric field generated by the high voltage source induces a voltage on these electrodes which causes a relative phase shift between the optical waves in each arm, resulting in an intensity modulation of the output beam. The voltage-in/optical power-out transfer function is given by [17.22]

$$P_{out} = \frac{\alpha P_{in}}{2}\left[1 + \gamma \cos\left(\frac{\pi V}{V_\pi} + \phi_i\right)\right],$$

(17.1.4)

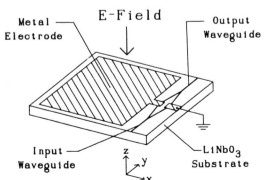

Fig. 17.8. IO high voltage sensor [17.22]

where P_{in} is the input power, ϕ_i is the intrinsic or zero voltage phase difference, V is the applied voltage and V_π is the half-wave phase shift voltage. The constants α and γ must be determined for a particular device. (For a perfect device $\alpha = \gamma = 1.0$.) Once the calibration curve has been determined for a particular sensor, the voltage can be accurately measured.

Since this sensor operates on the principle of an induced voltage it is not necessary to make electrical contact to the high voltage source, and input to, and output from the sensor can be via optical fiber. Thus good high voltage isolation can be maintained. The immunity to electrical noise provided by an optical fiber link is also a particularly important advantage when operating in a high voltage environment. This integrated optical high voltage sensor could be used, for example, for monitoring line voltages in SF_6-gas insulated bus ducts such as are used in power plants and switching stations.

17.2 Opto-Electronic Integrated Circuits

In the optical integrated circuits described in Sect. 17.1 all of the key elements were optical devices. However, there is another class of optical integrated circuits in which many of the devices are purely electronic and the signal is carried in parts of the circuit by electrical voltage or current waves rather than by an optical beam. Such circuits have come to be called opto-electronic integrated circuits (OEIC's). They are usually fabricated on substrates of semi-insulating GaAs or InP because both electronic and optical devices can be monolithically integrated on these materials.

17.2.1 An OEIC Transmitter

An OEIC four-channel optical transmitter is shown in Fig. 17.9 [17.23]. This circuit was fabricated on a semi-insulating GaAs substrate by molecular beam epitaxy. It features an array of four stripe-geometry, single-quantum-well GaAlAs lasers with microcleaved facets. Each laser is accompanied by a photodiode to monitor output power and a driver circuit containing three field effect transistors (FET's). The lasers have a relatively low threshold current of 15–29 mA and a differential quantum efficiency of 50–60%. The emission wavelength is 834 nm. The FET's are Schottky-barrier-gate devices. By monolithically integrating electronic and optical devices in an OEIC such as this, one can reduce parasitic capacitance and inductance to a minimum, thereby increasing the achievable maximum frequency of operation. In the circuit shown in Fig. 17.9 the transmitter is capable of operating at a data rate of 1.5 Gb/s. *Matsueda* and *Nakamura* [17.24] have produced a single-channel OEIC transmitter featuring a four FET driver circuit, a monitor photodiode and a GaAlAs laser diode, all monolithically integrated on a GaAs semi-

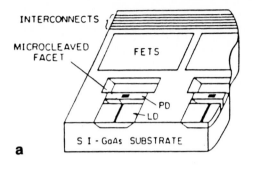

a

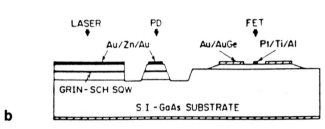

b

Fig. 17.9 a–c.
Schematic of four-channel OEIC transmitter **a** Overall view showing the layout. **b** Cross-section along the laser cavity showing the device structure. **c** Circuit diagram of OEIC transmitter. The *dashed line* shows the part of monolithic integration for a single channel.

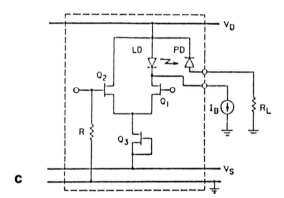

c

insulating substrate. That OEIC is capable of operating at a data rate of 2 Gb/s. Other OEIC transmitters are described in a review paper by *Matsueda* [17.25]

A four-channel laser transmitter OEIC for operation at 1.55 μm wavelength has been produced by *Woolnough* et al. [17.26]. It incorporates on an InP substrate, four single-mode, ridge-waveguide, InGaAsP, DFB lasers operating in the 1545 to 1560 nm window with 0.8 nm separation between adjacent channels. The drive circuits consist of lattice-matched, diffused-junction InGaAs-channel JFETs. Laser modulation at 155 Mbit/s has been achieved.

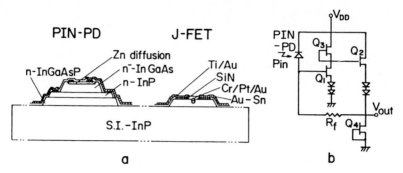

Fig. 17.10 a,b. OEIC photoreceiver **a**. Cross section of circuit chip; **b**. Circuit diagram [17.27]

17.2.2 An OEIC Receiver

The detection and amplification functions of a receiver can also be implemented in OEIC form. Figure 17.10 shows a typical OEIC receiver [17.27]. The circuit features a p-i-n photodiode detector and Schottky-barrier gate FET amplifier monolithically integrated on a semi-insulating InP substrate. It is designed to operate with a 5 volt power supply, which simplifies its interconnection with standard 5 volt logic IC's. The circuit has a 3 dB bandwidth of 240 MHz, with a 965 Ω transimpedance.

The increased speed of OEIC's as compared to that of discrete element circuits makes them attractive for use in lightwave communication systems, signal processing and sensing applications. Data rates in excess of 5 Gb/s have been achieved in a GaAs-MESFET/MSM photodiode receiver circuit [17.28], and an InP-based monolithically integrated balanced mixer receiver for coherent detection systems has exhibited a bandwidth of 14 GHz [17.29].

17.2.3 An OEIC Phased-Array Antenna Driver

Phased-array antennas have been used for many years in microwave applications in which a scanning microwave beam is required, but a moving antenna structure is impractical [17.30]. For example, the antenna for a RADAR transmitter in a supersonic aircraft can not be effectively scanned mechanically at a rapid enough rate. In this case, an electronically scanned phased-array antenna is used, consisting of a relatively large number of emitting antenna elements spaced many wavelengths apart, usually along the wings of the aircraft. If the relative phases of the waves transmitted by the various antenna elements are properly adjusted, a scanning microwave beam can be produced. To maintain phase coherence a frequency reference is provided by a stable master oscillator. This reference signal must be properly phase shifted and conveyed to each of the transmitting elements. If metallic microwave waveguide or coaxial cable is used to transport this signal, a great deal of

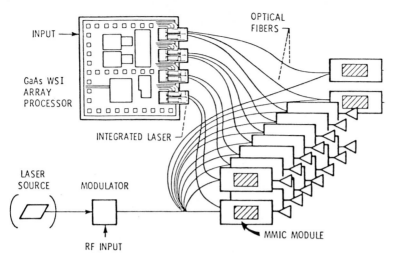

Fig. 17.11. OEIC phased array antenna driver [17.32]

undesirable weight is added to the aircraft. Obviously, this weight can be eliminated if the phase control signals are converted to optical signals and are distributed to the microwave emitters via optical fibers [17.31]

An OEIC which generates such phase control signals is shown in Fig. 17.11 [17.32]. The phase control signals are generated in a GaAs monolithic microwave integrated circuit (MMIC) with an integrated laser and driver circuit. These signals are then carried by optical fibers to the microwave generating MMIC modules which are the antenna elements. In the system shown in Fig. 17.11, which was designed for a data transmission application rather than RADAR, an information signal produced by modulating a laser was also carried to all of the antenna elements by optical fiber links.

17.3 Devices and Systems for Telecommunications

While optical integrated circuits are just emerging from the laboratory to become commercial products, optical fibers have already found widespread use in telecommunication systems. Almost from its beginning, the use of fiber waveguides in telecommunication systems has been an international phenomenon. Fiber links have been used in many countries to transmit audio and video signals, as well as digital data. There are presently so many examples of fiber telecommunication systems that have either been already implemented or are in the final planning stage that it is impractical to mention all of them in this chapter. Consequently, only a number of representative systems will be considered.

17.3.1 Trends in Optical Telecommunications

Over the past twenty-five years there has been remarkable improvement in the capabilities of optical fiber communications systems. The systems implemented before approximately 1980, listed in Table 17.3, generally had a length on the order of 10 km, operated at data rates less than 150 Mb/s and carried at most 30,000 telephone channels. By contrast, the systems which have become operational after about 1980, shown in Table 17.4, have lengths of thousands of kilometers, data rates ranging from hundred of megabits per second to gigabits per second and carry as many as 80,000 telephone channels. Also the spacing of repeaters has increased from a few kilometers in early systems to

Table 17.3. Commercial optical fiber communications links (pre-1980)

Company	Location	Length	Performance data
AT&T	Atlanta	10.9 km	44.7 Mbit/s 144 fibers, 6.2 dB/km
GTE	Long Beach	9 km	1.5 MHz/s 6 fibers, 6.2 dB/km
ITT-STL	Harlow	9 km	140 Mbit/s 4 fibers, 5 dB/km
Teleprompter	So. Cal.	240 m	CATV trunk 10K subscribers
Rediffusion Ltd.	Hastings	1.4 km	CATV trunk 34K subscribers
AT&T	Chicago	2.5 km	44.7 Mbit/s 12 fibers 8.5 dB/km
British Telcom	Brownhills-Walsall Croydon-Vauxhall London-Vauxhall	Total length 28 km	8 Mbit/s
GEC	London Subway	7 km	8 Mbit/s
Philips	Eindhoven to Helmond	14 km	140 Mbit/s 12 fibers, graded index, 1920 telephone channels/fiber
Siemens	Frankfurt/Main Oberursel	15.4 km	34 Mbit/s
Thomson-CSF	Paris	7 km	34 Mbit/s, 50 fibers, graded index 30,000 telephone channels
Martin Marietta Data Systems	Orlando, Fla.	9.2 km	45 Mbit/s
Israeli Post Office (Fibronics Fibers)	Tel-Aviv	2.7 km	8 fibers, 148 MHz, 6 dB/km loss at 0.82 μm

Table 17.4. Commercial optical fiber communications links (post-1980)

Company	Location	Length	Performance data
Pacific Telephone Co.	Sacramento, San Jose Stockton, Oakland San Francisco	Total length 270 km	
AT&T	Cambridge, MA Richmond, VA	1250 km	wavelength-division multiplexed 270 Mbit/s
AT&T	California Coast	830 km	wavelength-division multiplexed 270 Mbit/s
AT&T	Atlanta (1982)	65 km	432 Mbit/s, 6048 channels WDM (1335 & 1275 nm) single mode fiber, no repeater
AT&T	Atlanta (1984)	74 mi	420 Mbit/s, 6000 channels single mode laser, 1500 nm single mode fiber, no repeater
AT&T	5 intercity links Phila.-Pittsburgh Pittsburgh-Cleveland Dallas-Houston San Antonio-Seguia Atlanta-Charlotte		432 Mbit/s, 6048 channels single mode laser, 1300 nm two single mode fibers (no WDM) repeater spacing 46 ml
NTT (F400-M) commercial	Asahikawa to Kagoshima (main trunk)	4000 km	1.3 μm, single mode 40 Km repeater spacing 445 Mbit/s, Ge APD completion 1985
Field trial			1.55 μm, single mode InGaAs APD 80–120 km repeater spacing 5760 telephone channels
NTT (F1.6G) Field Trial		4000 km	1.6 Gbit/s DFB laser 1.3 μm-40 km re. sp. 1.55 μm-80–120 km re. sp. 23,040 telephone channels
NTT (FS-400 M)		1000 km (2 sections)	Submarine cable
Telecom Australia	Perth to Adelaide	2800 km	565 Mbit/s single mode 1.3 and 1.55 μm completion 1989
International Consortium- (TAT-8)	New Jersey to Widemouth, UK Penmarch, France	6687 km	Submarine cable 40,000 telephone channels 55 km repeater spacing single mode, 1.3 μm 274 Mbit/s completion 1988
International Consortium of 25 organizations (TAT-9)	North America to Europe & U.K.		submarine cable 80,000 telephone channels 565 Mbits/s completion 1991

over 100 km. These improvements in performance have come about because of a number of key technological breakthroughs.

Perhaps the most significant breakthrough was the development of efficient GaInAsP lasers and photodiodes operating at either 1.3-μm or 1.55-μm wavelength to take advantage of low loss, minimal dispersion optical fibers which were also developed for those wavelengths. As mentioned in Chap. 1, losses of less than 0.4 dB/km at 1.3-μm and 0.2 dB/km at 1.55-μm can be obtained. Minimal bulk dispersion can also be obtained either at 1.3-μm wavelength, or at 1.55-μm in "dispersion shifted" fibers.

The serious problem of modal dispersion which limited the distance-bandwidth product of early multimode systems has been completely eliminated in modern single-mode systems. The transition from multimode to single-mode systems required the development of new single-mode fibers, couplers and lasers, and did not happen overnight. There was a time, not very long ago, when many people felt the problems of single-mode coupling, with its submicrometer tolerances, could not be overcome. Fortunately, they were wrong. Single-mode couplers which can be installed by technicians and repairmen "in the field" are now commercially available from many different suppliers. For example, the AT&T Single-Mode ST^R connector, shown in Fig. 17.12, is a low-loss, low-reflection connector suitable for high-speed, wideband digital transmission [17.33]. The insertion loss of this connector averages 0.34 dB with a standard deviation of 0.28. Reflection averages −42.8 dB with a worst case of −34.1 dB. Practical single-mode optical fiber cables and lasers are also commercially available for use in telecommunications systems [17.34, 35].

In optical fiber communication systems operating at data rates greater than 1 Gb/s, generally DFB lasers have come to be used. The very narrow emission linewidth of these lasers (<1 Å) results in the least bulk dispersion effect and provides distance-bandwidth products in excess of 2000 GHz·km when the best fiber is also used. For example, 1.55 μm wavelength transmission

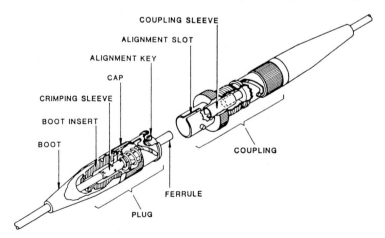

Fig. 17.12. AT&T SM ST connector [17.33]

has been reported in laboratory tests at 20 Gb/s over 109 km [17.36]. In general when high-power laser diodes are modulated at high data rates, frequency shift or "chirping" due to index change from carrier injection or temperature variation can be a problem. However, the inherent frequency selectivity of the DFB laser tends to limit chirping. Specially designed DFB lasers have exhibited 3 dB modulation bandwidth in excess of 13 GHz, with no-significant chirping being reported [17.37, 38].

The narrow linewidth and frequency stability of the DFB laser make it particularly useful in wavelength-division-multiplexed systems. A large number of information signal channels can be transmitted by using an array of DFB lasers, each operating at a slightly different wavelength. Typical wavelength spacing between channels is ten or twenty Angstroms; thus, many channels can be fit into the low-loss, low-dispersion passband of the optical fiber. Wavelength-division-multiplexing has been used in a number of the systems listed in Table 17.4 to achieve increased transmission capacity.

In summary, the key technological breakthroughs which have vastly expanded the capabilities of optical fiber telecommunication systems in recent years are: the development of efficient lasers and detectors for 1.3- or 1.5-μm wavelengths; practical single-mode fibers, couplers and lasers; frequency stable DFB lasers; and the introduction of wavelength-division-multiplexing techniques.

17.3.2 New Devices for Telecommunications

In addition to the major breakthroughs described in the previous section, there are also a number of new devices and techniques that are beginning to have an effect on telecommunications and offer great promise for the future. Certainly the quantum well devices described in Chap. 16, which are now emerging from the laboratory, will doubtless improve the performance of telecommunication systems. Also, optical amplifiers such as those described in Chap. 12 are finding implementation in newly installed systems. Such amplifiers can increase system bandwidth by eliminating the necessity of converting the optical signal to an electrical one for amplification in repeaters. An experimental test of a low-reflection (2.5%) optical amplifier as a midpoint repeater in a 200-km 1.5-μm wavelength system has been reported by *O'Mahoney* et al. [17.39]. The internal amplifier gain in this case was 27 dB, with input and output coupling losses of 5.5 and 6.5 dB respectively, yielding a net gain of 15 dB. A bit error rate of 10^{-9} was obtained at a data rate of 100 Mb/s. No significant problems were noted with regard to backscatter or reflections between fiber and amplifier.

A new type of optical amplifier that has been receiving a great deal of interest in recent years is the ERbium-DOped FIber Amplifier (EDFA). In this device a glass optical fiber is doped with erbium atoms which act as optically active centers to generate photons by stimulated emission and thereby amplify an optical signal as it passes through the fiber. The spectral characteristics of the emission from erbium atoms are such that they give efficient amplification in the 1520 to 1580 nm telecommunications window. EDFA's are generally pumped by high-power (>40 mW) semiconductor lasers operating at either 980 or 1480 nm to produce the required inverted population for emission at approximately 1550 nm. Optical-FIber Amplifiers (OFA) can also be made to amplify at 1300 nm by using other dopants, such as praseodymium. For a thorough discussion of the characteristics of OFA's see, for example, the review by *Fake* and *Parker* [17.40].

An example of the kind of performance improvement that can be obtained by using an EDFA in a long-distance telecommunication system is provided by *Wedding* and *Franz* [17.41]. They first transmitted 10 Gbit/s signals from a directly modulated DFB laser at 1.53 μm wavelength over 151 km of standard single mode fiber without any in-line amplification. When an EDFA was added to the system to provide in-line amplification they were able to transmit the same signal over a total distance of 204 km, achieving a Gbit × distance product of 2040 Gbit·km with a bit error rate of 10^{-9}.

Ultimately, probably the best system performance, in terms of signal-to-noise ratio and bandwidth-distance product, will be realized in coherent communication systems [17.42]. In such systems a highly stable, narrow line-width single-mode laser transmitter must be used, and phase-sensitive detection is employed at the receiver. Coherent communication systems are presently being developed and experimentally evaluated, but because of their complexity they will probably not find widespread use in installed systems for some years to come.

17.4 Opto-Microwave Applications

There are many opportunities for the use of integrated optic techniques in the generation, transmission, detection and processing of microwave signals. The RF spectrum analyzer described in Sect. 17.1 and the phased array antenna driver (Sect. 17.2) are two examples of such applications. There are a number of reasons why it is often advantageous to use optical signals to control microwave devices. First is the isolation of the optical control signal from undesired RF coupling encountered in the use of an electronic signal to control a microwave device or signal. This eliminates the problem of the control signal directly feeding through, or coupling, in an unwanted fashion to other channels. Optical control will significantly reduce the weight of the control signal transmission lines by replacing metallic waveguides with optical fibers. Also, optical control generally provides faster response because the optical signals are not slowed down by a parasitic capacitance and the electronic signal delays of coax or metallic waveguide. A final feature of optical control which will make it very useful in the long run is that the III-V substrate materials used for making optical integrated circuits are also used for making microwave monolithic integrated circuits, so that the combination of the two can produce an opto-microwave monolithic integrated circuit (OMMIC). Some examples of optical signals being used to control microwave devices include RF switching [17.43], phase shifting [17.44], optical injection locking of microwave oscillators [17.45] and optically pumped microwave mixer diodes [17.46]. A detailed discussion of these techniques is beyond the scope of this text but they have been reviewed elsewhere [17.47, 48].

In addition to being used to control microwave devices, integrated optic techniques can also be used to analyze microwave signals. The IO RF spectrum analyzer is an example of this being done in the frequency domain. An IO time domain RF voltage wave sampler has been produced by *Ridgway* and *Davis* [17.49], as shown in Fig. 17.13. This device consists of a series of electrooptic dual-channel directional couplers spaced along the length of an overlaying coplanar RF stripline which forms traveling wave electrodes for the couplers. In operation, an optical pulse is propagated along the main waveguide in the direction opposite to the propagation of the RF voltage wave. Depending on the magnitude of the RF voltage wave at a given coupler when the optical pulse passes, a certain fraction of the light is coupled into the auxiliary waveguide of that coupler. Thus the RF voltage waveform is sampled, and its magnitude and shape can be determined. Since the velocity of the optical pulse is much greater than that of the voltage wave, good time-domain resolution can be obtained. An optimum temporal resolution of a few picoseconds at a rate in excess of 4×10^{10} samples per second has been estimated for the OIC of Fig. 17.13, which was fabricated on a $LiNbO_3$ substrate with Ti diffused waveguides. The light source used was a GaAlAs laser emitting at a wavelength of 0.84 μm.

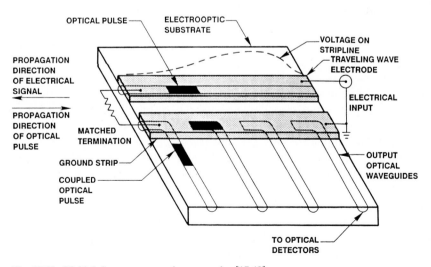

Fig. 17.13. IO high frequency waveform sampler [17.49]

The most widespread opto-microwave application is, of course, the transmission of signals modulated at microwave frequencies over high-speed fiber optic links. Many examples of such transmission in telecommunication systems have been given in this text. However, such transmission is not limited to just telecommunications. Fiber-optic links are being considered for use as millimeter-wave subcarrier transmission links for personal radio communication systems [17.50]. In this case fiber-optic links were used to transmit 25 GHz FM and also 70 MHz quadrature phase-shift keyed (QPSK) signals. Optical links have also been used for transmission of broadcast AM video signals [17.51]. An 18 mW, 1310 nm DFB laser was used as the source. Using 62 channel loading, a carrier to noise ratio of 59.5 dB was achieved for a 20 km link with 9 dB optical loss. For a detailed description of high-speed fiber-optic links for microwave-signal transmission see, for example, [17.52] by *Herczfeld*.

17.5 Future Projections

The technical feasibility and economic viability of optical fiber interconnectors has clearly been demonstrated in a wide variety of applications. The use of optical fiber cable transmission systems is now widespread for interexchange links and undersea cables. Digital data transmission rates on the order of gigabits per second are commonplace, and rates up to several tens of gigabits per second will become common within the next few years. Optical fiber cables have not yet replaced copper wire cables in the subscriber loop in many locations but the plans for that conversion are already being made. When it occurs, the public telephone network will be transformed into a vastly more flexible medium for the delivery of all sorts of new services including interactive high-speed data transmission, high-definition TV cable channels, video telephone, high-speed FAX, and security and medical alarm systems. Many households are already enjoying the benefits of data communication over the Internet and the World Wide Web.

With high-data-rate optical-fiber-cable transmission lines becoming well established, the need for optical integrated circuits has substantially grown. As a result, optical integrated circuits have ceased to be merely laboratory curiosities and have entered the marketplace. A number of companies are now marketing optical integrated circuits as off-the-shelf commercial products, and at least one company is offering to produce user-designed OIC's in a "foundry" type of service such as has been available for many years for electronic IC chips. Presently the commercially available OIC's are relatively limited – 2 × 2 optical switches, electro-optic phase modulators, Mach-Zehnder inteferometer intensity modulators, multiplexers, demultiplexers, couplers and splitters. However, these products have only been on the market for a relatively short time compared to electronic IC's. As systems engineers become more familiar with their use and their advantages, the demand for more sophisticated OIC's will surely grow. As optical fiber cables are extended into the subscriber loops of telecommunications systems the demand for large quantities of high-speed optical switches and couplers can best be met by producing them in OIC form.

The use of OIC's and fibers as sensors is also a rapidly expanding field of application. So far most of the work has been confined to laboratory experiments and military applications, but a growing consumer market is evident. For example, the sensing and control functions in a modern automobile are becoming extremely complex and have been highly automated. Integrated optic sensors and signal processors with optical fiber interconnects can offer improved performance and reduced weight. One luxury car is already advertizing a fiber optic instrumentation system.

Finally, the combination of integrated optic technology with microwave devices and systems, which has only just begun, has the potential of providing another expanding market, with applications in data transmission, avionics, security systems and television distribution.

Now that optical integrated circuits are emerging from the laboratory and some have become commercially available products which are being installed in systems "in the field", a further impetus will be provided to continuing research and development of more sophisticated OIC's.

References

Chapter 1

1.1   T. Tamir (ed.): *Integrated Optics*, 2nd edn., Topics Appl. Phys., Vol. 7 (Springer, Berlin, Heidelberg 1979)
1.2   S.E. Miller, A.G. Chynoweth: *Optical Fiber Communications* (Academic, New York 1979) Chap. 1
1.3   OSA Topical Meeting on Integrated Optics – Guided Waves, Materials and Devices, Las Vegas, NV (1972)
1.4   OSA Topical Meeting on Integrated Optics – Guided Waves, Materials and Devices, Las Vegas, NV (1974)
1.5   R.D. Maurer: Properties of research fibers for optical communications. OSA Topical Meeeting on Integrated Optics, New Orleans, LA (1974)
1.6   S.E. Miller: Optical-fiber transmission research. OSA Topical Meeting on Integrated Optics, New Orleans, LA (1974)
1.7   B. Wedding, B. Franz: Electron. Lett. **29**, 402 (1993)
1.8   J. Tamura, S. Nakamura, E. Kinoshita, A. Iino, M. Ogai: Low-loss fully fluorine-doped dispersion-shifted fibers. OSA/IEEE Optical Fiber Communications Conf., New Orleans, LA (1988)
1.9   L.G. Cohen, P. Kaiser, C. Lin: IEEE Proc. **68**, 1203 (1980)
1.10 P.N. Woolnough, P. Birdsall, P.J. O'Sullivan, A.J. Cockburn, M.J. Horlow: Electron. Lett. **29**, 1388 (1993)
1.11 S. Somekh, E. Garmire, A. Yariv, H. Garvin, R.G. Husperger: Appl. Opt. **13**, 327 (1974)
1.12 J.F. Dalgleish: IEEE Proc. **68**, 1226 (1980)
1.13 M. Papachon: Appl. Phys. Lett. **27**, 289 (1975)
1.14 R.A. Becker: Commercially available integrated optics products and services. SPIE Proc. **993**, 246 (1988)
1.15 M.C. Hamilton, D.A. Wille, M.J. Miceli: Opt. Eng. **16**, 475 (1977)
1.16 D. Mergerian, E.C. Malarkey: Microwave J. **23**, 27 (September 1980)
      D. Mergerian, E.C. Malarkey, R.P. Pautienus, J.C. Bradly, M. Mill, C.W. Baugh, A.L. Kellner, M. Mentzer: SPIE Proc. **321**, 149 (1982)
1.17 H. Kressel, M. Ettenberg: J. Appl. Phys. **47**, 3533 (1976)
1.18 C.S. Tsai: IEEE Trans. CAS-**26**, 1072 (1980)
1.19 C.L. Chen, J.T. Boyd: Channel waveguide array coupled to an integrated charge-coupled device (CCD). OSA Topical Meeting on Integrated Optics, Salt Lake City, UT (1978)
1.20 G. Giesecke: Lattice constants. *Semiconductors and Semimetals* **2**, 63-75 (Academic, New York 1976), in particular, ps.68 and 69
1.21 K. Nakajimi, K. Nakajima, A. Yamaguchi, K. Akita, T. Kotani: J. Appl. Phys. **49**, 5944 (1979)

1.22 K. Sato, S. Sekine, Y. Kondo, M. Yamamoto: IEEE J. QE-**29**, 180 (1993)

1.23 P. Hvertas, G. Mier, M. Dotor, J. Anguita, D. Golmayo, F. Briones: Sensors and Actuators A: Physical **37**, 512 (1993)

1.24 S. Ura, M. Shinohara, T. Suhara, N. Nishihara: IEEE Photon. Techn. Lett. **6**, 239 (1994)

1.25 H. Matsueda, M. Nakamura: Optical integrated circuit for optical communications. Appl. Opt. **23**, 779 (1984)

1.26 S. Sakano, H. Inove, H. Nakamura, T. Katsuyama, M. Matsumura: Electron. Lett. **22**, 594 (1966)

1.27 T.Q. Vu, C.S. Tsai: Integrated optics RF signal processor module in a GaAs waveguide. OSA/IEEE Integrated Photonics Research Conf., New Orleans, LA (1992)

1.28 H. Yano: IEEE J. LT-**10**, 933 (1992)

1.29 J. Wang, C. Shih, W. Chang, J. Middleton, P. Apostolakis, M. Feng: IEEE Photonics Tech. Lett. **5**, 316 (1993)

1.30 W. Hong, G. Chang, R. Bhat, J. Hayes: Proc. 1993 IEEE 5th Int'l Conf. on InP and Related Mater., Paris, France (1993) p. 275

1.31 K. Jackson, E. Flint, M. Cina, D. Lacey, Y. Kwark, J. Trewlella, T. Caulfield, P. Buchmann, Ch. Harder, P. Vettiger: IEEE J. LT-**12**, 1185 (1994)

1.32 D. Nichols, J. Lopata, W. Hobson, P. Berger, P. Smith, N. Dutta, D. Sivco, A. Cho: Integrated optical receivers and transmitters for use in wide-bandwidth optical transmission systems. Optical Fiber Commun. Conf., San Jose, CA (1994) p. 34

1.33 D. Trommer, G. Unterborsch, F. Reier: Proc. 6th IEEE Int'l Conf. on InP and Rel. Mater., Santa Barbara, CA (1994) p. 251

1.34 C. Shih, D. Barlage, J. Wang. M. Feng: Proc. 1994 IEEE MTT-S Int'l Microwave Symp., San Diego, CA (1994) p. 1379

1.35 H. Toda, M. Haruna, N. Nishihara: IEEE J. LT-**5**, 901 (1987)

1.36 R.W. Ridgeway, D.T. Davis: IEEE J. LT-**4**, 1514 (1986)

1.37 L.M. Johnson, F.J. Leonberger, G.W. Pratt, Jr.: Appl. Phys. Lett. **41**, 134 (1982)

1.38 B.L. Booth: Optical interconnection polymers, in *Polymers for Lightwave and Integrated Optics: Technology and Applications*, ed. by L.A. Hornak (Dekker, New York 1992)

1.39 R.K. Watts: Lithography, in *VLSI Technology*, ed. by S.M. Sze, 2nd edn. (McGraw Hill, New York 1988)

1.40 R.R. Krchnavek, G.R. Lalk, D.H. Hartman: J. Appl. Phys. **66**, 5156 (1989)

1.41 D.H. Hartman, G.R. Lalk, J.W. Howse, R.R. Krchnavek: Appl. Opt. **28**, 40 (1989)

1.42 J. Hagerhorst-Trewhella, J. Gelorme, B. Fan, A. Speth, D. Flagello, M. Oprysko: SPIE Proc. **1777**, 379 (1989)

1.43 A. Guha, J. Bristow, C. Sullivan, A. Husain: Appl. Opt. **29**, 1077 (1990)

1.44 R. Selvaraj: IEEE J. LT-**6**, 1034 (1988)

1.45 A. Neyer, T. Knoche, L. Müller: Electron. Lett. **29**, 399 (1993)

1.46 D.P. Prakash, DV. Plant, D. Zhang, H.R. Fetterman: SPIE Proc. **1774**, 118 (1993)

1.47 J.E. Thomson, H. Levesque, E. Savov, F. Horowitz, B.L. Booth, J.E. Marchegiano: Opt. Eng. **33**, 939 (1994)

1.48 M. Oguma, T. Kitagawa, K. Hattori, M. Horiguchi: IEEE Photonics Tech. Lett. **6**, 586 (1994)

1.49 K. Hattori, T. Kitigawa, M. Oguma, Y. Ohmori, M. Horiguchi: 1994 IEEE Optical Fiber Communications Conf., San Jose, CA (1994) Digest, p. 280
1.50 E. Mwarania, D. Murphy, M. Hempstead, L. Reekie, J. Wilkenson: IEEE Photon. Tech. Lett. **4**, 235 (1992)
1.51 R. Chen, M. Lee, S. Natarajan, C. Lin, Z. Ho, D. Robinson: IEEE Photon. Tech. Lett. **5**, 1328 (1993)

Chapter 2

2.1 A. McWhorter: Solid State Electron. **6**, 417 (1963)
2.2 J. McKenna: Bell Syst. Techn. J. **46**, 1491 (1967)
2.3 P.K. Tien: Appl. Opt. **10**, 2395 (1971)
2.4 D. Marcuse: *Theory of Dielectric Optical Waveguides* (Academic, New York 1974)
2.5 H.F. Taylor, A. Yariv: IEEE Proc. **62**, 1044 (1974)
2.6 H. Kogelnik: Theory of dielectric waveguides, in *Integrated Optics*, ed. by T. Tamir, 2nd edn., Topics Appl. Phys., Vol. 7 (Springer, Berlin, Heidelberg 1979) Chap. 2
2.7 A. Yariv: *Introduction to Optical Electronics*, 2nd edn. (Holt, Rinehart and Winston, New York 1976) Chap. 13
2.8 D.T.F. Marple: J. Appl. Phys. **35**, 1241 (1964)
2.9 F. Zernike: Fabrication and measurement of passive components, in *Integrated Optics*, ed. by T. Tamir, 2nd. edn., Topics Appl. Phys., Vol. 7 (Springer, Berlin, Heidelberg 1979) Chap. 5
2.10 E. Garmire, H. Stoll, A. Yariv, R.G. Hunsperger: Appl. Phys. Lett. **21**, 87 (1972)
2.11 J. Goell: Bell Syst. Tech. J. **48**, 2133 (1969)
2.12 H.M. Stoll, A. Yariv, R.G. Hunsperger, E. Garmire: Proton-implanted waveguides and integrated optical detectors. OSA Topical Meeting on Integrated Optics, New Orleans, LA (1974)
2.13 P.K. Tien, G. Smolinsky, R.J. Martin: Appl. Opt. **11**, 637 (1972)
2.14 D.H. Hensler, J. Cuthbert, R.J. Martin, P.K. Tien: Appl. Opt. **10**, 1037 (1971)
2.15 R.G. Hunsperger, A. Yariv, A. Lee: Appl. Opt. **16**, 1026 (1977)
2.16 P.K. Tien, R. Ulrich, R.J. Martin: Appl. Phys. Lett. **14**, 291 (1969)
2.17 R. Ulrich, R.J. Martin: Appl. Opt. **10**, 2077 (1971)
2.18 S.J. Maurer, L.B. Felsen: IEEE Proc. **55**, 1718 (1967)
2.19 H.K.V. Lotsch: Optik **27**, 239 (1968)
2.20 E.U. Condon: Electromagnetic waves, in *Handbook of Physics*, ed. by E.U. Condon, H. Odishaw (McGraw-Hill, New York 1967) pp. 6-8
2.21 B.H. Billings: Optics, in *American Institute of Physics Handbook*, ed. by D.E. Gray, 3rd edn. (McGraw-Hill, New York 1972) pp. 6-9
2.22 H.K.V. Lotsch: Optik **32**, 116-137, 189-204, 299-319, 553-569 (1970/71)
2.24 M. Born, E. Wolf: *Pricinples of Optics*, 3rd edn. (Pergamon, New York 1970) p. 49

Chapter 3

3.1   W. Hayt, Jr.: *Engineering Electromagnetics*, 4th edn. (McGraw-Hill, New York 1981) p. 151 and p. 317

3.2   A. Yariv: *Introduction to Optical Electronics*, 2nd edn. (Holt, Rinehart and Winston, New York 1976) p. 364

3.3   D. Hall, A. Yariv, E. Garmire: Opt. Commun. **1**, 403 (1970)

3.4   M. Barnoski: *Introduction to Integrated Optics* (Plenunm, New York 1974) p. 61

3.5   D. Mergerian, E. Malarkey: Microwave J. **23**, 37 (1980)

3.6   H. Kressel, M. Ettenberg, J. Wittke, I. Laddany: Laser diodes and LEDs for fiber optical communications, in *Semiconductor Devices*, ed. by H. Kressel, 2nd edn., Topics Appl. Phys., Vol. 39 (Springer, Berlin, Heidelberg 1982) pp. 23-25

3.7   S. Somekh, E. Germaire, A. Yariv, H. Garvin, R.G. Hunsperger: Appl. Opt. **13**, 327 (1974)

3.8   J. Goell: Bell Syst. Tech. J. **48**, 2133 (1969)

3.9   W. Schlosser, H.G. Unger: In *Advances in Microwaves*, ed. by L. Young (Academic, New York 1966) pp. 319-387

3.10   A.A.J. Marcatilli: Bell Syst. Tech. J. **48**, 2071 (1969)

3.11   H. Furuta, H. Noda, A. Ihaya: Appl. Opt. **13**, 322 (1974)

3.12   V. Ramaswamy: Bell Syst. Tech. J. **53**, 697 (1974)

3.13   H. Kogelnik: Theory of dielectric waveguides, in *Integrated Optics*, ed. by T. Tamir, 2nd edn., Topics Appl. Phys., Vol. 7 (Springer, Berlin, Heidelberg 1979) Chap. 2

3.14   J. Campbell, F. Blum, D. Shaw, K. Shaw, K. Lawley: Appl. Phys. Lett. **27**, 202 (1975)

3.15   F. Blum, D. Shaw, W. Holton: Appl. Phys. Lett. **25**, 116 (1974)

3.16   F. Reinhart, R. Logan, T. Lee: Appl. Phys. Lett. **24**, 270 (1974)

3.17   H.G. Unger: *Planar Optical Waveguides and Fibers* (Oxford Univ. Press, Oxford 1978)

3.18   H. Kogelnik: Theory of optical waveguides, in *Guided Wave Optoelectronics*, ed. by T. Tamir, 2nd edn., Springer Ser. Electron. Photon., Vol. 26 (Springer, Berlin, Heidelberg 1990)

3.19   J.T. Chilwell, I.J. Hodgkinson: Thin-film matrix description of optical multilayer planar waveguides. J. Opt. Soc. Am. **72**, 1821 (1982)

3.20   L.G. Ferreira, M. Pudensi: Waveguiding in a dielectric medium varying slowly in one transverse direction. J. Opt. Soc. Am. **71**, 1377 (1981)

3.21   T. Tamir: Microwave modeling of periodic waveguides. IEEE Trans. MTT-**29**, 979 (1981)

3.22   T. Tamir: Guided-wave methods for optical configurations. Appl. Phys. **25**, 201 (1981)

3.23   E.A. Kolosovsky, D.V. Petrov, A.V. Tsarev, I.B. Yakovkin: An exact method for analyzing light propagation in anisotropic inhomogeneous optical waveguides. Opt. Commun. **43**, 21 (1982)

3.24   K. Yasumoto, Y. Oishi: A new evaluation of the Goos Hänchen shift and associated time delay. J. Appl. Phys. **54**, 2170 (1983)

3.25   F.P. Payne: A new theory of rectangular optical waveguides. Opt. Quant. Electron. **14**, 525 (1982)

3.26   H. Yajima: Coupled-mode analysis of anisotropic dielectric planar branching waveguides. IEEE J. LT-**1**, 273 (1983)

3.27 S.A. Shakir, A.F. Turner: Method of poles for multiyer thin film waveguides. Appl. Phys. A **29**, 151 (1982)

3.28 W.H. Southwell: Ray tracing in gradient-index media. J. Opt. Soc. Am. **72**, 909 (1982)

3.29 J. Van Roey, J. VanderDonk, P.E. Lagasse: Beam-propagation method: Analysis and assessment. J. Opt. Soc. Am. **71**, 803 (1981)

3.30 J. Nezval: WKB approximation for optical modes in a periodic planar waveguide. Opt. Commun. **42**, 320 (1982)

3.31 Ch. Pichot: Exact numerical solution for the diffused channel waveguide. Opt. Commun. **41**, 169 (1982)

3.32 V. Ramaswamy, R.K. Lagu: Numerical field solution for an arbitary asymmetrical graded-index planar waveguide. IEEE J. LT-**1**, 408 (1983)

3.33 Y. Li: Method of successive approximations for calculating the eigenvalues of optical thin-film waveguides. Appl. Opt. **20**, 2595 (1981)

3.34 J.P. Meunier, J. Piggeon, J.N. Massot: A numerical technique for the determination of propagation characteristics of inhomogeneous planar optical waveguides. Opt. Quant. Electron. **15**, 77 (1983)

3.35 M. Belanger, G.L. Yip: Mode conversion analysis in a single-mode planar taper optical waveguide. J. Opt. Soc. Am. **72**, 1822 (1982)

3.36 E. Khular, A. Kumar, A. Sharma, I.C. Goyal, A.K. Ghatak: Modes in buried planar optical waveguide with graded-index profiles. Opt. Quant. Electron. **13**, 109 (1981)
A.K. Ghatak: Exact modal analysis for buried planar optical waveguides with asymmetric graded refractive index. Opt. Quant. Electron. **13**, 429 (1981)

3.37 A. Hardy, E. Kapon, A. Katzir: Expression for the number of guided TE modes in periodic multilayer waveguides. J. Opt. Soc. Am. **71**, 1283 (1981)

3.38 L. Eyges, P. Wintersteiner: Modes in an array of dielectric waveguides. J. Opt. Soc. Am. **71**, 1351 (1981)
L. Eyges, P.D. Gianino: Modes of cladded guides of arbitrary cross-sectional shape. J. Opt. Soc. Am. **72**, 1606 (1982)

3.39 P.M. Rodhe: On radiation in the time-dependent coupled power theory for optical waveguides. Opt. Quant. Electron. **15**, 71 (1983)

3.40 L. McCaughan, E.E. Bergmann: Index distribution of optical waveguides from their mode profile. IEEE J. LT-**1**, 241 (1983)

3.41 H. Kogelnik: Devices for lightwave communications, in *Lasers and Applications*, ed. by W.O.N. Guimaraes, C.-T. Lin, A. Mooradian, Springer Ser. Opt. Sci., Vol. 26 (Springer, Berlin, Heidelberg 1981)

Chapter 4

4.1 P.K. Tien: Appl. Opt. **10**, 2395

4.2 K.E. Wilson, E. Garmire, R.M. Silva, W.K. Stowell: J. Opt. Soc. Am. **71**, 1560 (1981)

4.3 R. Behrisch (ed.): *Sputtering by Particle Bombardment I*, Topics Appl. Phys., Vol. 47 (Springer, Berlin, Heidelberg 1981)

4.4 F. Zernike: Fabrication and measurement of passive components, in *Integrated Optics*, ed. by T. Tamir, 2nd edn., Topics Appl. Phys., Vol. 7 (Springer, Berlin, Heidelberg 1979)

4.5 L. Maissel, M. Francombe: *An Introduction to Thin Films* (Gordon and Breach, New York 1973)

4.6   A.B. Glaser, G.E. Subak-Sharpe: *Integrated Circuit Engineering* (Addison-Wesley, Reading, MA 1977) pp. 169-181

4.7   R.Agard: Appl. Phys. Lett: **27**, 607 (1975)

4.8   F. Hinkernell: Low loss zinc oxide optical waveguides on amorphous substrates. OSA Topical Meeting on Integrated Optics, Incline Village, NV (1980)

4.9   R. McGraw, F. Zernike: Annual Meeting, Am. Ceramic Soc., Chicago, IL (1974)

4.10   D. Hensler, J. Cuthbert, R. Martin, P.K. Tien: Appl. Opt. **10**, 1037 (1971)

4.11   M. Fujimori, M. Sasaki, M. Honda: Optical transmission characteristics in Ta_2O_5 film. OSA Topical Meeting on Integrated Optics, Las Vegas, NV (1972)

4.12   D. Ostrowsky, A. Jaques: Appl. Phys. Lett. **18**, 556 (1971)

4.13   R. Ulrich, H. Weber: Appl. Opt. **11**, 428 (1972)

4.14   T. Sosnowski, H. Weber: Appl. Phys. Lett. **21**, 310 (1972)

4.15   D.A. Ramey, J.T. Boyd: IEEE Trans. CAS-**26**, 1041 (1979)

4.16   R. Reuter, H. Franke, C. Feger: Appl. Opt. **27**, 4565 (1988)

4.17   C.T. Sullivan: SPIE Proc. **994**, 92 (1988)

4.18   A. Neyer, T. Knoche, L. Müller: Electron. Lett. **29**, 399 (1993)

4.19   D.P. Prakash, D.v. Plant, D. Zhang, H.R. Fetterman: SPIE Proc. **1774**, 118 (1993)

4.20   P.K. Tien, G. Smolinsky, R. Martin: Appl. Opt. **11**, 637 (1972)

4.21   D.J. Hamilton, W.G. Howard: *Basic Integrated Circuit Engineering* (McGraw-Hill, New York 1975) pp. 32-52

4.22   J.L. Jackel, V. Ramaswamy, S.P. Lyman: Appl. Phys. Lett. **38**, 509 (1981)

4.23   M.N. Armenese, M. DeSario: J. Opt. Soc. Am. **72**, 1514 (1982)

4.24   L. McCaughan, E.J. Murphy: J. Opt. Soc. Am. **72**, 1821 (1982)

4.25   B. Chen: A new technique for waveguide formation in $LiNbO_3$. OSA Topical Meeting on Integrated Optics, Salt Lake City, UT (1978)

4.26   H.F. Taylor, W.E. Martin, D.B. Hall, V.N. Smiley: Appl. Phys. Lett. **21**, 325 (1972)

4.27   W.E. Martin, D.B. Hall: Appl. Phys. Lett. **21**, 325 (1972)

4.28   E.M. Zolatov, V.A. Kiselyov, A.M. Prokhorov, E.A. Sacherbakov: Determination of characteristics of diffused optical waveguides. OSA Topical Meeting on Integrated Optics, Salt Lake City, UT (1978)

4.29   B.L. Booth: Optical interconnection polymers, in *Polymers for Lightwave and Integrated Optics: Techniques and Applications*, ed. by L.A. Hornak (Dekker, New York 1992) p. 232

4.30   T. Izawa, H. Nakagome: Appl. Phys. Lett. **21**, 584 (1972)

4.31   R.C. Alfernes: Titanium-diffused lithium niobate waveguide devices, in *Guided-Wave Optoelectronics*, ed. by T. Tamir, 2nd edn., Springer Ser. Electron. Photon., Vol. 26 (Springer, Berlin, Heidelberg 1990) pp. 145-206, in particular, p. 148

4.32   A. Tervonen, S. Honkanen, P. Poyhonen, M. Tahkokorpi: SPIE Proc. **1794**, 264 (1993)
       P. Masalkar, V. Rao, R. Sirohi: SPIE Proc. **1794**, 271 (1993)

4.33   H. Ryssel, H. Glawisching (eds.): *Ion Implantation*, Springer Ser. Electrophys., Vols. 10 and 11 (Springer, Berlin, Heidelberg 1982 and 1983)

4.34   W.S. Johnson, J.F. Gibbons: *Projected Range Statistics in Semiconductors* (Stanford Univ. Press, Stanford, CA 1969)

4.35   F.J. Leonberger, J.P. Donnelly, C.O. Bozler: Low-loss GaAs p^+-n^--n^+ optical striplines fabricated using Be^+ ion implantation. OSA Topical Meeting on Integrated Optics, Salt Lake City, UT (1976)

4.36 E.V.K. Rao: J. Appl. Phys. **46**, 955 (1975)
4.37 R. Standley, W.M. Gibson, J.W. Rodgers: Appl. Opt. **11**, 1313 (1972)
4.38 E. Garmire, H. Stoll, A. Yariv, R.G. Hunsperger: Appl. Phys. Lett. **21**, 87 (1972)
4.39 M.A. Mentzer, R.G. Hunsperger, S. Sriram, J. Bartko, M.S. Wlodawski, J.M. Zavada, H.A. Jenkinson: Opt. Eng. **24**, 225 (1985)
4.40 D. Hall, A. Yariv, E. Garmire: Opt. Commun. **1**, 403 (1970)
4.41 M. Barnoski, R.G. Hunsperger, R. Wilson, G. Tangonan: J. Appl. Phys. **44**, 1925 (1973)
4.42 J. Zavada, H. Jenkinson, T. Gavanis, R.G. Hunsperger, M. Mentzer, D. Larson, J. Comas: SPIE Proc. **239**, 157 (1980)
4.43 J.P. Donnelley, A.G. Foyt, W.T. Lindley, G.W. Iseler: Solid State Electron. **13**, 755 (1970)
4.44 R.M. Allen, British Embassy, Washinton, DC 20008: Priv. Commun. (1976)
4.45 P. Bhattacharya: *Semiconductor Optoelectronic Devices* (Prentice-Hall, Englewood Cliffs, NJ 1944) pp. 133-137, 294-299
4.46 E. Garmire: Semiconductor components for monolithic applications, in *Integrated Optics*, ed. by T. Tamir, 2nd edn., Topics Appl. Phys., Vol. 7 (Springer, Berlin, Heidelberg 1979) Chap. 6, in particular, pp. 293-301
4.47 T. Moss, G. Hawkins: Infrared Phys. **1**, 111 (1961)
4.48 D. Hill: Phys. Rev. A **133**, 866 (1963)
4.49 J. Shah, B.I. Miller, A.E. DiGiovanni: J. Appl. Phys. **43**, 3436 (1972)
4.50 J.T. Boyd: IEEE J. QE-**8**, 788 (1972)
4.51 H.C. Casey Jr., D.D. Sell, M.B. Panish: Appl. Phys. Lett. **24**, 633 (1974)
4.52 V. Evtuhov, A. Yariv: IEEE Trans. MTT-**23**, 44 (1975)
4.53 Zn.I. Alferov, V.M. Andreev, E.L. Portnoi, M.K. Trukn: Sov. Phys. – Semiconductors **3**, 1107 (1970)
4.54 J.H. McFee, R.H. Nahory, M.A. Pollack, R.A. Logan: Appl. Phys. Lett. **23**, 571 (1973)
4.55 F.K. Reinhart, R.A. Logan, T.P. Lee: Appl. Phys. Lett. **24**, 270 (1974)
4.56 S.M. Jensen, M.K. Barnoski, R.G. Hunsperger, G.S. Kamath: J. Appl. Phys. **46**, 3547 (1975)
4.57 M.G. Craford, W.O. Groves: IEEE Proc. **61**, 862 (1973)
4.58 S. Kamath: Epitaxial GaAs-(GaAl)As layers for integrated optics. OSA Topical Meeting on Integrated Optics, New Orleans, LA (1974)
5.59 H. Kressel (ed.): *Semiconductor Devices for Optical Communication*, 2nd edn., Topics Appl. Phys., Vol. 39 (Springer, Berlin, Heidelberg 1982)
4.60 C.M. Wolfe, G.E. Stillman, M. Melngallis: Epitaxial growth of InGaAs-GaAs for integrated optics. OSA Topical Meeting on Integrated Optics, New Orleans, LA (1974)
4.61 M.Kawabe: Appl. Phys. Lett. **26**, 46 (1975)
4.62 K. Nakajimi, A. Yamaguch, K. Akita, T. Kotani: J. Appl. Phys. **49**, 5944 (1979)
4.63 R.U. Martinelli: LEOS'88, Santa Clara, CA (1988) Digest p. 55
4.64 A.A. Chernov (ed.): *Modern Crystallography III, Crystal Growth*, Springer Ser. Solid-State Sci., Vol. 36 (Springer, Berlin, Heidelberg 1984)
4.65 K. Ploog, K. Graf: *Molecular Beam Epitaxy of III-V compounds, a Comprehensive Bibliography 1958-1983* (Springer, Berlin, Heidelberg 1984)
      M.A. Herman, H. Sitter: *Molecular Beam Epitaxy*, 2nd edn., Springer Ser. Mater. Sci., Vol. 7 (Springer, Berlin, Heidelberg 1996)
4.66 W.T. Tsang: Appl. Phys. Lett. **38**, 587 (1981)

4.67   J.C.M. Hwang, J.V. DiLorenzo, P.E. Luscher, W.S. Knodle: Solid State Techn. **25**, 166 (1982)

4.68   C. Goldstein, C. Stark, J. Emory, F. Gaberit, D. Bonnevie, F. Poingt, M. Lambert: J. Crystal Growth **120**, 157 (1992)

4.69   R.G. Walker, R.C. Goodfellow: Electron. Lett. **19**, 590 (1983)

4.70   T. Matsumoto, P. Bhattacharya, M.J. Ludowise: Appl. Phys. Lett. **42**, 52 (1983)

4.71   H. Ishiguro, T. Kawabata, S. Koike: Appl. Phys. Lett. **51**, 12 (1987)

4.72   H. Jvergensen: Microelectron. Eng. **18**, 119 (1992)

4.73   A. Yariv: *Optical Electronics*, 4th edn. (Holt, Rinehart and Winston, New York 1991) pp. 309-316

4.74   B.G. Streetman: *Solid State Electronic Devices*, 3rd edn. (Prentice-Hall, Englewood Cliffs, NJ 1990) p. 147

4.75   E. Spiller, R. Feder: X-ray lithography, in *X-Ray Optics*, ed. by H.-J. Queiser, Topics Appl. Phys., Vol. 22 (Springer, Berlin, Heidelberg 1977)

4.76   R.A. Bartolini: Photoresits, in *Holographic Recording Materials*, ed. by H.M. Smith, Topics Appl. Phys., Vol. 20 (Springer, Berlin, Heidelberg 1977)
       H.I. Bjelkhagen: *Silver-Halide Recording Materials for Holography and Their Processing*, 2nd edn., Springer Ser. Opt. Sci., Vol. 66 (Springer, Berlin, Heidelberg 1995)

4.77   M.C. Rowland: The preparation and properties of gallium arsenide, in *Gallium Arsenide Lasers*, ed. by C.H. Gooch (Wiley-Interscience, New York 1969) p. 166

4.78   S. Somekh, E. Garmire, A. Yariv, H. Garvin, R.G. Hunsperger: Appl. Opt. **12**, 455 (1973)

4.79   A.R. Goodwin, D.H. Lovelace, P.R. Selway: Opto-Electron. **4**, 311 (1972)

4.80   M. Kawabe, S. Hirata, S. Namba: IEEE Trans. CAS-**26**, 1109 (1979)

4.81   F.A. Blum, D.W. Shaw, W.C. Holton: Appl. Phys. Lett. **25**, 116 (1974)

4.82   H.F. Taylor, W.E. Martin, D.B. Hall, V.N. Smiley: Appl. Phys. Lett. **21**, 95 (1972)

4.83   S. Somek, E. Garmire, A. Yariv, H. Garvin, R.G. Hunsperger: Appl. Opt. **13**, 327 (1974)

4.84   J. Heibe, E. Voges: Fabrication of strip waveguides in $LiNbO_3$ by combined metal diffusion and ion implantation. OSA Topical Meeting on Integrated Optics, Incline Village (1980)

4.85   S.M. Sze: *VLSI Technology*, 2nd edn. (McGraw-Hill, New York 1988)

4.86   D.F. Barbe: *Very Large Scale Integration (VLSI)*, 2nd edn, Springer Ser. Electrophys., Vol. 5 (Springer, Berlin, Heidelberg 1982)

4.87   J.L. Jackel, R.E. Howard, E.L. Hu, S.P. Lyman: Appl. Phys. Lett. **38**, 907 (1981)

4.88   M.A. Bosch, L.A. Coldren, E. Good: Appl. Phys. Lett. **38**, 264 (1981)

4.89   K. Gamo: Mater. Sci. Eng. B: Solid-State Mater. for Adv. Techn. B **9**, 307 (1991)

Chapter 5

5.1   See, for example, E.U. Condon: Electromagnetic waves, in *Handbook of Physics*, ed. by E.U. Condon, H. Odishaw (McGraw-Hill, New York 1967) Sect. 6.1

5.2   P.K. Tien: Appl. Opt. **10**, 2395 (1971)

5.3   D.H. Hensler, J.D. Cuthbert, R.J. Martin, P.K. Tien: Appl. Opt. **10**, 1037 (1971)

5.4   D. Marcuse: Bell Syst. Techn. J. **48**, 3187, 3233 (1969); ibid. **49**, 273 (1970); ibid. **51**, 429 (1972)
5.5   Y. Suematsu, K. Furuya: Electron. and Commun. Jpn. **56**-C, 62 (1973)
5.6   S. Miyanaga, M. Imai, T. Asakura: IEEE J. QE-**14**, 30 (1978)
5.7   M. Gottlieb, G. Brandt, J. Conroy: IEEE Trans. CAS-**26**, 1029 (1979)
5.8   D.G. Hall, G.H. Ames, R.W. Modavis: J. Opt. Soc. Am. **72**, 1821 (1982)
5.9   D.D. North: IEEE J. QE-**15**, 17 (1979)
5.10   N.K. Uzunoglu, J.G. Fikioris: J. Opt. Soc. Am. **72**, 628 (1982)
5.11   A. Bostrom, P. Olsson: J. Appl. Phys. **52**, 1187 (1981)
5.12   T. Moss, G. Hawkins: Infrared Phys. **1**, 111 (1961)
       M.D. Sturge: Phys. Rev. **127**, 768 (1962)
       H. Stoll, A. Yariv, R.G. Hunsperger, E. Garmire: Proton-implanted waveguides and integrated optical detectors in GaAs. OSA Topical Meeting on Integrated Optics, New Orleons, LA (1974)
5.13   V. Evtuhov, A. Yariv: IEEE Trans. MTT-**23**, 44 (1975)
5.14   D.L. Spears, A.J. Strauss, S.R. Chinn, I. Melngailis, P. Vohl: CdTe waveguide devices and HgCdTe epitaxial layers for integrated optics. OSA Topical Meeting on Integrated Optics, Salt Lake City, UT (1976)
5.15   P.K. Cheo, J.M. Berak, W. Oshinsky, J.L. Swindal: Appl. Opt. **12**, 500 (1973)
5.16   M. Barnoski, R.G. Hunsperger, R. Wislon, G. Tangonan: J. Appl. Phys. **44**, 1925 (1973)
5.17   A. Yariv: *Introduction to Optical Electronics*, 2nd edn. (Holt, Rinehart and Winston, New York 1976) p. 100
5.18   M.A. Mentzer, R.G. Hunsperger, S. Sriram, J. Bartko, M.S. Wlodawski, J.M. Zavada, H.A. Jenkenson: Opt. Eng. **24**, 225 (1985)
5.19   J.I. Pankove: *Optical Processes in Semiconductors* (Prentice-Hall, Englewood Cliffs, NJ 1971) p. 75
5.20   H.Y. Fan: Effects of free carriers on the optical properties. *Semiconductors and Semimetals* **3**, 409 (Academic, New York 1967)
5.21   D. Marcuse: Bell Syst. Tech. J. **48**, 3187 (1969)
5.22   E.A.J. Marcatili, S.E. Miller: Bell Syst. Tech. J. **48**, 2161 (1969)
5.23   S.E. Miller: Bell Syst. Tech. J. **43**, 1727 (1964)
5.24   J.E. Goell: Loss mechanisms in dielectric waveguides, in *Introduction to Integrated Optics*, ed. by M.K. Barnoski (Plenum, New York 1974) p. 118
       E. Neumann, W. Richter: Appl. Opt. **22**, 1016 (1983)
5.25   P.K. Tien, R. Ulrich, R.J. Martin: Appl. Phys. Lett. **14**, 291 (1969)
5.26   H.P. Weber, F.A. Dunn, W.N. Leibolt: Appl. Opt. **12**, 755 (1973)
5.27   H. Osterberg, L.W. Smith: J. Opt. Soc. Am. **54**, 1078 (1964)

Chapter 6

6.1   A. Yariv: IEEE J. QE-**9**, 919 (1973)
6.2   R.G. Hunsperger, A. Yariv, A. Lee: Appl. Opt. **16**, 1026 (1977)
6.3   J.M. Hammer, D. Botez, C.C. Neil, J.C. Connoly: J. Appl. Phys. **39**, 943 (1981)
6.4   S. Enoch: Optical fiber interconnect to a single mode laser. OSA/IEEE Meeting on Integrated and Guided Wave Optics, Atlanta, GA (1986)
6.5   M. Yanagisawa, H. Teroi, K. Shuto, T. Miya, M. Kobayashi: Photon. Techn. Lett. **4**, 21 (1992)
6.6   M. Aoki, M. Suzuki, T. Taniwatari, H. Sano, T. Kawano: Microwave Opt. Techn. Lett. **7**, 132 (1994)

6.7   T. Tamir: Beam and waveguide couplers, in *Integrated Optics*, ed. by T. Tamir, 2nd edn., Topics Appl. Phys., Vol. 7 (Springer, Berlin, Heidelberg 1979) Chap. 3, in particular, pp. 102-107

6.8   W.A. Pasmooli, P.A. Mandersloot, M.K. Smit: J. LT-7, 175 (1989)

6.9   M.S. Klimov, V.A. Sychugov, A. Tishchenko, O. Parriaux: Fiber Integr. Opt. **11**, 85 (1992)

6.10   P.K. Tien: Appl. Opt. **10**, 2395 (1971)

6.11   T. Tamir, S.T. Peng: Appl. Phys. **14**, 235 (1977)

6.12   M. Shams, D. Botez, S. Wang: Opt. Lett. **4**, 96 (1979)

6.13   A. Gruss, K.T. Tam, T. Tamir: Appl. Phys. Lett. **36**, 523 (1980)

6.14   K. Rokushima, J. Yamakita: J. Opt. Soc. Am. **73**, 901 (1983)

6.15   K.C. Chang, T. Tamir: Appl. Opt. **19**, 282 (1980)

6.16   K.C. Chang, V. Shah, T. Tamir: J. Opt. Soc. Am. **70**, 804 (1980)

6.17   L. Comerford, P. Zory: Appl. Phys. Lett. **25**, 208 (1974)

6.18   S. Somekh, E. Garmire, A. Yariv, H. Garvin, R.G. Hunsperger: Appl. Opt. **12**, 455 (1973)

6.19   M. Dakss, L. Kuhn, P.F. Heidrich, B.A. Scott: Appl. Phys. Lett. **16**, 523 (1970)

6.20   E. Kapon, A. Katzir: J. Appl. Phys. **53**, 1387 (1982)

6.21   Yu.I. Ostrovsky, M.M. Butusov, G.V. Ostrovskaya: *Interferometry by Holography*, Springer Ser. Opt. Sci., Vol. 20 (Springer, Berlin, Heidelberg 1980)
Yu.I. Ostrovsky, V.P. Shchepinov, V.V. Yakovlev: *Holographic Interferometry in Experimental Mechanics*, Springer Ser. Opt. Sci., Vol. 60 (Springer, Berlin, Heidelberg 1991)

6.22   H. Yen, M. Nakamura, E. Garmire, S. Somekh, A. Yariv: Opt. Commun. **9**, 35 (1973)

6.23   S. Hava, H.B. Sequeira, R.G. Hunsperger: Fabrication of monolithic Peltier cooling structures for semiconductor laser Diodes. Joint Meeting, Nat'l Sci. Foundation, Grantee-User Group in Opt. Commun. and the Opt. Nat'l Telecommun. and Inform. Administr. Task Force on Opt. Commun., St. Louis, MO (1981) Proc. pp. 46-51

6.24   P.K. Tien, R.J. Martin: Appl. Phys. Lett. **18**, 398 (1974)

6.25   W.L. Emkey: IEEE J. LT-1, 436 (1983)

6.26   A.B. Glaser, G.E. Subak-Sharpe: *Integrated Circuit Engineering* (Addison-Wesly, Reading, MA 1977) pp. 263-265 and 267-268

6.27   H. Hsu, A. Milton, W. Burns: Appl. Phys. Lett. **33**, 603 (1978)

6.28   S.K. Sheem, C.H. Bulmer, R.P. Moeller, W.K. Burns: High efficiency single-mode fiber/channel waveguide flip-chip coupling. OSA Topical Meeting on Integrated Optics, Incline Village, NV (1980)

6.29   E.J. Murphy, T.C. Rice: IEEE J. QE-**22**, 928 (1986)

6.30   A. Sugita, K. Onosa, Y. Ohnori, M. Yasu: Fiber Inegr. Opt. **12**, 347 (1993)

6.31   B.L. Booth: Optical interconnection polymers. in *Polymers for Lightwave and Integrated Optics: Technology and Applications*, ed. by L.A. Hornak (Dekker, New York 1992)

6.32   A.T. Andreev, K.P. Panajotov, B.S. Zatirova, J.B. Koprinarova: SPIE Proc. **1973**, 72 (1993)

6.33   J. Lee, H. Lee, C. Lee: SPIE Proc. **1813**, 76 (1991)

6.34   T. Brenner, H. Melchior: IEEE Photon. Techn. Lett. **5**, 1059 (1993)

6.35   M.K. Barnoski: Fiber couplers, in *Semiconductor Devices for Optical Communications*, 2nd edn., ed. by H. Kressel, Topics Appl. Phys., Vol. 39 (Springer, Berlin, Heidelberg 1982) pp. 201-211

6.36 P.K. Tien, G. Smolinsky, R.J. Martin: Theory and experiment on a new film-fiber coupler. OSA Topical Meeting on Integrated Optics. New Orleans, LA (1974)

6.37 G.A. Teh, G.I. Stegeman: Appl. Opt. **17**, 2483 (1978)

Chapter 7

7.1   W.K. Burns, A.F. Milton: Waveguide transitions and junctions, in *Guided-Wave Optoelectronics*, 2nd edn., ed. by T. Tamir, Springer Ser. Electron. Photon., Vol. 26 (Springer, Berlin, Heidelberg 1990) pp. 89-144

7.2   G.A. Vawter, J.L. Merz, L.A. Coldren : IEEE J. QE-**25**, 154 (1989)

7.3   K. Utaka, Y. Suematsu, K. Kobayashi, H. Kawanishi: Room-temperature operation of GaInAsP/InP integrated twin-guide lasers with first-order distributed Bragg reflectors. OSA Topical Meeting on Integrated Optics, Incline Village, NV (1980)

7.4   Y. Suematsu, M. Yamada, K. Hayashi: IEEE Proc. **63**, 208 (1975)

7.5   A.J. Baden Fuller: *Microwaves* (Pergamon, Oxford 1979) pp. 237-238

7.6   S. Somekh, E. Garmire, A. Yariv, R.G. Hunsperger: Appl. Opt. **13**, 327 (1974)

7.7   A. Jervenen, S. Horikanen, S. Najafi: Opt. Eng. **32**, 2083 (1993)

7.8   A. Yariv: IEEE J QE-**9**, 919 (1975)

7.9   A. Yariv: *Quantum Electronics*, 3rd edn. (Wiley, New York 1989) pp. 623-631

7.10 R.G. Peall, R.R.A. Syms: IEEE J. QE-**7**, 540 (1989)

7.11 A. von Hippel: Dielectrics, in *Handbook of Physics*, 2nd edn., ed. by E.U. Condon, H. Odishaw (McGraw-Hill, New York 1967) pp. 4.110-112

7.12 S. Somekh: Theory, fabrication and performance of some integrated optical devices. PhD Thesis, California Institute of Technology (University Microfilms, Ann Arbor, MI 1974) p. 46

7.13 H. Garvin, E. Garmire, S. Somekh, H. Stoll, A. Yariv: Appl. Opt. **12**, 455 (1973)

7.14 M.A. Mentzer, R.G. Hunsperger, S. Sriram, J. Bartko, M.S. Wlodowski, J.M. Zavada, H.A. Jenkinson: Appl. Eng. **24**, 225 (1985)

7 15 S. Somekh, E. Garmire, A. Yariv, H. Garvin, R.G. Hunsperger: Appl. Phys. Lett. **22**, 46 (1973)

7.16 E. Garmire, D. Lovelace, G.H.B. Thompson: Appl. Phys. Lett. **26**, 329 (1975)

7.17 N. Schulz, K. Bierwirth, F. Arndt: IEEE Trans. MTT-**38**, 722 (1990)

7.18 W.E. Martin, D.B. Hall: Appl. Phys. Lett. **21**, 325 (1972)

7.19 I.P. Kaminow, L.W. Stulz, E.H. Turner: Appl. Phys. Lett. **27**, 555 (1975)

7.20 L. Riviere, A. Carenco, A. Yi-Yan, R. Guglielmi: Normalized diagrams for diffused waveguides optical propterties: Applications to Ti:LiNbO$_3$ electrooptic directional coupler design, in *Integrated Optics*, ed. by H.P. Nolting, R. Ulrich, Springer Ser. Opt. Sci., Vol. 48 (Springer, Berlin, Heidelberg 1985) pp. 53-57

7.21 J.C. Campbell, F.A. Blum, D.W. Shaw, K.L. Lawley: Appl. Phys. Lett. **27**, 202 (1975)

7.22 I.P. Kaminov, V. Ramaswamy, R.V. Schmidt, H. Turner: Appl. Phys. Lett. **24**, 622 (1974)

7.23 M. Kawabe, S. Hirata, S. Namba: IEEE Trans. CAS-**26**, 1109 (1978)

7.24 M. Kawabe, M. Kubota, K. Masuda, S. Namba: J. Vac. Sci. Technol. **15**, 1096 (1978)

7.25 M. Koshiba, H. Saitoh, M. Eguch, K. Hirayama: IEE Part J Optoelectron. **139**, 166 (1992)

7.26 B.L. Booth: Optical interconection polymers, in *Polymers for Lightwave and Integrated Optics: Technology and Applications*, ed. by A. Hornak (Dekker, New York 1992) p. 291

7.27 W.K. Burns, A.F. Milton: Waveguide transitions and junctions, in *Guided-Wave Optoelectronics*, 2nd edn., ed. by T. Tamir, Springer Ser. Electron. Photon., Vol. 26 (Springer, Berlin, Heidelberg 1990) Chap. 3, in particular, pp. 102-125

7.28 M. Haruna, J. Koyama: IEEE J. LT-**1**, 223 (1983)

7.29 A. Beguin, T. Dumas, M.J. Hackert: IEEE J. LT-**6**, 1483 (1988)

7.30 P.B. Keck, A.J. Morrow, D.A. Nolan, D.A. Thompson: IEEE J. LT-**7**, 1623 (1989)

Chapter 8

8.1 D.A. Pinnow: IEEE J. QE-**6**, 223 (1970)
J.M. Hammer: Modulation and switching of light in dielectric waveguides, in *Integrated Optics*, 2nd edn., ed. by T. Tamir, Topics Appl. Phys., Vol. 7 (Springer, Berlin, Heidelberg 1979) p. 142

8.3 See, for example, A. Yariv: *Qunatum Electronics*, 3rd edn. (Wiley, New York 1989) pp. 298-307

8.4 See, for example, J.F. Nye: *Physical Properties of Crystals* (Oxford Univ. Press, New York 1957) p. 123
L.A. Shuvalov (ed.): *Modern Crystallography IV*, Springer Ser. Solid-State Sci., Vol. 37 (Springer, Berlin, Heidelberg 1988)

8.5 S. Namba: J. Opt. Soc. Am. **51**, 76 (1961)

8.6 R.C. Alferness: Titanium-diffused lithium niobate waveguide devices, in *Guided-Wave Optoelectronics*, 2nd edn., ed. by T. Tamir, Springer Ser. Electron. Photon., Vol. 26 (Springer, Berlin, Heidelberg 1990) Chap. 4, in particular, pp. 155-157

8.7 D. Hall, A. Yariv, E. Garmire: Appl. Phys. Lett. **17**, 127 (1970)

8.8 I.P. Kaminow, V. Ramaswamy, R.V. Schmidt, F.H. Turner: Appl. Phys. Lett. **27**, 555 (1975)

8.9 I.P. Kaminov, L.W. Stultz, E.H. Turner: Appl. Phys. Lett. **27**, 555 (1975)

8.10 Y. Shuto, M. Amano, T. Kaino: IEEE Photon. Tech. Lett. **3**, 1003 (1991)

8.11 J.C. Campbell, F.A. Blum, D.W. Shaw: Appl. Phys. Lett. **26**, 640 (1975)

8.12 M. Kawabe, S. Hirata, S. Namba: IEEE Trans. CAS-**26**, 1109 (1979)

8.13 See, for example, J.I. Pamkove: *Optical Processes in Semiconductors* (Prentice Hall, Englewood Cliffs, NJ 1971) p. 29

8.14 B.G. Streetman: *Solid State Electronic Devices*, 4th edn. (Prentice Hall, Englewood Cliffs, NJ 1995) pp. 301-307

8.15 V.S. Vavilov: Sov. Phys. – Uspekhi **4**, 761 (1962)

8.16 F.K. Reinhart: Appl. Phys. Lett. **22**, 372 (1973)

8.17 Y. Node: IEEE J LT-**4**, 1445 (1986)

8.18 H. Soda, K. Nakai, H. Ishikawa: High-speed and low-chirp GaInAsP/InP optical intensity modulator. *Integrated and Guided Wave Optics*, 1988 Techn. Digest Ser., Vol. 5 (Opt. Soc. Am., Washington, DC 1988) p. 28

8.19 A. Yariv: IEEE J. QE-**9**, 919 (1975)

8.20 E.A.J. Marcatili: Bell Syst. Techn. J. **48**, 2130 (1969)

8.21 S. Somekh, E. Garmire, A. Yariv, H.L. Garvin, R.G Hunsperger: Appl. Phys. Lett. **22**, 46 (1973)

8.22 S. Somekh, E. Garmire, A. Yariv, H.L. Garvin, R.G. Hunsperger: Appl. Opt. **13**, 327 (1974)

8.23 H.F. Taylor: J. Appl. Phys. **44**, 3257 (1973)

8.24 J.C. Campbell, F.A. Blum, D.W. Shaw, K.I. Lawley: Appl. Phys. Lett. **27**, 202 (1975)

8.25 M. Papuchon, Y. Combernale, X. Mathieu, D.B. Ostrowsky, L. Reiber, A.M. Roy, B. Sejourne, M. Werner: Appl. Phys. Lett. **27**, 289 (1975)

8.26 M. Papuchon: Switching and coupling in Ti:LiNbO₃ waveguides. OSA Topical Meeting on Integrated Optics, Salt Lake City, UT (1976)

8.27 H. Kogelnik, R.V. Schmidt: IEEE J. QE-**12**, 396 (1976)

8.28 Y. Zhou, W. Qiu, Y. Chen: Optica Sinica **14**, 264 (1994)

8.29 R.C. Alferness: Titanium diffused lithium niobate devices, in *Guided-Wave Optoelectronics*, ed. by T. Tamir, 2nd edn., Springer Ser. Electron. Photon., Vol. 26 (Springer, Berlin, Heidelberg 1990) Chap. 4, in particular, ps.179,180

8.30 J.J. Veselka, D.A. Herr, T.O. Murphy, L.L. Buhl, S.K. Korotky: IEEE J. LT-**7**, 908 (1989)

8.31 R.A. Steinberg, T.G. Giallorenzi: IEEE J. QE-**13**, 122 (1977)

8.32 R.A. Steinberg, T.G. Giallorenzi, R.G. Priest: Appl. Opt. **16**, 2166 (1977)

8.33 F. Zernike: Integrated optic switc. OSA Topical Meeting on Integrated Optics, Ne Orleons, LA (1974)

8.34 W.E. Martin: Appl. Phys. Lett. **26**, 562 (1975)

8.35 R.A. Becker: Appl. Phys. Lett. **43**, 131 (1983)

8.36 V. Ramaswami, M.D. Divino, R.D. Standley: Appl. Phys. Lett. **32**, 644 (1978)

8.37 See, for example, K. Iizuka: *Engineering Optics*, 2nd edn., Springer Ser. Opt. Sci., Vol. 35 (Springer, Berlin, Heidelberg 1985) p. 395

8.38 J.M. Hammer: Modulation and switching of light in dielectric waveguides, in *Integrated Optics*, 2nd edn., ed. by T. Tamir (Springer, Berlin, Heidelberg 1979) p. 182

8.39 J.M. Hammer: Appl. Phys. Lett. **18**, 147 (1971)

8.40 D.P. Giarusso, J.H. Harris: Appl. Opt. **10**, 27861 (1971)

8.41 J.M. Hammer, D.J. Channin, M.T. Duffy: Appl. Phys. Lett. **23**, 176 (1973)

8.42 J.M. Hammer, W. Phillips: Appl. Phys. Lett. **24**, 545 (1974)

8.43 Y. Lee, S. Wang: Appl. Opt. **15**, 1565 (1976)

8.44 J.C. An, Y. Cho, Y. Matsuo: IEEE J. QE-**13**, 206 (1977)

8.45 G.L. Tangonan, L. Persechini, J.F. Lotspeich, M.K. Barnoski: Appl. Opt. **17**, 3259 (1978)

8.46 C.S. Tsai, B. Kim, F.R. El-Akkari: IEEE J. QE-**14**, 513 (1978)

8.47 S.K. Sheem: Appl. Opt. **17**, 3679 (1978)

8.48 K. Oh, K. Park, D. Oh, H. Kim, H. Park, K. Lee: IEEE Photon. Tech. Lett. **6**, 65 (1994)

8.49 G.K. Gopalakrishnan, C.H. Bulmar, W.R. Burns, R.W. McElhanon, A.S. Greenble: OSA/IEEE Integrated Photonics Research Conf., New Orleans, LA (1992)

8.50 R. Spickermann, N. Dagli: Electron. Lett. **29**, 774 (1993)

8.51 F. Kappe, G. Mekonnen, C. Bernholdt, P. Reier, D. Hoffman: Electron. Lett. **30**, 1049 (1994)

Chapter 9

9.1   K. Izuka: *Engineering Optics*, 2nd edn., Springer Ser. Opt. Sci., Vol. 35 (Springer, Berlin, Heidelberg 1985) pp. 394-401

9.2   D. Pinnow: IEEE J. QE-**6**, 223 (1970)

9.3   C.S. Ih, R.G. Hunsperger, J.J. Kramer, R. Tian, X. Wang, K. Kissa, J. Butler: SPIE Proc. **876**, 30 (1988)

9.4   A.A. Oliner (ed.): *Acoustic Surface Waves*, Topics Appl. Phys., Vol. 24 (Springer, Berlin, Heidelberg 1978)
       S.V. Biryukov, Yu.V. Gulyaev, V.V. Krylov, V.P. Plessky: *Surface Acoustic Waves in Inhomogeneous Media*, Springer Ser. Wave Phenom., Vol. 20 (Springer, Berlin, Heidelberg 1995)

9.5   J.M. Hammer: Modulation and switching of light in dielectric waveguides, in *Integrated Optics*, 2nd edn., ed. by T. Tamir, Topics Appl. Phys., Vol. 7 (Springer, Berlin, Heidelberg 1982) p. 155

9.6   R. Adler: IEEE Spectrum **4**, 42 (May 1967)

9.7   R.W. Dixon: J. Appl. Phys. **38**, 5149 (1967)

9.8   G. Wade: Bulk-wave acousto-optic Bragg diffraction, in *Guided-wave Acousto-Optic Interactions, Devices and Applications*, ed. by C. Tsai, Springer Ser. Electron. Photon., Vol. 23 (Springer, Berlin, Heidelberg 1990) Chap. 2

9.9   L. Kuhn, M.L. Dakss, P.F. Heidrich, B.A. Scott: Appl. Phys. Lett. **17**, 265 (1970)

9.10   T.G. Gialorenzi, A.F. Milton: J. Appl. Phys. **45**, 1762 (1974)

9.11   D.A. Wille, M.C. Hamilton: Appl. Phys. Lett. **24**, 159 (1974)

9.12   Y. Omachi: J. Appl. Phys. **44**, 3928 (1973)

9.13   T. Gryba, J. Lefebvre, J. Gazalen: *Proc. Ultrasonics Int'l Conf. 1993* (Butterworth-Heinemann, London 1993) pp. 73-76

9.14   N. Chubachi, J, Kushibiki, Y. Kikuchi: Electron. Lett. **9**, 193 (1973)

9.15   R.V. Schmidt, I.P. Kaminow, J.R. Carruthers: Appl. Phys. Lett. **23**, 417 (1973)

9.16   K.W. Loh, W.S.C. Wang, W.R. Smith, T. Grudkowski: Appl. Phys. **15**, 156 (1976)

9.17   R.S. Chu, T. Tamir: IEEE Trans. MTT-**18**, 486 (1970)

9.18   L. Kuhn, P. Heidrich, E. Lean: Appl. Phys. Lett. **19**, 428 (1971)

9.19   M.L. Shah: Appl. Phys. Lett. **23**, 75 (1973)

9.20   G.B. Brandt, M. Gottlieb, J.S. Conroy: Appl. Phys. Lett. **23**, 53 (1973)

9.21   R. Karapinar, E. Gunduz: Opt. Commun. **105**, 29 (1994)

9.22   T. Gryba, J. Lefebvre: IEE Proc. Optoelectronics **141**, 62 (1994)

9.23   See, e.g., A. Nussbaum, R.A. Phillips: *Contemporary Optics for Scientists and Engineers* (Prentice Hall, Englewood Cliffs, NJ 1976) pp. 248-255

9.24   E.G.H. Lean: Acousto-optical interactions, in *Introduction to Integrated Optics*, ed. by M.K. Barnoski (Plenum, New York 1974) p. 458

9.25   E.G.H. Lean, C.G. Powell: IEEE Proc. **58**, 1939 (1970)

9.26   S.M. Sze: *VLSI Technology*, 2nd edn. (McGraw-Hill, New York 1988) pp. 450-457

9.27   C.S. Tsai: IEEE Trans. CAS-**26**, 1072 (1979)

9.28   R.H. Tancrell, M.G. Holland: IEEE Proc. **59**, 393 (1971)

9.29   W.R. Smith, H.M. Gerard, W.R. Jones: IEEE Trans. MTT-**20**, 458 (1972)

9.30   C.S. Tsai, M.A. Alhaider, Le T. Nguyen, B. Kim: IEEE Proc. **64**, 318 (1976)

9.31   C.C. Lee, B. Kim, C.S. Tsai: IEEE J. QE-**15**, 1166 (1979)

9.32   Le T. Nguyen, C.S. Tsai: Appl. Opt. **16**, 1297 (1977)

9.33   D. Mergerian, E.C. Malarkey: Microwave J. **23**, 37 (1980)

9.34 R. Stone, F. Zoise, P. Guifoyle: SPIE Proc. **1563**, 267 (1990)
9.35 K. Kissa, X. Wang, C.S. Ih, R.G. Hunsperger: Novel integrated-optic modulator for optical communications, OSA Opticon'88 (Santa Clara, CA 1988) Paper MK3

Chapter 10

10.1 See, e.g., O. Madelung: *Introduction to Solid-State Theory*, Springer Ser. Solid-State Sci., Vol. 2 (Springer, Berlin, Heidelberg 1978) p. 57
10.2 See, e.g., E.L. Hill: The theory of relativity, in *Handbook of Physics*, ed. by E.U. Condon, H. Odishaw (McGraw-Hill, New York 1967) pp. 2-44
10.3 E.W. Williams, H.B. Bebb: Gallium arsenide photoluminescence, in *Transport and Optical Phenomena*, ed. by R.K. Willardson, A.C. Beer, Semiconductor and Semimetals, Vol. 8 (Academic, New York 1972) Chap. 5
10.4 J.I. Pankove (ed.): *Electroluminescence*, Topics Appl. Phys. Phys., Vol. 17 (Springer, Berlin, Heidelberg 1977) Chap. 3
10.5 J.I. Pankove (ed.): *Display Devices*, Topics Appl. Phys., Vol. 40 (Springer, Berlin, Heidelberg 1980) Chap. 2
10.6 D.A. Cusano: Solid State Commun. **2**, 353 (1964)
10.7 D.A. Cusano: Appl. Phys. Lett. **7**, 151 (1964)
10.8 G. Lucovsky, A. Varga: J. Appl. Phys. **35**, 3419 (1964)
10.9 B.G. Streetman: *Solid State Electronics*, 4th edn. (Printice-Hall, Englewood Cliffs, NJ 1995) pp. 377-380
10.10 See, e.g., M. Shur: *Physics of Semiconductor Devices* (Prentice-Hall, Englewood Cliffs, NJ 1990) pp. 92-102
10.11 R. Bonifacio (ed.): *Dissipative Systems in Quantum Optics*, Topics Curr. Phys., Vol. 27 (Springer, Berlin, Heidelberg 1982)

Chapter 11

11.1 B.N. Hall, G.E. Fenner, J.D. Kingsley, T.J. Soltys, R.O. Carlson: Phys. Rev. Lett. **9**, 366 (1962)
11.2 M.I. Nathan, W.P. Dumke, G. Burns, F.H. Dill Jr., G. Lasher: Appl. Phys. Lett. **1**, 62 (1962)
11.3 T.M. Quist, R.H. Rediker, R.J. Keyes, W.E. Krag, B. Lax, A.L. McWhorter, H.J. Zeiger: Appl. Phys. Lett. **1**, 91 (1962)
11.4 N. Holonyak Jr., S.F. Bevacqua: Appl. Phys. Lett. **1**, 82 (1962)
11.5 See, e.g., A. Yariv: *Optical Electronics*, 4th edn. (Holt, Rinehart and Winston, New York 1991) p. 116
11.6 G. Wade, C.A. Wheeler, R.G. Hunsperger, T.O. Caroll: 5th Int'l Cong. on Microwave Tubes, Paris, France (1964)
11.7 G.J. Lasher: IBM J. **7**, 58 (1963)
11.8 G. Wade, C.A. Wheeler, R.G. Hunsperger: IEEE Proc. **53**, 98 (1965)
11.9 G. Diemer, B. Bölger: Physica **29**, 600 (1963)
11.10 T. Pecany. Phys. Stat. Sol. **6**, 651 (1964)
11.11 H. Kroemer: IEEE Proc. **51**, 1782 (1963)
11.12 Zh.I. Alferov: Sov. Phys. – Solid State **7**, 1919 (1966)
11.13 H. Kressel, N. Nelson: RCA Rev. **30**, 106 (1969)
11.14 R.H. Fowler, L. Nordheim: Proc. Roy. Soc. (London) A **119**, 173 (1928)
11.15 A.G. Chynoweth: Progr. Semiconduct. **4**, 97 (1960)

Supplementary Reading on Semiconductor-Laser Fundamentals

Carlson N.W.: *Monolithic Diode-Laser Arrays*, Springer Ser. Electron. Photon., Vol. 33 (Springer, Berlin, Heidelberg 1994)

Chow W.W., S.W. Koch, M. Sargent III: *Semiconductor-Laser Physics* (Springer, Berlin, Heidelberg 1994)

Klingshirn C.F.: *Semiconductor Optics* (Springer, Berlin, Heidelberg 1995)

Kressel H., M. Ettenberg, J.P. Wittke, I. Ladany: Laser diodes and LEDs for fiber optical communications, in *Semiconductor Devices for Optical Communications*, 2nd edn., ed. by H. Kressel (Springer, Berlin, Heidelberg 1982) pp. 9-62

Pankove J.: *Optical Processes in Semiconductors* (Printice-Hall, Reading, MA 1971) Chap. 10

Yariv A.: *Quantum Electronics*, 3rd edn. (Wiley, New York 1989) pp. 232-263

Yu P., M. Cardona: *Fundamentals of Semiconductors: Physics and Material Properties* (Springer, Berlin, Heidelberg 1995)

Chapter 12

12.1 G. Diemer, B. Bölger: Physica **29**, 600 (1963)

12.2 T. Pecany: Phys. Stat. Sol. **6**, 651 (1964)

12.3 H. Kroemer: IEEE Proc. **51**, 1782 (1963)

12.4 Zh.I. Alferov: Sov. Phys. – Solid State **7**, 1919 (1966)

12.5 I. Hayashi, M.B. Panish, P. Foy: IEEE J. QE-**5**, 211 (1969)

12.6 H. Kressel, H. Nelson: RCA Rev. **30**, 106 (1969)

12.7 Zh.I. Alferov, V. Andreev, E. Portnoi, M. Trukhan: Sov. Phys. – Semicond. **3**, 1107 (1969)

12.8 K. Sheger, A. Milnes, D. Feught: Proc. Int'l Conf. on Chem. Semicond. Heterojunction Layer Structures, Budapest (Hung. Acad. Sci., Budapest 1970) Vol. 1, p. 73
P.H. Holloway, T.J. Anderson: *Compound Semiconductors: Growth, Processing and Devices* (CRC Press, Boca Raton, FL 1989) p. 115

12.9 Q.H.F. Vrehen: J. Phys. Chem. Solids **29**, 129 (1968)

12.10 H. Yonezu, I. Sakuma, Y. Nannich: Jpn. J. Appl. Phys. **9**, 231 (1970)

12.11 A. Yariv, R.C.C. Leite: Appl. Phys. Lett. **2**, 173 (1963)

12.12 H.C. Casey Jr., M.B. Panish: *Heterostructure Lasers*, Pt.B: Materials and Operating Characterizations (Academic, New York 1978) pp. 109-132

12.13 H. Kroemer: IEEE Trans. ED-**39**, 2635 (1992)

12.14 K. Iga, S. Kinoshita: *Semiconductor Lasers and Related Epitaxies*, Springer Ser. Mater. Sci., Vol. 30 (Springer, Berlin, Heidelberg 1995)

12.15 A. McWhorter: Solid State Electron. **6**, 417 (1963)

12.16 H.C. Casey Jr., M.B. Panish: *Heterostructure Lasers*, Pt.A: Fundamental Principles (Academic, New York 1978) pp. 54-57

12.17 L. Goldstein, C. Starck, J. Emery, F. Gaborit, O. Bonnevie, F. Poingt, M. Lambert: J. Crystal Growth **120**, 157 (1992)

12.18 S. Oshiba, Y. Kawai: OSA/IEEE OFC'90, San Francisco (1990)

12.19 P.K. Cheo: *Fiber Optics and Optoelectronics* (Prentice-Hall, Englewood Cliffs, NJ 1990) p. 239

12.20 D. Greenaway, G. Harbeke: *Optical Properties and Band Structure of Semiconductors* (Pergamon, Oxford 1968) p. 67

12.21 D.N. Payne, W.A. Gambling: Electron. Lett. **11**, 176 (1975)

12.22 See, e.g., T. Izawa: *Optical Fibers: Materials and Fabrication* (KTK Scientific Pub., Tokyo 1986)

12.23 H. Kressel (ed.): *Semiconductor Devices for Optical Communications*, 2nd edn., Topics Appl. Phys., Vol.39 (Springer, Berlin, Heidelberg 1982) pp. 285-289

12.24 M. Kawahara, S. Oshiga, A. Matoba, Y. Kawai, Y. Tamura: OSA/IEEE OFC'87, Reno, NV, Paper ME-1

12.25 E. Rezek: OSA Topical Meeting on Semiconductor Lasers, Albuquerque, NM (1987)

12.26 H. Mani, A. Joullie, G. Boissier, E. Tournie, F. Pitard, C.A. Ailibert: Electron. Lett. **24**, 1542 (1988)

12.27 R.V. Martinelli: LEOS'88, Santa Clara, CA, Digest p.55

12.28 H. Ebe, Y. Nishijima, K. Shinohara: 11th IEEE Int'l Conf. on Semicond. Lasers, Boston, MA (1988) Digest p.68

12.29 D.L. Partin: IEEE J. QE-**24**, 1716 (1988)

12.30 J.C. Dyment: Appl. Phys. Lett. **10**, 84 (1967)

12.31 L.A. D'Asaro: J. Lumin. **7**, 310 (1973)

12.32 H. Yonezu, I. Sakuma, K. Kobayashi, T. Kamejima, M. Ueno, Y. Nannicki: Jpn. J. Appl. Phys. **12**, 1585 (1973)

12.33 T. Tsukada: J. Appl. Phys. **45**, 4899 (1974)

12.34 M. Nakamura: IEEE Trans. CAS-**26**, 1055 (1979)

12.35 N. Chinone: J. Appl. Phys. **48**, 3237 (1977)

12.36 K. Seki, T. Kamiya, H. Yanai: Trans. IECE (Jpn.) E-**62**, 73 (1979)

12.37 W.O. Schlosser: Bell. Syst. Tech. J. **52**, 887 (1973)

12.38 W.T. Tsang, R.A. Logan, M. Ilegems: Appl. Phys. Lett. **32**, 311 (1978)

12.39 T. Kobayashi, H. Kawaguchi, Y. Furukawa: Jpn. J. Appl. Phys. **16**, 601 (1977)

12.40 H. Namizaki: IEEE J QE-**11**, 427 (1975)

12.41 K. Aiki, M. Nakamura, T. Kuroda, J. Umeda: Appl. Phys. Lett. **48**, 649 (1977)

12.42 H. Yonezu: Jpn. J. Appl. Phys. **16**, 209 (1977)

12.43 D. Botez: Appl. Phys. Lett. **33**, 872 (1978)

12.44 C.S. Hong, Y.Z. Liu, J.J. Coleman, P.D. Dapkus: OSA/IEEE Topical Meeting on Integrated Optics, Asilomar, CA (1982) Paper FC2

12.45 I.P. Kaninow, R.S. Tucker: Mode-controlled semiconductor lasers, in *Guided-Wave Optoelectronics*, 2nd edn., ed. by T. Tamir, Springer Ser. Electron. Photon., Vol.26 (Springer, Berlin, Heidelberg 1990) pp.211-263

12.46 R. Baets: Solid State Electron. **30**, 1175 (1987)

12.47 C.E. Hurwitz, J.A. Rossi, J.J. Hsieh, C.M. Wolfe: Appl. Phys. Lett. **27**, 241 (1975)

12.48 L.A. Koszi, A.K. Chin, B.P. Segner, T.M. Shen, N.K. Dutta: Electron. Lett. **21**, 1209 (1985)

12.49 A. Antreasyn, C.Y. Chen, R.A. Logan: Electron. Lett. **21**, 405 (1985)

12.50 A. Antreasyn, S.G. Napholtz, D.P. Wilt, P.A. Garbinski: IEEE J. QE-**22**, 1064 (1986)

12.51 M. Ishii, K. Kamon, M. Shimazu, M. Mihara, H. Kumabe, K. Isshiki: Electron. Lett. **23**, 179 (1987)

12.52 M. Ishii, K. Kamon, M. Shimazu, M. Mihara, H. Kumabe, K. Isshiki: Optoelectronics – Devices and Technologies 2, 83 (1987)

12.53 K. Oe, Y. Noguchi, C. Canea: IEEE Photon. Tech. Lett. **6**, 479 (1994)

12.54 M. Kawabe, H. Kotani, K. Masuda, S. Namba: Appl. Phys. Lett. **26**, 46 (1975)

12.55 K. Aiki, M. Nakamura, J. Umeda: Appl. Phys. Lett. **29**, 506 (1976)

12.56 M. Krakowski, R. Blondeau, J. Ricciardi, J. Hirtz, M. Razeghi, B. de Cremoux: OSA/IEEE OFC/IGWO'86, Atlanta, GA, Paper TU33

12.57 P. Petroff, R.L. Hartman: Appl. Phys. Lett. **23**, 469 (1973)

12.58 J.R. Baird, G.E. Pittman, J.F. Leezer: *Proc. 1966 Symp. on GaAs* (IoP, London 1967) p. 113

12.59 H. Yonezu, I. Sakuma, T. Kamejima, M. Ueno, K. Nishida, Y. Nannicki, I. Hayashi: Appl. Phys. Lett. **24**, 18 (1974)

12.60 R. Ito, H. Nakashima, S. Kishino, O. Nakada: IEEE J. QE-**11**, 551 (1975)

12.61 B.C. DeLoach, B.W. Haki, R.L. Hartman, L.A. D'Asaro: IEEE Proc. **61**, 1042 (1973)

12.62 R.L. Hartman, N.E. Schumaker, R.W. Dixon: Appl. Phys. Lett. **31**, 756 (1977)

12.63 H. Kressel, M. Ettenberg, I. Ladany: Appl. Phys. Lett. **32**, 305 (1978)

12.64 S. Oshiba, Y. Kawai: Electron. Lett. **23**, 843 (1987)

12.65 J. Buus, R. Plastow: IEEE J QE-**21**, 614 (1985)

12.66 S. Kobayashi, T. Kimura: IEEE Spectrum **21**, 26-33 (May 1984)

12.67 Y.C. Zhang, Z.X. Qin, S.L. Wu, L.J. Wu, L.J. Wang, D.E. Lee: Fiber and Integrated Optics **8**, 99 (1989)

12.68 T. Brenner, R. Dall, A. Holtmann, P. Besse, N. Melchior: IEEE 5th Int'l Conf. on InP and Related Materials, Paris (1993) Digest p. 88

Supplementary Reading on Heterojunction Lasers

Butler J.K. (ed.): *Semiconductor Injection Lasers* (IEEE Press, New York 1980)

Casey Jr H.C., M.B. Panish: *Heterostructure Lasers* (Academic, New York 1978)

Kressel H. (ed.): *Semiconductor Devices for Optical Communication*, 2nd edn., Topics Appl. Phys., Vol. 39 (Springer, Berlin, Heidelberg 1982) Chap. 2

Kressel H., J.K. Butler: *Semiconductor Lasers and Heterojunction LEDs* (Academic, New York 1977)

Tamir T. (ed.): *Guided-Wave Optoelectronics*, 2nd edn., Springer Ser. Electron. Photon., Vol. 26 (Springer, Berlin, Heidelberg 1990) Chap. 5

Chapter 13

13.1 See, e.g., Z.G. Pinsker: *Dynamical Scattering of X Rays in Crystals*, Springer Ser. Solid-State Sci., Vol. 3 (Springer, Berlin, Heidelberg 191978)
B.K. Agarwal: *X-Ray Spectroscopy*, 2nd edn., Springer Ser. Opt. Sci., Vol. 15 (Springer, Berlin, Heidelberg 1991)

13.2 H. Kogelnik, C.V. Shank: J. Appl. Phys. **43**, 2327 (1972)

13.3 A. Yariv: IEEE J. QE-**9**, 919 (1973)

13.4 S. Wang: IEEE J. QE-**10**, 413 (1974)

13.5 D.R. Scifres, R.D. Burnham, W. Streifer: Appl. Phys. Lett. **26**, 48 (1975)

13.6 W. Streifer, D.R. Scifres, R.D. Burnham: IEEE J. QE-**11**, 867 (1975)

13.7 A. Yariv: *Optical Electronics*, 4th edn. (Holt, Rinhart, Winston, New York 1991) p. 503

13.8 R. Buchmann, H. Dietrich, G. Sasso, P. Vettiger: Microelectron. Eng. **9**, 485 (1989)

13.9 D.R. Scifres, R. Burnham, W. Streifer: Appl. Phys. Lett. **25**, 203 (1974)

13.10 M. Nakamura, K. Aiki, J. Umeda, A. Yariv: Appl. Phys. Lett. **27**, 403 (1975)
K. Aiki, M. Nakamura, J. Umeda: IEEE J. QE-**12**, 597 (1976)

13.11 S. Tsuji, A. Ohishi, N. Nakamura, M. Hirao, N. Chinone, H. Matsumura: IEEE J. LT-**5**, 822 (1987)

13.12 P. Bhattacharya: *Semiconductor Optoelectronic Devices* (Prentice-Hall, Engle-wood Cliffs, NJ 1994) pp. 286-292

13.13 S. Tsuji, K. Mizuishi, M. Hirao, M. Nakamura: IEEE Int'l Conf. on Communi-cations, Amsterdam, The Netherlands (1984) Proc. p. 1123

13.14 R. Martin, S. Forouhar, S. Keo, R. Lang, R.G. Hunsperger, R. Tiberio, P. Chapman: Electron. Lett. **30**, 1058 (1994)

13.15 S. Wang: IEEE J. QE-**10**, 413 (1974)

13.16 H.M. Stoll: IEEE Trans. CAS-**26**, 1065 (1979)

13.17 K. Aiki, M. Nakamura, J. Umeda: IEEE J. QE-**13**, 220 (1977)

13.18 R. Kaiser, F. Fidorra, H. Heidrich, P. Albrecht, W. Rehbein, S. Malchow, H. Schroeter-Janssen, D. Franke, G. Sztefka: 6th Int'l Conf. on InP and Related Materials, Santa Barbara, CA (1994) Proc. p. 474

13.19 W. Tsang, M. Wu, Y. Chen, F. Choa, R. Logan, S. Chu, A. Sergent, P. Magill, K. Reichmann, C. Burrus: IEEE J. QE-**30**, 1370 (1994)

13.20 M. Nakamura, A. Yariv, H.W. Yen, S. Somekh, H.L. Garvin: Appl. Phys. Lett. **22**, 515 (1973)

13.21 M. Nakamura, H.W. Yen, A. Yariv, E. Garmire, S. Somekh, H.L. Garvin: Appl. Phys. Lett. **23**, 224 (1973)

13.22 H. Yusuka, J. Nakano, M. Fukuda, Y. Nakano, Y. Itaya: IEEE Photon. Tech. Lett. **1**, 74 (1989)

13.23 H.A. Haus, C.V. Shank: IEEE J. QE-**12**, 532 (1976)

13.24 S. Akiba, Y. Matsushima, M. Usami, K. Utaka: Electron. Lett. **23**, 316 (1987)

13.25 W. Tsang, R. Kapre, R. Logan, T. Tanbun-Ek: IEEE Photon. Tech. Lett. **5**, 978 (1993)

13.26 M. Okai, T. Tsuchiya: Electron. Lett. **29**, 349 (1993)

13.27 H. Mawatari, F. Kano, N. Yamomoto, Y. Kondo, Y. Tohmori, Y. Yoshikun: Jpn. J. Appl. Phys. Pt.1 **33**, 811 (1994)

13.28 M. Okai, A. Tsuchiya, A. Takai, N. Chinone: IEEE Photon. Tech. Lett. **4**, 526 (1992)

13.29 S. Takigawa, T. Uno, M. Kume, K. Hamada, N. Yoshikawa, H. Shimizu, G. Kano: IEEE J. QE-**25**, 1489 (1989)

13.30 B. Wedding, B. Franz: Electron. Lett. **29**, 402 (1993)

Chapter 14

14.1   G.J. Lasher: Solid State Electron. **7**, 707 (1964)

14.2   T.L. Paoli, J.E. Ripper: IEEE Proc. **58**, 1457 (1970)

14.3   T. Ikegami, Y. Suematsu: Electron. Commun. (Jpn.) B **51**, 51 (1968)

14.4   See, e.g., T.P. Lee, R.M. Derosier: IEEE Proc. **62**, 1176 (1974)

14.5   K. Konnerth, C. Lanza: Appl. Phys. Lett. **4**, 120 (1964)

14.6   M. Ross: IEEE Trans. AES-**3**, 324 (1967)

14.7   G.E. Fenner: Pulse width modulated laser. US Patent no. 3,478,280 (Nov. 1969)

14.8   L.A. D'Asaro, J.M. Cherlow, T.L. Paoli: IEEE J. QE-**4**, 164 (1968)

14.9   J.E. Ripper, T.L. Paoli: Appl. Phys. Lett. **15**, 203 (1969)

14.10 T.L. Paoli, J.E. Ripper: IEEE J. QE-**6**, 335 (1970)

14.11 Yu P. Zakharov, V.V. Nikitin, V.D. Samoilov: Sov. Phys. – Semicond. **2**, 895 (1969)

14.12 G.E. Fenner: Appl. Phys. Lett. **5**, 198 (1964)

14.13 W.T. Tsang, N.A. Olsson, R.A. Logan: Appl. Phys. Lett. **42**, 650 (1983)

14.14 L.A. Coldren, D.I. Miller, K. Iga, A. Rentschler: Appl. Phys. Lett. **38**, 315 (1981)

14.15 W.T. Tang, N.A. Olsson, R.A. Logan: IEEE J. QE-**19**, 1621 (1983)

14.16 W. Streifer, D. Yevick, T.L. Paoli, R.D. Burnham: IEEE J. QE-**20**, 754 (1984)

14.17 J.E. Ripper, G.W. Pratt, C.G. Whitney: IEEE J. QE-**2**, 603 (1966)

14.18 J.E. Ripper: IEEE J. QE-**6**, 129 (1970)

14.19 C.G. Whitney, G.W. Pratt: IEEE J. QE-**6**, 352 (1970)

14.20 H. Reick: Solid State Electron. **8**, 83 (1965)

14.21 D. Kleinman: Bell Syst. Tech. J. **43**, 1505 (1964)

14.22 B. Goldstein, R. Wigand: IEEE Proc. **53**, 195 (1965)

14.23 R. Myers, P. Pershan: J. Appl. Phys. **36**, 22 (1965)

14.24 T. Nakano, T. Oku: Jpn. J. Appl. Phys. **6**, 1212 (1967)

14.25 T. Ikegami, Y. Suematsu: IEEE Proc. **55**, 122 (1967)

14.26 T. Ikegami, Y. Suematsu: Electron. Commun. (Jpn.) B **51**, 51 (1968)

14.27 J. Nishizawa: IEEE J. QE-**4**, 143 (1968)

14.28 T. Ikegami, Y. Suematsu: IEEE J. QE-**4**, 148 (1968)

14.29 S. Takamiga, F. Kitasawa, J. Nishizawa: IEEE Proc. **56**, 135 (1968)

14.30 Zh.I. Alferov: Sov. Phys. – Semicond. **3**, 1170 (1970)

14.31 M.B. Panish, I. Hayashi, S. Sumski: Appl. Phys. Lett. **17**, 109 (1970)

14.32 M. Lakshminarayana, R.G. Hunsperger, L. Partain: Electron. Lett. **14**, 640 (1978)

14.33 M. Maeda, K. Nagano, I. Ikushima, M. Tanaka, K. Saito, R. Ito: 3rd Europ. Conf. Opt. Commun., NTG Fachberichte **59**, 120 (1977)

14.34 H. Yania, M. Yano, T. Kamiya: IEEE J. QE-**11**, 519 (1975)

14.35 J. Caroll, J. Farrington: Electron. Lett. **9**, 166 (1973)

14.36 P. Russer, S. Schultz: Arch. Elektr. Übertrag. **27**, 193 (1973)

14.37 T. Ozeki, T. Ito: IEEE J. QE-**9**, 388 (1973)

14.38 A.J. Seeds, J.R. Forrest: Electron. Lett. **14**, 829 (1978)

14.39 H.W. Yen: OSA/IEEE Conf. on Laser Engineering and Applications, Washington, DC (1979) Digest p. 9D

14.40 F. Mengel, V. Ostoich: IEEE J. QE-**13**, 359 (1977)

14.41 D.J. Channin, M. Ettenberg, H. Kressel: J. Appl. Phys. **50**, 6700 (1979)

14.42 D.J. Channin: SPIE Proc. **224**, 128 (1980)

14.43 K.Y. Lau, A. Yariv: Appl. Phys. Lett. **40**, 452 (1982)

14.44 K.Y. Lau, A. Yariv: Appl. Phys. Lett. **46**, 326 (1985)

14.45 R. Olshansky, V. Lanzisera, C. Bsu, W. Powazinik, R.B. Lauer: Appl. Phys. Lett. **49**, 128 (1986)

14.46 J.E. Bowers: Solid State Electron. **30**, 1 (1987)

14.47 T. Chen, P. Chen, J. Ungar, N. Bar-Chain: Electron. Lett. **30**, 1055 (1994)

14.48 H. Lipsanen, D. Coblentz, R. Logan, R. Yadvish, P. Morton, H. Temkin: IEEE Photon. Tech. Lett. **4**, 673 (1992)

14.49 J. Ralston, S. Weisser, K. Eisele, R. Sah, E. Larkins, J. Rosenzweig, J. Fleissner, K. Bender: IEEE Photon. Tech. Lett. **6**, 1076 (1994)

14.50 B. Tromberg, H. Lassen, H. Oleseni: IEEE J. QE-**30**, 939 (1994)

14.51 See, e.g., P.A. Rizzi: *Microwave Engineering Passive Circuits* (Prentice-Hall, Englewood Cliffs, NJ 1988) pp. 248-299

14.52 R.G. Hunsperger, N. Hirsch: Solid State Electron. **18**, 349 (1975)

14.53 S. Dods, M. Ogura, M. Watanabe: IEEE J. QE-**29**, 2631 (1993)

14.54 S. Margalit, N. Bar-Chaïm, I. Ury, D. Wilt, M. Yust, A. Yariv: Monolithic integration of optical and electronic devices on semi-insulating GaAs substrates. OSA/IEEE Topical Meeting on Integrated and Guided-Wave Optics, Incline Village, NV (1980)

14.55 I. Ury, S. Margalit, M. Yust, A. Yariv: Appl. Phys. Lett. **34**, 430 (1979)

14.56 T. Fukuzama, M, Nakamura, M. Hirao, T. Kuroda, J. Umeda: Appl. Phys. Lett. **36**, 181 (1980)

14.57 H. Nakano, S. Yamashita, T. Tanaka; N. Hirao, N. Naeda: IEEE J. LT-**4**, 574 (1986)

14.58 J. Carney, M. Helix, R. Kolbas: GaAs IC Symp., Phoenix, AZ (1983) Digest p. 48

14.59 A. Suzuki, K. Kasahara, M. Shikada: IEEE J. LT-**5**, 1479 (1987)

14.60 K. Dretting, W. Idler, P. Wiedermann: Electron. Lett. **29**, 2195 (1993)

14.61 P. Woolnough, P. Birdsall, P. O'Sullivan, A. Cockburn, M. Harlow: Electron. Lett. **29**, 1388 (1993)

14.62 N. Suzuki, H. Furuyama, Y. Hirayama, M. Morinaga, E. Eguchi, M. Kushibe, M. Funamizu, M. Nakamura: Electron. Lett. **24**, 467 (1988)

14.63 J. Wang, C. Shih, N. Chang, J. Middleton, P. Apostolakis, M. Feng: IEEE Photon. Tech. Lett. **5**, 316 (1993)

14.64 J. Wang, C. Shih, W. Chang, J. Middleton, P. Apostolakis, M. Feng: IEEE MTT-S Int'l Symp. on Circuits and Systems, Atlanta, GA (1993) Digest Vol. 2, p. 1047

14.65 W. Hong, G. Chang, R. Bhat, J. Hayes: 5th Int'l Conf. on InP and Related Materials, Paris (1993) Proc. p. 275

14.66 D. Trommer, Unterborsch, F. Reier: 5th Int'l Conf. on InP and Related Materials, Paris (1993) Proc. p. 251

14.67 C. Shih, D. Barlage, J. Wang, M. Feng: IEEE MTT-S Int'l Microwave Symp., San Diego, CA (1994) Digest, Vol. 3, p. 1379

14.68 M. Yust, N. Bar-Chaim, S. Izapanah, S. Margalit, I. Ury, D. Wilt, A. Yariv: Appl. Phys. Lett. **35**, 796 (1979)

14.69 K. Runge: CLEO'93, Anaheim, CA, p. 374

14.70 K. Jackson, E. Flint, M. Cina, D. Lacey, Y. Kwark, J. Trewhella, T. Caulfield, P. Buchmann, Ch. Harder, P. Vettiger: IEEE J. LT-**12**, 1185 (1994)

Supplementary Reading on Modulation of Laser Diodes

Arnold G., P. Russer, K. Petermann: Modulation of laser diodes, in *Semiconductor Devices for Optical Communication*, 2nd edn., ed, by H. Kressel, Topics Appl. Phys., Vol. 39 (Springer, Berlin, Heidelberg 1982) Chap. 7

Butler J.K (ed.): *Semiconductor Injection Lasers* (IEEE Press, New York 1980) pp. 332-389

Casey Jr. H.C, M.B. Panish: *Heterostructure Lasers* (Academic, New York 1978) pp. 258-264

Chapter 15

15.1 D.P. Schinke, R.G. Smith, A.R. Hartmann: Photodetectors, in *Semiconductor Devices for Optical Communication*, 2nd edn., ed. by H. Kressel, Topics Appl. Phys., Vol. 39 (Springer, Berlin, Heidelberg 1982) Chap. 3

15.2 R.J. Keyes (ed.): *Optical and Infrared Detectors*, 2nd edn., Topics Appl. Phys., Vol. 19 (Springer, Berlin, Heidelberg 1980)

15.3 S.M. Sze: *Physics of Semiconductor Devices*, 2nd edn. (Wiley, New York 1981) p. 665

15.4 D. Bossi, R. Ade, R. Basilica, J. Berak: IEEE Photon. Tech. Lett. **5**, 166 (1993)

15.5 M. Erman, Ph. Riglet, Ph. Jarry, B. Martin, M. Renaud, J. Vinchant, J. Cavailles: IEE Proc. Pt. J Optoelectron. **138**, 101 (1991)

15.6 D. Parker: IEE Coll. on Optical Control and Generation of Microwave Signals, London (1986) Digest p. 9

15.7 B.G. Streetman: *Solid State Electronic Devices*, 4th edn. (Prentice-Hall, Englewood Cliffs, NJ 1994) pp. 183-190

15.8 S. Miller: Phys. Rev. **99**, 1234 (1955)

15.9 H. Melchior, W.T. Lynch: IEEE Trans. ED-**13**, 829 (1966)

15.10 See, e.g., M. Lakshminarayana, R.G. Hunsperger, L. Partain: Electron. Lett. **14**, 640 (1978)

15.11 H. Melchior, A.R. Hartman, D.P. Schinke, T.E. Seidel: Bell Syst. Tech. J. **57**, 1791 (1978)

15.12 K.Kato, A. Kozen, Y. Muramoto, Y. Itaya, T. Nagatsuma, M. Yaita: IEEE Photon. Tech. Lett. **6**, 719 (1994)

15.13 T. Hsiang, S. Alexandrou, C. Wang, M. Liu, S. Chou: SPIE Proc. **2022**, 76 (1993)

15.14 I.R. Mactaggart, M. Bendett, S. Tayler: IEEE J. SSC-**28**, 1018 (1993)

15.15 V. Hurm, M. Ludwig, J. Rosenzweig, W. Benz, M. Berroth, R. Bosch, W. Bronner, A. Hulsmann, K. Kohler, B. Raynor, J. Schneider: Electron. Lett. **29**, 9 (1993)

15.16 M. Leary, J. Ballantyne: IEEE Cornell Conf. on Advanced Concept High Speed Semiconductor Devices and Circuits, Ithaca, NY (1993) Proc. p. 383

15.17 A. Iliadis, N. Papanicolaou, A. Christou, G. Anderson: 6th Int'l SAMPLE Electronics Conf., Baltimore, MD (1992) Proc. p. 816

15.18 W. Gao, A. Khan, P. Berger, R.G. Hunsperger, G. Zydzik, H. O'Bryan, D. Sivco, A. Cho: Appl. Phys. Lett. **65**, 1930 (1994)

15.19 J. Seo, C. Caneau, R. Bhat, I. Adesida: IEEE Photon. Tech. Lett. **5**, 1313 (1993)

15.20 D. Ostrowsky, R. Poirier, L. Reiber, C. Deverdun: Appl. Phys. Lett. **22**, 463 (1973)

15.21 S. Koike, H. Takahara, K. Katsura: Electron. and Commun. Jpn., Pt.II: Electron. **75**, 41 (1922)

15.22 G. Stillman, C.M. Wolfe, I. Melngailis: Appl. Phys. Lett. **25**, 36 (1974)

15.23 T. Kimura, K. Daikoku: Opt. and Quantum Electron. **9**, 33 (1977)

15.24 S. Miura, H. Kuwatsuka, T. Mikawa, O. Wada: IEEE J. LT-**6**, 399 (1988)

15.25 S. Chandrasakhar, J.C. Campbell, A.G. Dentai, C.H. Joyner, G.J. Qua, W.W. Snell: IEEE Electron Dev. Lett. **8**, 512 (1987)

15.26 S. Chandrasekhar, J.C. Campbell, A.G. Dentai, G.J. Qua: Electron. Lett. **23**, 501 (1987)

15.27 M. Erman, P. Jarry, R. Gamonal, J. Gentner, P. Stephan, C. Guedon: IEEE J. LT-**6**, 399 (1988)

15.28 H. Stoll, A. Yariv, R.G. Hunsperger, G. Tangonan: Appl. Phys. Lett. **23**, 664 (1973)

15.29 H.J. Stein: In *Int'l Conf. on Ion Implantation of Semiconductors and Other Materials*, Yorktown Heights, NY, 1972 (Plenum, New York 1973)

15.30 M.K. Barnoski, R.G. Hunsperger, R.G. Wilson, G. Tangonan: J. Appl. Phys. **44**, 1925 (1973)

15.31 S. Valete, G. Labrunie, J. Deutsch, J. Lizet: Appl. Phys. **16**, 1289 (1977)

15.32 D.L. Spears, A.J. Strauss, S.R. Chinn, I. Melngailis, P. Vohl: OSA/IEEE Topical Meeting on Integrated Optics, Salt Lake City, UT (1976) Digest Paper TUD3-1

15.33 A. Carenco, L. Menigaux: Appl. Phys. Lett. **37**, 880 (1980)

15.34 See, e.g., J.I. Pankove: *Optical Processes in Semiconductors* (Prentice-Hall, Englewood Cliffs, NJ 1971) p. 29

15.35 K.H. Nichols, W.S.C. Wang, C.M. Wolfe, G.E. Stillman: Appl. Phys. Lett. **31**, 631 (1977)

15.36 R.G. Hunsperger: Monolithic dual mode emitter-detector terminal for optical waveguide transmission lines. US Patent No.3,952,265 (issued 20 April 1976)

15.37 J. Park, S. Wadekar, R.G. Hunsperger: SPIE Proc. **835**, 283 (1987)

15.38 S. Wadekar, E. Armour, M. Donhowe, R.G. Hunsperger: SPIE Proc. **994**, 133 (1988)

15.39 R. Link, K. Reichmann, T. Koch, V. Koren: IEEE Phot. Tech. Lett. **1**, 278 (1989)

15.40 S. Sakano, H. Inoue, H. Nakamura, T. Matsumura: Electron. Lett. **22**, 594 (1986)

15.41 H. Inoue, S. Tsuji: Appl. Phys. Lett. **51**, 1577 (1987)

15.42 J.E. Bowers, C.A. Burrus Jr.; IEEE J. LT-5, 1339 (1987)

15.43 D. Wolf (ed.): *Noise in Physical Systems*, Springer Ser. Electrophys., Vol. 2 (Springer, Berlin, Heidelberg 1978)

15.44 M. DiDomerrico Jr., O. Svelto: IEEE Proc. **52**, 136 (1964)

15.45 B. Saleh: *Photoelectron Statistics*, Springer Ser. Opt. Sci., Vol. 6 (Springer, Berlin, Heidelberg 1978)

15.46 J.W. Goodman, E. Rawson: Speckle phenomena in optical communication, in *Laser Speckle and Related Phenomena*, 2nd edn., ed. by J.C. Dainty, Topics Appl. Phys., Vol. 9 (Springer, Berlin, Heidelberg 1982)

Chapter 16

16.1 See, e.g., K. Seeger: *Semiconductor Physics*, 5th edn., Springer Ser. Solid-State Sci., Vol. 40 (Springer, Berlin, Heidelberg 1991)

16.2 R. Dingle: *Proc. 13th Int'l Conf. on Physics of Semiconductors* ed. by F.G. Fumi (North-Holland, Amsterdam 1976)

16.3 A. Yariv: *Quantum Electronics*, 3rd edn. (Wiley York, New York 1988) pp. 267-268

16.4 A. Yariv: *Quantum Electronics*, 3rd edn. (Wiley York, New York 1988) p. 270

16.5 A. Yariv: *Quantum Electronics*, 3rd edn. (Wiley York, New York 1988) p. 272

16.6 A. Larsson, P. Andrekson, B. Jonsson, C. Lindstrom: IEEE J. QE-25, 2013 (1989)

16.7 J. Feldmann, G. Peter, E.O. Göbel, K. Leo, H. Polland, K. Ploog, K. Fujiware, T. Nakayama: Appl. Phys. Lett. **54**, 226 (1987)

16.8 W.T. Tsang: Appl. Phys. Lett. **40**, 217 (1982)

16.9 L.M. Walpita: J. Opt. Soc. Am. A **2**, 592 (1985)

16.10 P.L. Derry, A. Yariv, K.Y. Lau, N. Bar-Chaim, K. Lee, J. Rosenberg: Appl. Phys. Lett. **50**, 1773 (1987)

16.11 See, e.g., 16.3 A. Yariv: *Quantum Electronics*, 3rd edn. (Wiley York, New York 1988) p. 274

16.12 W.T. Tsang: Appl. Phys. Lett. **39**, 786 (1981)

16.13 R. Chin, N. Hollonyak Jr., B. Vojak, K. Hess, R. Dupuis, P. Dapkus: Appl. Phys. Lett. **36**, 19 (1980)

16.14 Z.Y. Yu, V. Kreismanis, C.L. Tung: Appl. Phys. Lett. **44**, 136 (1984)

16.15 Y. Sasai, J. Ohya, M. Ogura, T. Kajiwara: Electron. Lett. **23**, 232 (1987)

358   References

16.16 Y. Sasai, N. Huse, M. Ogura, T. Kajiwara: J. Appl. Phys. **59**, 28 (1986)
16.17 Y. Sasai, M. Ogura, T. Kajiwara: J. Cryst. Growth **78**, 461 (1986)
16.18 Y. Arakawa, A. Yariv: IEEE J. QE-**21**, 1666 (1985)
16.19 M. Okai, T. Tsuchiya, A. Takai, N. Chinone: IEEE Photon. Tech. Lett. **4**, 526 (1992)
16.20 A. Ghiti: SPIE Proc. **861**, 96 (1987)
16.21 A. Cavicchii, O. Lang, D. Gershoni, A. Sergent, J. Vandenberg, S.N.G. Chu, M.B. Panish: Appl. Phys. Lett. **54**, 739 (1989)
16.22 P.J.A. Thijs, T. vanDongen, B.H. Verbeek: OFC'90, San Francisco, Paper WJ2
16.23 J.J. Coleman: OFC'90, San Francisco, Paper WJ1
16.24 Y. Arakawa, K. Vahala, A. Yariv: Appl. Phys. Lett. **45**, 950 (1984)
16.25 H. Temkin, G. Dolan, M.B. Panish, S.N.G. Chu: Appl. Phys. Lett. **50**, 413 (1987)
16.26 J. Cibert, P.M. Petroff, G.J. Dolan, C.J. Pearton, A.C. Gossard, J.H. English: Appl. Phys. Lett. **49**, 1275 (1986)
16.27 M.E. Hoenk, C.W. Nieh, H. Chen, K.J. Vahala: Appl. Phys. Lett. **55**, 53 (1989)
16.28 W. Tsang, M. Wu, Y. Chen, F. Chou, R. Logan, S. Chu, A. Sergant, P. Magill, K. Reichmann, C. Burrus: IEEE J. QE-**30**, 1370 (1994)
16.29 D.A.B. Miller, D.S. Chemla, T.C. Damen, A.C. Gossard, W. Wiegmann, T.H. Wood, L.A. Burrus: Phys. Rev. Lett. **53**, 2173 (1984)
16.30 D.S. Chemla, T.C. Damen, D.A.B. Miller, A.C. Gossard, W. Wiegmann: Appl. Phys. Lett. **42**, 864 (1983)
16.31 T.H. Wood, C.A. Burrus, D.A.B. Miller, D.S. Chemia, T.C. Damen, A.C. Gossard, W. Wiegemann: IEEE J. QE-**21**, 117 (1985)
16.32 D.A.B. Miler, D.S. Chemia, T.C. Damen, A.C. Gossard, W. Wiegmann, T.H. Wood, C.A. Burrus: Phys. Rev. B **32**, 1043 (1985)
16.33 T.H. Wood: IEEE J LT-**6**, 743 (1988)
16.34 P.J. Stevens, G. Parry: IEEE J. LT-7, 1101 (1989)
16.35 T.H. Wood, C.A. Burrus, R.S. Tucker, J.S. Weiner, D.A.B. Miller, D.S. Chemla, T.C. Damen, A.C. Gossard, Wiegmann: Electron. Lett. **21**, 693 (1985)
16.36 T.H. Wood, R.W. Tkach, A.R. Chraplyvy: Appl. Phys. Lett. **50**, 798 (1987)
16.37 T.H. Wood: Appl. Phys. Lett. **48**, 1413 (1986)
16.38 K. Wakita, K. Sato, I. Kotaka, M. Yamamoto, T. Kataoka: Electron. Lett. **30**, 302 (1994)
16.39 F. Koyama, K. Liou, A. Dentai, R. Gnall, G. Aapon, H. Presby, C. Burrus: Lasers and Electro-optics Soc., Annual Meeting, Anaheim, CA (1994) Proc p. 240
16.40 K. Wakita, I. Kotaka: Microwave and Opt. Tech. Lett. **7**, 120 (1994)
16.41 See, e.g., J.I. Pankove: *Optical Processes in Semiconductors* (Prentice-Hall, Englewood Cliffs, NJ 1971) pp. 89-90
16.42 T. Hiroshima: Appl. Phys. Lett. **50**, 968 (1987)
16.43 H. Nagai, M. Yamanishi, Y. Kam, I. Suemune: Electron. Lett. **22**, 888 (1986)
16.44 T.H. Wood, R.W. Tkach, A.R. Chraplyvy: Appl. Phys. Lett. **50**, 798 (1987)
16.45 I.D. Hennings, J.V. Collins: Electron. Lett. **19**, 927 (1983)
16.46 J.S. Weiner, D.A.B. Miller, D.S. Chemla: Appl. Phys. Lett. **50**, 842 (1987)
16.47 P. Li Kam Wa, J.E. Sitch, N.J. Mason, J.S. Roberts, P.N. Robson: Electron. Lett. **21**, 26 (1985)
16.48 A. Ajisawa, M. Fujiware, J. Shimizu, M. Sugimoto, M. Uchida, Y. Ohta: Electron. Lett. **23**, 1121 (1987)

16.49 T.H. Wood, J.Z. Pastalan, C.A. Burrus, B.I. Miller, J.L. de Miguel, U. Koren, M. Young: OSA/IEEE Conf. on Integrated Photonics Research, Hilton Head, SC (1990) Paper TuG4

16.50 F. Capasso, K. Mohammed, A.Y. Cho, R. Hull, L. Hutchinson: Appl. Phys. Lett. **47**, 420 (1995)

16.51 F. Capasso, K. Mohammed, A.Y. Cho, R. Hull, L. Hutchinson: Phys. Rev. Lett. **55**, 1152 (1985)

16.52 K. Mohammed, F. Capasso, J. Allam, A.Y. Cho, L. Hutchinson: Appl. Phys. Lett. **47**, 597 (1985)

16.53 F. Capasso: OFC/IGWO'86m Atlanta, GA, Paper WCC1

16.54 For a review of SEED devices see, e.g., D.A.B. Miller: Optics and Photon. News **1**, 7 (April 1990)

16.55 M. Cao, H. Li, A. Jun, F. Luo, X. Jun, L. Nu, W. Gao: Optics and Laser Tech. **26**, 271 (1994)

16.56 A.L. Lentine, H.S. Hinton, D.A.B. Miller, J.E. Henry, J.E. Cunningham, L.M.F. Chirovsky: IEEE J. QE-**25**, 1928 (1989)

16.57 L.M.F. Chirovsky, L.A. D'Asaro, R.F. Kopf, J.M. Kuo, A.L. Lentine, F.B. McCormick, R.A. Novotny, G.D. Boyd: OSA Annual Meeting, Orlando, FL (1989) Paper PD28

16.58 S. Tarucha, H. Okamoto: Appl. Phys. Lett. **48**, 1 (1986)

16.59 Y. Kawamura, K. Wakita, Y.I. Taya, Y. Yoshikuni, H. Asahi: Electron. Lett. **22**, 242 (1986)

16.60 Y. Kawamura, K. Wakita, Y. Yoshikuni, Y. Itaya, H. Asahi: IEEE J. QE-**23**, 915 (1987)

16.61 H. Nakano, S. Yamashita, T. Tanaka, H. Hirao, N. Naeda: IEEE J. LT-**4**, 574 (1986)

16.62 Y. Zebda, R. Lipa, M. Tutt, D. Pavlidis, P. Bhattacharya, J. Pamulapati, J. Oh: IEEE Trans. ED-**35**, 2435 (1988)

16.63 O. Wada, N. Nobuhara, T. Sanada, M. Kuno, M. Makiuchi, T. Fujii, T. Sakarai: IEEE J. LT-**7**, 186 (1989)

Supplementary Reading on Quantum Wells

P. Bhattacharya: *Semiconductor Optoelectronic Devices* (Prentice-Hall, Englewood Cliffs, NJ 1944) pp. 133-137, 294-299

Miller D.A.B., D.S. Chemla, S. Schmitt-Rink: Electric field dependence of optical properties of semiconductor quantum wells, in *Semiconductorts*, ed. by H. Haug (Academic, New York 1988) pp. 325-360

Chapter 17

17.1 M.C. Hamilton, D.A. Wille, W.J. Micelli: Opt. Eng. **16**, 475 (1977)

17.2 D. Mergerian, E.C. Malarkey: Microwave J. **23**, 37 (May 1980)
  D. Mergerian, E.C. Malarkey, R.P. Pautienus, J.C. Bradley, M. Mill, C.W. Baugh, A.L. Kellner, M. Mentzer: SPIE Proc. **321**, 149 (1982)

17.3 Milton: Charge transfer devices for infrared imaging, in *Optical and Infrared Detectors*, 2nd edn., ed. by R.J. Keyes, Topics Appl. Phys., Vol. 19 (Springer, Berlin, Heidelberg 1980)

17.4 D.F. Barbe (ed.): *Charge-Coupled Devices*, Topics Appl. Phys., Vol. 38 (Springer, Berlin, Heidelberg 1980)

17.5 M.E. Pedinoff, T.R. Ranganath, T.R. Joseph, J.Y. Lee: NASA Conf. on Optical Information Processing for Aerospace Applications, Houston, TX (1981) Proc. p. 173
B. Chen, T.R. Joseph, J.Y. Lee: IGWO'80, Incline Village, NV Paper ME-3

17.6 K. Aiki, M. Nakamura, J. Umeda: Appl. Phys. Lett. **29**, 506 (1976)

17.7 K. Aiki, M. Nakamura, J. Umeda: IEEE J. QE-**13**, 220 (1977)

17.8 K. Aiki, M. Nakamura, J. Umeda: IEEE J. QE-**13**, 597 (1977)

17.9 C. Zah, R. Bhat, B. Pathak, L. Curtis, F. Favire, P. Lin, C. Caneny, A. Gozdz, W. Lin, N. Andreadakis, D. Mahoney, M. Koza, W. Young, T. Lee: IEEE 5th Int'l Conf. on InP and Related Materials, Paris (1993) Proc. p. 77

17.10 K. Sato, S. Sakine, Y. Kondo, M. Yamamoto: IEEE J. QE-**29**, 1805 (1993)

17.11 H.F. Taylor: IEEE J. **15**, 210 (1979)

17.12 H.F. Taylor, M.J. Taylor, P.W. Bauer: Appl. Phys. Lett. **32**, 559 (1978)

17.13 S. Yamada, M. Minakota, J. Noda: Appl. Phys. Lett. **39**, 124 (1981)

17.14 H. Toda, M. Haruna, N. Nishihara: IEEE J. LT-**5**, 901 (1987)

17.15 H. Toda, K. Kasazum, M. Haruna, N. Nishihara: IEEE J. LT-**7**, 364 (1989)

17.16 S. Ura, T. Suhara, H. Nishihara: IEEE J. LT-**4**, 913 (1986)

17.17 M. Miller, M. Skalsky: Opt. Commun. **33**, 13 (1980)

17.18 T. Suhara, H. Nishihara, K. Koyama: IGWO'84, Kissimmee, FL, Paper ThD4

1719 S. Ura, T. Suhara, N. Nishihara: IEEE J. LT-**7**, 270 (1989)

17.20 S. Ura, M. Shimohara, T. Suhara, N. Nishihara: IEEE Photon. Tech. Lett. **6**, 239 (1994)

17.21 L. Johnson, F. Leonberger, G. Pratt: Appl. Phys. Lett. **41**, 134 (1982)

17.22 N.A.F. Jaeger, L. Young: IEEE J. LT-**7**, 229 (1989)

17.23 O. Wada, H. Nobuhara, T. Sanada, M. Kuno, M. Makiuchi, T. Fujii, T. Sakurai: IEEE J. LT-**7**, 186 (1989)

17.24 H. Matsueda, M. Nakamura: Appl. Opt. **23**, 779 (1984)

17.25 H. Matsueda: IEEE J. LT-**5**, 1382 (1987)

17.26 P. Woolnough, P. Birdsall, P. O'Sullivan, A. Cockburn, M. Harlow: Electron. Lett. **29**, 1388 (1993)

17.27 T. Horimatsu, M. Sasaki: IEEE J. LT-**7**, 1612 (1989)

17.28 C. Shih, D. Barlage, J. Wang, M. Feng: IEEE MTT-S Int'l Microwave Symp., San Diego, CA (1994) Proc. p. 1379

17.29 D. Trommer, G. Unterborsch, F. Reier: 6th Int'l Conf. on InP and Related Materials, Santa Barbara, CA (1994) Proc. p. 251

17.30 J.-F. Luy, P. Russer (eds.): *Silicon-Based Millimeter-Wave Secives*, Springer Ser. Electron. Photon., Vol. 32 (Springer, Berlin, Heidelberg 1994)

17.31 R.G. Hunsperger, M.K. Barnoski, H.W. Yen: A system for optical injection locking and switching of microwave oscillators. US Patent No. 4,264,857 (issued April 1981)

17.32 R.R. Kunath, K.B. Bhasin: IEEE AP-S/VRSI Symp., Philadelphia, PA (1986)

17.33 G.M. Alameel, A.W. Carlisle: Fiber and Integ. Opt. **8**, 45 (1989)

17.34 C.H. Gartside, C.F. Cottingham: OSA/IEEE OFC/IGWO'86, Atlanta, GA, Paper W12

17.35 See, e.g., Laser Focus World Buyers Guide '95 (PennWell Publ., Tulsa, OK 1995)

17.36 D. Mathoorasing, C. Kazmirierski, M. Blez, Y. Sorel, J. Kerdiles, M. Henry, C. Thebault: Electron. Lett. **30**, 507 (1994)

17.37 Y. Hirayama, H. Furuyama, M. Moringa, N. Suzuki, M. Kushibe, K. Eguchi, M. Nakamura: IEEE J. QE-**25**, 1320 (1989)

17.38 H. Ishikawa, H. Soda, K. Wakao, K. Kirhara, K. Kamite, K. Kotaki, M. Matsuda, H. Sudo, S. Yamakoshi, S. Isozum, H. Imai: IEEE J. LT-**5**, 848 (1987)

17.39 M. O'Mahoney, I. Marshall, H. Westlake, W. Devlin, J. Regnault: OFC'86, Atlanta, GA, Paper WE5

17.40 M. Fake, D. Parker: Optical amplifiers, in *Photonic Devices and Systems*, ed. by R.G. Hunsperger (Dekker, New York 1994) Chap. 5

17.41 B. Wedding, B. Franz: Electron. Lett. **29**, 402 (1993)

17.42 T. Okoshi: OFC'86, Atlanta, GA, Paper TuK1

17.43 A. Rosen, P. Stabile, W. Janton, A. Gombar, P. Basile, J. Delmaster, R. Hurwitz: IEEE Trans. MTT-**37**, 1255 (1989)

17.44 A. Vaucher, W. Streiffer, C.H. Lee: IEEE Trans. MTT-**31**, 209 (1983)

17.45 P.R. Herczfeld, A.S. Durousch, A. Rosen, A.K. Sharma, V.M. Contarino: IEEE Trans. MTT-**34**, 1371 (1986)

17.46 N.J. Gomes, A.J. Seeds: Electron. Lett. **23**, 1084 (1987)

17.47 R.G. Hunsperger, M.A. Mentzer: SPIE Proc. **993**, 204 (1988)

17.48 R.G. Hunsperger: Proc. SBMO Int'l Microwave Symp., Sao Paulo, Brazil (1989), IEEE Cat. No.89TH0260-0, p. 743

17.49 R.W. Ridgway, D.T. Davis: IEEE J. LT-**4**, 1514 (1986)

17.50 H. Ogawa, D. Polifko: IEEE MTT-S Int'l Microwave Symp., Albuquerque, NM (1992) Digest Pt.2, p. 558

17.51 H. Blauvelt, P. Chen, N. Bar-Chaim, P. Jiang, C. Lii, I. Jav: IEEE J. LT-**10**, 1766 (1992)

17.52 P. Herczfeld: Application of photonics to microwave devices and systems, in *Photonic Devices and Systems*, ed. by R.G. Hunsperger (Dekker, New York 1994) Chap. 8

Subject Index

Printing: Druckerei Zechner, Speyer
Binding: Buchbinderei Schäffer, Grünstadt